F.H. MONTGOMERY, Professor Emeritus of Botany and Genetics at
the University of Guelph, is the author of several books on the plants
and trees of Canada and the northeastern United States.

Hitherto, there has been no way to identify Canadian and US seed
samples except to compare them with known specimens. Identification has been a process of elimination depending on the experience,
knowledge, and memory of the observer. *The Seeds and Fruits of
Plants of Eastern Canada and Northeastern United States* de-
scribes and illustrates the seeds of about 1100 species of native wild
and introduced weedy plants from some 118 families, and provides
keys for their identification based on their geometric shapes. The
primary tool used in describing the shape of the seeds and keying
them is the chart of symmetric plane figures developed by the Systematics Association Committee of the International Association for
Plant Taxonomy.

The seeds, which were obtained from herbaria or the author's own
collection, were thus sorted and classified according to the exact shape
of their longitudinal and cross sections to form the major sections of
the key. Morphological details, observed under magnifications up to
30 x, were then used both for minor separations in the key and in
describing species. Photography was chosen as the most accurate
means of illustration.

While primarily intended for taxonomists and those teaching seed
identification, as a text and for verification of identifications, this
book will also be an important reference for students of archaeology
studying seed remains, ecologists studying bog formations, and seed
analysts both in Canada and the US.

Seeds and fruits of plants of eastern Canada and northeastern United States

F.H. MONTGOMERY

UNIVERSITY OF TORONTO PRESS

TORONTO AND BUFFALO

© University of Toronto Press 1977
Toronto and Buffalo
Printed in Canada
Reprinted in 2018
ISBN 978-1-4875-8187-9 (paper)
This book has been published during the
Sesquicentennial year of the University of Toronto

Library of Congress Cataloging in Publication Data

Montgomery, Frederick Howard, 1902–

Seeds and fruits of plants of eastern Canada and
northeastern United States.
Bibliography: p.
Includes index.
1. Seeds – Identification. 2. Fruit – Identification.
3. Botany – Canada. 4. Botany – Northeastern States.
I. Title.
QK660.M66   582'.0467'09713   76-23241
ISBN 0-8020-5341-6

To my wife

Contents

Preface

Seed* morphology is one of the most stable features of a plant and for this reason it is a necessary study for several sciences which may depend on it directly or indirectly. A knowledge of seed forms and marking is a valuable asset to the plant taxonomist, and the seed analyst is dependent to a large extent upon seed morphology for his identifications. The wild-life specialist finds it important in determining the food preferences of many animals and birds. An awareness of the nature of seed form and sculpture is essential to the archaeologist examining food relics at encampments or waste heaps of early man. The literature of palaeobotany and palaeoecology contains many references to, and illustrations of, seed remains (see, for example, references 18, 29, 58, 60, 68).

Because of their economic importance, the seeds of agricultural plants and weeds have been well described and illustrated. However, floras, and even monographic works, seldom give adequate descriptions or illustrations of the seeds of indigenous species. This is particularly true for the flora of northeastern North America. There is no one reference work in which one may find a means of identifying the seeds of even the commonest of our native plants. This deficiency led the author to collect, photograph, and describe the seeds of a number of the plants that are considered native or naturalized to eastern Canada, which, of course, are also a part of the flora of the northeastern United States. Over 1100 species have been included in this project.

The materials were obtained either from specimens collected by the author or from herbarium specimens. Voucher specimens of the plants from which the seeds originated are in the herbaria of the University of Guelph, Guelph, Ontario, or the University of Waterloo, Waterloo, Ontario. The seed collection itself has been placed in the

*The term seed is used here in its broadest sense to include seed-like structures which are fruits in botanical terminology.

Department of Geology of the Royal Ontario Museum, Toronto.

How to describe the form or shape of a seed was a primary problem. It was soon found that the descriptive terms in common use did not have the same meaning to everyone. For example, there was no standard concept for the term oval or ovate, and just as important, there was no standard for expressing the degree of ovateness. This difficulty was avoided by using the system proposed by the Systematics Association, Committee for Descriptive Biological Terminology, which uses mathematical ratios to describe the shapes of plane figures (65). This system has been applied wherever possible in describing the longitudinal sections and cross sections of the seeds. Although this system was published in 1962, its use in describing seeds has been limited. Berggren (10) has used it to describe the achenes of the Cyperaceae. It is to be expected that its comparative newness will initially cause some difficulties and misunderstandings. It is believed, however, that with perseverance it will prove less difficult, and certainly more accurate, than a system based on generalized descriptive terms.

With the kind permission of the International Association for Plant Taxonomy, the parts of the chart of symmetric plane figures and the language equivalents that are applicable to this work are reproduced on pages 2 and 3.

Since all seeds are not symmetrical, or are of a special form that cannot be measured or described mathematically, it has been necessary to employ three additional series (X, Y, and Z), which are not shown on the chart. The technique of using the system is explained in the Introduction to 'The Key.' For the convenience of users of the key, the chart and accompanying material have been placed before this explanation.

Line drawings have been used to illustrate seeds in the majority of publications, although W.H. Wright (73) produced an excellent series of paintings of weed seeds for the Canada Department of Agriculture, Berggren (10) used photographs to illustrate the fruits of the Cyperaceae, Martin and Barkley (45) also used this technique in illustrating their seed manual, and more recently R.J. Delorit (17) has used colour photography in a manual of weed seeds. Photography was chosen to illustrate the seeds in this publication, first because few artists are botanists and an artist views an object and draws it from a different point of view than would a botanist, and secondly, because photographs show more clearly the actual appearance of an object as seen under a hand lens or a binocular microscope.

Seeds are like human beings; some are photogenic and others are not. Green specimens such as the florets of the grasses and dull, black seeds which reflect little light are particularly difficult subjects to photograph. In small seeds, the curvature of the surface often restricted the magnification at which they could be photographed. For such seeds the scanning electron microscope would be a most helpful instrument even though the overall appearance might be masked by the fine surface detail revealed at the greater magnification. The photographs are oriented generally so that the hilum is either basal or to the left side of the photograph, with some obvious variations to the rule made for photographic purposes.

Magnifications of up to 30 x were used in observing surface features. The descriptive terms used have, as far as possible, been limited to those in *Taxonomic Terminology of the Higher Plants* by H.I. Featherly (20). Even though the author is aware that the manuals for this geographical area may not include the latest nomenclatural revisions, for the sake of consistency he has followed the nomenclature of *Manual of Vascular Plants of Northeastern United States and Adjacent Canada* by Gleason and Cronquist (28). The arrangement of families, genera, and species is, however, alphabetical.

The author has received assistance from many individuals during

the course of the work and he expresses his sincere appreciation to
all. He is particularly grateful to Dr Erica Gaertner for examining an
early draft of the manuscript and making many helpful suggestions;
to Mr W.J. Cody of the Biosystematic Research Institute, Ottawa,
who read and suggested a number of revisions in a later manuscript;
and to Dr J.K. Morton of the University of Waterloo, whose aid and
personal interest have been invaluable. A few seed collections were
obtained from a number of herbaria, and to the curators of these he
expresses his sincere thanks.

Publication of this book has been made possible by grants from
the National Research Council of Canada and the publications fund
of the University of Toronto Press.

F.H.M.

The chart

Symmetric plane figures

SERIES	SUB-SERIES 1	2		3		4	X	5	6		7		8
	12:1	6:1	3:1	2:1	3:2	6:5	1:1	5:6	2:3	1:2	1:3	1:6	1:12
A Elliptic		1	2	3	4	5	6	7	8	9	10	11	
B Oblong	12	13	14	15	16	17	18	19	20	21	22	23	24
C Rhombic		25	26	27	28	29	30	31	32	33	34	35	
D Ovate		36	37	38	39	40	41	42	43	44			
E Obovate		45	46	47	48	49	50	51	52	53			
H Triangular	72	73	74	75	76	77	78	79	80	81	82	83	
I Obtriangular	84	85	86	87	88	89	90	91	92	93	94	95	

3

Language Equivalents for Shapes in Chart

SERIES A: ELLIPTIC
1-2 narrowly elliptic
3-4 elliptic
5 broadly elliptic
6 circular
7 transversely broadly elliptic
8-9 transversely elliptic
10-11 transversely narrowly
elliptic

SERIES B: OBLONG
12 linear
13-14 narrowly oblong
15-16 oblong
17 broadly oblong
18 square
19 transversely broadly oblong
20-21 transversely oblong
22-23 transversely narrowly
oblong
24 transversely linear

SERIES C: RHOMBIC
25-26 narrowly rhomic
27-28 rhombic
29 broadly rhombic
30 quadrate rhombic
31 transversely broadly rhombic
32-33 transversely rhombic
34-35 transversely narrowly
rhombic

SERIES D: OVATE
36-37 narrowly ovate
38-39 ovate
40-41 broadly ovate
41-42 very broadly ovate
43-44 depressed ovate

SERIES E: OBOVATE
45-46 narrowly obovate
47-48 obovate
49-50 broadly obovate
50-51 very broadly obovate
52-53 depressed obovate

SERIES H: TRIANGULAR
72 linear-triangular
73-74 narrowly triangular
75-76 triangular
77-78 broadly triangular
78-79 very broadly triangular
80-81 shallowly triangular
82-83 very shallowly triangular

SERIES I: OBTRIANGULAR
84 linear-obtriangular
85-86 narrowly obtriangular
87-88 obtriangular
89-90 broadly obtriangular
90-91 very broadly obtriangular
92-93 shallowly obtriangular
94-95 very shallowly obtriangular

Correlation of Ratios with Shapes in Chart

Ratios of length to width or thickness to width for the elliptic series, and the corresponding shape number from the chart. The equivalent numbers for the other series may be determined by reading downward on the chart to the desired series.

Ratio	Shape
1 : 1	6
1 : 2	9
1 : 3	10
1 : 4–5	10–11
1 : 6	11
2 : 1	3
2 : 3	8
2 : 5	9–10
2 : 7–11	10–11
3 : 1	2
3 : 2	4
3 : 4	7–8
3 : 5	8–9
3 : 7–8	9–10
3 : 10–17	10–11
4 : 1	1–2
4 : 3	4–5
4 : 5	7–8
4 : 7	8–9
4 : 9–11	9–10
4 : 13–23	10–11
5 : 1	1–2
5 : 2	2–3
5 : 3	3–4

Ratio	Shape
5 : 4	4–5
5 : 6	7
5 : 7	7–8
5 : 8–9	8–9
5 : 11–14	9–10
5 : 16–19	10–11
6 : 1	1
6 : 5	5
6 : 7	6–7
6 : 10–11	8–9
6 : 13–17	9–10
6 : 19–35	10–11
7 : 2	1–2
7 : 3	2–3
7 : 4	3–4
7 : 5	4–5
7 : 6	5–6
7 : 8	6–7
7 : 9–10	7–8
7 : 11–13	8–9
7 : 15–20	9–10
7 : 22–41	10–11
8 : 3	2–3
8 : 5	3–4
8 : 7	5–6

Ratio	Shape
8 : 9	6–7
8 : 10–11	7–8
8 : 13–15	8–9
8 : 17–23	9–10
8 : 25–47	10–11
9 : 2	1–2
9 : 4	2–3
9 : 5	3–4
9 : 7	4–5
9 : 8	5–6
9 : 10	6–7
9 : 11–13	7–8
9 : 14–17	8–9
9 : 19–26	9–10
9 : 28–53	10–11
10 : 2–3	1–2
10 : 7–8	4–5
10 : 9	5–6
10 : 11	6–7
10 : 13–14	7–8
10 : 16–19	8–9
10 : 21–29	9–10
10 : 31–59	10–11
11 : 2–3	1–2
11 : 4–5	2–3
11 : 6–7	3–4
11 : 8–9	4–5
11 : 10	5–6
11 : 12–13	6–7
11 : 14–16	7–8
11 : 17–21	8–9
11 : 23–32	9–10
11 : 34–65	10–11

Ratio	Shape
12 : 5	2–3
12 : 7	3–4
12 : 11	5–6
12 : 13–14	6–7
12 : 16–17	7–8
12 : 19–23	8–9
12 : 25–35	9–10
12 : 37–71	10–11
13 : 3–4	1–2
13 : 5–6	2–3
13 : 7–8	3–4
13 : 9–10	4–5
13 : 11–12	5–6
13 : 14–15	6–7
13 : 16–19	7–8
13 : 20–25	8–9
13 : 27–38	9–10
14 : 3–4	1–2
14 : 5–6	2–3
14 : 8–9	3–4
14 : 10–11	4–5
14 : 12–13	5–6
14 : 15–16	6–7
14 : 17–20	7–8
14 : 22–27	8–9
14 : 29–41	9–10
15 : 3–4	1–2
15 : 6–7	2–3
15 : 8–9	3–4
15 : 11–12	4–5
15 : 13–14	5–6
15 : 16–17	6–7

Ratio	Shape	Ratio	Shape	Ratio	Shape	Ratio	Shape
15 : 19–22	7–8	19 : 10–12	3–4	23 : 4–7	1–2	26 : 32–38	7–8
15 : 23–29	8–9	19 : 13–15	4–5	23 : 8–11	2–3	26 : 40–51	8–9
15 : 31–44	9–10	19 : 16–18	5–6	23 : 12–15	3–4	26 : 53–77	9–10
		19 : 20–22	6–7	23 : 16–19	4–5		
16 : 3–5	1–2	19 : 23–28	7–8	23 : 20–22	5–6	27 : 5–9	1–2
16 : 6–7	2–3	19 : 29–37	8–9	23 : 24–27	6–7	27 : 10–13	2–3
16 : 9–10	3–4	19 : 39–56	9–10	23 : 28–34	7–8	27 : 14–17	3–4
16 : 11–13	4–5			23 : 37–45	8–9	27 : 19–22	4–5
16 : 14–15	5–6	20 : 4–6	1–2	23 : 47–68	9–10	27 : 23–26	5–6
16 : 17–19	6–7	20 : 7–9	2–3			27 : 28–32	6–7
16 : 20–23	7–8	20 : 11–13	3–4	24 : 5–7	1–2	27 : 33–40	7–8
16 : 25–31	8–9	20 : 14–16	4–5	24 : 9–11	2–3	27 : 41–53	8–9
16 : 33–47	9–10	20 : 17–19	5–6	24 : 13–15	3–4	27 : 55–80	9–10
		20 : 21–23	6–7	24 : 17–19	4–5		
17 : 3–5	1–2	20 : 25–29	7–8	24 : 21–23	5–6	28 : 5–9	1–2
17 : 6–8	2–3	20 : 31–39	8–9	24 : 25–28	6–7	28 : 10–13	2–3
17 : 9–11	3–4	20 : 41–59	9–10	24 : 29–35	7–8	28 : 15–18	3–4
17 : 12–14	4–5			24 : 37–47	8–9	28 : 19–23	4–5
17 : 15–16	5–6	21 : 4–6	1–2	24 : 49–71	9–10	28 : 24–27	5–6
17: 18–20	6–7	21 : 8–10	2–3			28 : 29–33	6–7
17 : 21–25	7–8	21 : 11–13	3–4	25 : 5–8	1–2	28 : 34–41	7–8
17 : 26–33	8–9	21 : 15–17	4–5	25 : 9–12	2–3	28 : 43–55	8–9
17 : 35–50	9–10	21 : 18–20	5–6	25 : 13–16	3–4	28 : 57–83	9–10
		21 : 22–25	6–7	25 : 17–20	4–5		
18 : 4–5	1–2	21 : 26–31	7–8	25 : 21–24	5–6	29 : 5–9	1–2
18 : 7–8	2–3	21 : 32–41	8–9	25 : 26–29	6–7	29 : 10–14	2–3
18 : 10–11	3–4	21 : 43–62	9–10	25 : 31–37	7–8	29 : 15–19	3–4
18 : 13–14	4–5			25 : 38–49	8–9	29 : 20–24	4–5
18 : 16–17	5–6	22 : 4–7	1–2	25 : 51–74	9–10	29 : 25–28	5–6
18 : 19–21	6–7	22 : 8–10	2–3			29 : 30–34	6–7
18 : 22–26	7–8	22 : 12–14	3–4	26 : 5–8	1–2	29 : 35–43	7–8
18 : 28–35	8–9	22 : 15–18	4–5	26 : 9–12	2–3	29 : 44–57	8–9
18 : 37–53	9–10	22 : 19–21	5–6	26 : 14–17	3–4	29 : 59–86	9–10
		22 : 23–26	6–7	26 : 18–21	4–5		
19 : 4–6	1–2	22 : 27–32	7–8	26 : 22–25	5–6	30 : 6–9	1–2
19 : 7–9	2–3	22 : 34–43	8–9	26 : 27–31	6–7	30 : 11–14	2–3
		22 : 45–65	9–10				

Ratio	Shape	Ratio	Shape	Ratio	Shape	Ratio	Shape
30 : 16–19	3–4	34 : 6–11	1–2	38 : 7–12	1–2	42 : 8–13	1–2
30 : 21–24	4–5	34 : 12–16	2–3	38 : 13–18	2–3	42 : 15–20	2–3
30 : 26–29	5–6	34 : 18–22	3–4	38 : 20–25	3–4	42 : 22–27	3–4
30 : 31–35	6–7	34 : 23–28	4–5	38 : 26–31	4–5	42 : 29–35	4–5
30 : 37–44	7–8	34 : 29–33	5–6	38 : 32–37	5–6	42 : 36–41	5–6
30 : 46–59	8–9	34 : 35–40	6–7	38 : 39–45	6–7	42 : 43–50	6–7
30 : 61–89	9–10	34 : 41–50	7–8	38 : 46–57	7–8	42 : 51–63	7–8
		34 : 52–67	8–9	38 : 58–75	8–9	42 : 64–83	8–9
31 : 6–10	1–2						
31 : 11–15	2–3	35 : 6–11	1–2	39 : 7–12	1–2	43 : 8–14	1–2
31 : 16–20	3–4	35 : 12–17	2–3	39 : 14–19	2–3	43 : 15–21	2–3
31 : 21–25	4–5	35 : 18–23	3–4	39 : 20–25	3–4	43 : 22–28	3–4
31 : 26–30	5–6	35 : 24–29	4–5	39 : 27–32	4–5	43 : 29–35	4–5
31 : 32–37	6–7	35 : 30–34	5–6	39 : 33–38	5–6	43 : 36–42	5–6
31 : 38–46	7–8	35 : 36–42	6–7	39 : 40–46	6–7	43 : 44–51	6–7
31 : 47–61	8–9	35 : 43–52	7–8	39 : 47–58	7–8	43 : 52–64	7–8
		35 : 53–69	8–9	39 : 59–77	8–9	43 : 65–85	8–9
32 : 6–10	1–2						
32 : 11–15	2–3	36 : 7–11	1–2	40 : 7–13	1–2	44 : 8–14	1–2
32 : 17–21	3–4	36 : 13–17	2–3	40 : 14–19	2–3	44 : 15–21	2–3
32 : 22–26	4–5	36 : 19–23	3–4	40 : 21–26	3–4	44 : 23–29	3–4
32 : 27–31	5–6	36 : 25–29	4–5	40 : 27–33	4–5	44 : 30–36	4–5
32 : 33–38	6–7	36 : 31–35	5–6	40 : 34–39	5–6	44 : 37–43	5–6
32 : 39–47	7–8	36 : 37–43	6–7	40 : 41–47	6–7	44 : 45–52	6–7
32 : 49–63	8–9	36 : 44–53	7–8	40 : 49–59	7–8	44 : 53–65	7–8
		36 : 55–71	8–9	40 : 61–79	8–9	44 : 67–87	8–9
33 : 6–10	1–2						
33 : 12–16	2–3	37 : 7–12	1–2	41 : 7–13	1–2	45 : 8–14	1–2
33 : 17–21	3–4	37 : 13–18	2–3	41 : 14–20	2–3	45 : 16–22	2–3
33 : 23–27	4–5	37 : 19–24	3–4	41 : 21–27	3–4	45 : 23–29	3–4
33 : 28–32	5–6	37 : 25–30	4–5	41 : 28–34	4–5	45 : 31–37	4–5
33 : 34–39	6–7	37 : 31–36	5–6	41 : 35–40	5–6	45 : 38–44	5–6
33 : 40–49	7–8	37 : 38–44	6–7	41 : 42–49	6–7	45 : 46–53	6–7
33 : 50–65	8–9	37 : 45–55	7–8	41 : 50–61	7–8	45 : 55–67	7–8
		37 : 56–73	8–9	41 : 62–81	8–9	45 : 68–89	8–9

Ratio	Shape	Ratio	Shape	Ratio	Shape	Ratio	Shape
46 : 8–15	1–2	50 : 9–16	1–2	54 : 10–17	1–2	58 : 39–48	4–5
46 : 16–22	2–3	50 : 17–24	2–3	54 : 19–26	2–3	58 : 49–57	5–6
46 : 24–30	3–4	50 : 26–33	3–4	54 : 28–35	3–4	58 : 59–69	6–7
46 : 31–38	4–5	50 : 34–41	4–5	54 : 37–44	4–5	58 : 70–87	7–8
46 : 39–45	5–6	50 : 41–49	5–6	54 : 46–53	5–6		
46 : 47–55	6–7	50 : 51–59	6–7	54 : 55–64	6–7	59 : 10–19	1–2
46 : 56–69	7–8	50 : 61–74	7–8	54 : 65–81	7–8	59 : 20–29	2–3
46 : 70–91	8–9	50 : 76–99	8–9			59 : 30–39	3–4
				55 : 10–18	1–2	59 : 40–49	4–5
47 : 8–15	1–2	51 : 9–16	1–2	55 : 19–27	2–3	59 : 50–58	5–6
47 : 16–23	2–3	51 : 18–25	2–3	55 : 28–36	3–4	59 : 60–71	6–7
47 : 24–31	3–4	51 : 26–34	3–4	55 : 37–45	4–5	59 : 72–88	7–8
47 : 32–39	4–5	51 : 35–42	4–5	55 : 46–54	5–6		
47 : 40–46	5–6	51 : 43–50	5–6	55 : 56–65	6–7	60 : 11–19	1–2
47 : 48–56	6–7	51 : 52–61	6–7	55 : 67–82	7–8	60 : 21–29	2–3
47 : 57–70	7–8	51 : 62–76	7–8			60 : 31–39	3–4
47 : 71–93	8–9			56 : 10–18	1–2	60 : 41–49	4–5
		52 : 9–17	1–2	56 : 19–27	2–3	60 : 51–59	5–6
48 : 9–15	1–2	52 : 18–25	2–3	56 : 29–37	3–4	60 : 61–71	6–7
48 : 17–23	2–3	52 : 27–34	3–4	56 : 38–46	4–5	60 : 73–89	7–8
48 : 25–32	3–4	52 : 35–43	4–5	56 : 47–55	5–6		
48 : 33–39	4–5	52 : 44–51	5–6	56 : 57–67	6–7		
48 : 41–47	5–6	52 : 53–62	6–7	56 : 68–83	7–8		
48 : 49–57	6–7	52 : 63–77	7–8				
48 : 58–71	7–8			57 : 10–18	1–2		
48 : 73–95	8–9	53 : 9–17	1–2	57 : 20–28	2–3		
		53 : 18–26	2–3	57 : 29–38	3–4		
49 : 9–16	1–2	53 : 27–35	3–4	57 : 39–47	4–5		
49 : 17–24	2–3	53 : 36–44	4–5	57 : 48–56	5–6		
49 : 25–32	3–4	53 : 45–52	5–6	57 : 58–68	6–7		
49 : 33–40	4–5	53 : 54–63	6–7	57 : 69–85	7–8		
49 : 41–48	5–6	53 : 64–79	7–8				
49 : 50–58	6–7			58 : 10–19	1–2		
49 : 59–73	7–8			58 : 20–28	2–3		
49 : 74–97	8–9			58 : 30–38	3–4		

The key

The primary basis for describing the shape of seeds and keying them is the chart, Symmetric Plane Figures, on page 2. This chart in its entirety illustrates and delimits mathematically the form of 95 plane figures, but only those series and figures which apply to this work have been illustrated here. This system can be applied to seeds by considering a longitudinal section or a cross section to be a plane figure. Because some seeds are not symmetrical, or for other reasons do not fit into any of the original series, three additional series have been added which do not appear on the chart. Series X includes seeds that are winged and for that reason may be asymmetrical; series Y comprises the grasses, a very specialized family; and series Z includes seeds that cannot be placed in any other series.

The key is based on first a separation into series and sub-series according to the exact shape of a longitudinal section and then a separation according to the shape of a cross section. Since mathematical ratios are required to specify these shapes, it is essential to obtain seed dimensions. The position of the hilum is considered the base of the seed, and the length is measured from the hilum to the apex or, where a style is present, to the base of the style. The width is the horizontal measurement at right angles to the length and does not include wings that may be present. The thickness is the perpendicular measurement at right angles to the length and the width. In the text the measurements are in the above order of length x width x thickness. Whenever possible, the figures are averages of three separate measurements of 10 seeds each and comparisons have been made with those given in other publications. Seeds may vary somewhat in size depending on the growing conditions, but the ratios are usually constant.

The first step in describing and keying the seed is the determination of its longitudinal shape so that it may be placed in the proper vertical series A to I of the chart. The opinions of observers will differ in many border-line cases, and an effort has been made to anticipate some of these by placing such genera or species in more than one series or sub-series of the key. The term oblique is used to distinguish those seeds that have the overall appearance of being elliptic or oblong, etc., but are not symmetrical because of their curved axes (e.g., *Sagittaria* spp.). The overall form must be given primary consideration and minor form variations are considered secondary. The form of the bean is usually described as reniform or kidney-shaped, but it is basically elliptic with a depression in the area of the hilum.

The second step is to determine the length to width ratio and, from this, the longitudinal section sub-series, as indicated across the top of the chart, as well as the descriptive number for the particular shape (note that the language equivalents by which the seed shapes may be described are given after the chart). Steps one and two must then be repeated to determine the descriptive number for the cross section of the seed.

To facilitate the determination of the descriptive number from a set of measurements, a table of ratios with an initial figure of 1 to 60 is given on pages 4–7 for the elliptic series. The corresponding equivalent in any other series can be determined by reading downward on the chart. For example, if the ratio of length to width is 3:2, the seed would fit descriptive number 4 for the elliptic series, 16 for the oblong series, 28 for the rhombic series, and 39, 48, 76, and 88 for the remaining series. In all cases the ratio should be reduced to its lowest terms, e.g. 9:6 should be reduced to 3:2. If the first number in a ratio is greater than 60, the descriptive number can, of course, still be determined from the table by dividing both numbers in the ratio by a suitable number.

When the cross-sectional form of a seed is distorted by compression from adjacent seeds, or the seed is the product of a divided ovary, as in the fruits of the Boraginaceae, Verbenaceae, and Labiatae, the broader portion is considered to be the apex for the cross section. Thus, some seeds of the above families will be considered as obovate or obtriangular in cross section.

The remaining parts of the key are dichotomous and follow the familiar pattern of floral keys.

EXAMPLE OF USE OF THE KEY

Suppose we have a seed which measures 0.9 x 0.9 x 0.5 mm and which, by observation, fits within the elliptic series A in longitudinal section. From the length and width dimensions 0.9 x 0.9 mm the length-width ratio is 1:1, and it falls in longitudinal section sub-series X and fits descriptive number 6, taking one to page 21. Suppose the seed is also elliptic in cross section. From the thickness and width dimensions 0.5 x 0.9, the thickness-width ratio is 5:9 and the descriptive number 8–9, which is keyed out in section C on page 22. Further details will assist in identifying the seed beyond this point, and reference to the individual descriptions in the text may be required to complete the identification where differences between species are slight.

ABBREVIATIONS USED IN THE KEY
AND IN THE DESCRIPTIONS

l-w length to width
t-w thickness to width
l.s. longitudinal section
c.s. cross section

Series Separation Based on Gross Morphology

Series X: Seeds obviously winged, page 41
Series Y: Seeds with glumes and/or with a lemma and palea,
The Gramineae, page 42
Series Z: Seeds very irregular in form and not readily placed in any
of the series above or below, page 44

Series Separation Based on Shape of the Longitudinal Section

Series A: Seeds elliptic in l.s., page 11
Series B: Seeds oblong in l.s., page 26
Series D: Seeds ovate in l.s., page 28
Series E: Seeds obovate in l.s., page 32
Series I: Seeds obtriangular in l.s., page 40

Series A (seeds elliptic in l.s.)

Sub-series 2: Seeds elliptic 1–2 in l.s., page 11
Sub-series 3: Seeds elliptic 3–4 in l.s., page 13
Sub-series 4: Seeds elliptic 5 in l.s., page 18
Sub-series X: Seeds elliptic 6 in l.s., page 21
Sub-series 5: Seeds elliptic 7 in l.s., page 23
Sub-series 6: Seeds elliptic 8–9 in l.s., page 25
Sub-series 7: Seeds elliptic 10–11 in l.s., page 25

KEY TO SUB-SERIES 2 (seeds elliptic 1–2 in l.s.)

A Seeds elliptic 6 in c.s., page 11
B Seeds elliptic 7 in c.s., page 12
C Seeds elliptic 8–9 in c.s., page 12
D Seeds elliptic 10–11 in c.s., page 13
E Seeds oblong 18 in c.s., *Tragopogon* spp.
F Seeds oblong 22–23 or 24–25 in c.s., *Lactuca canadensis*,
Pedicularis lanceolata

G Seeds triangular 78–79 in c.s., *Cyperus* spp.
H Seeds obtriangular 90–91 or 92–93 in c.s., *Ilex* sp.,
Lechea intermedia

A / *Seeds elliptic 6 in c.s.*

1 Seeds separating at maturity into 3 or more carpels,
Triglochin spp.
1 Seeds remaining intact at maturity   2

2 Seeds averaging 3.0 mm or more in length   3
2 Seeds averaging less than 3.0 mm in length   6

3 Surface with several longitudinal ridges or sulci   4
3 Surface with a single ridge or sulcus   5

4 Surface coarsely ridged and the ridges smooth or slightly
pubescent, *Anemonella* sp., *Thalictrum* spp.
4 Surface finely ridged or ribbed and the ribs transversely
rugulose, *Picris* sp., *Tragopogon* spp.

5 Seeds with a marginal ridge and rugulose, caruncle present and
large, *Jeffersonia* sp.
5 Seed surface with a sulcus, rugulose, caruncle absent, *Vinca* sp.

6 Surface muricate or papillose, *Penthorum* sp.
6 Surface not muricate or papillose   7

7 Surface longitudinally ribbed and transversely rugulose,
reticulate, *Odontites* sp.
7 Surface merely striate, reticulate, areolate, or smooth   8

8 Surface striate and reticulate, *Epifagus* sp., *Lobelia* spp.,
Pinguicula sp.
8 Surface striate, areolate, or smooth   9

9 Surface striate, areolate, apex or base or both caudate, *Juncus* spp.
9 Surface smooth or obscurely areolate, *Najas* sp.

B / *Seeds elliptic 7 in c.s.*

1 Seeds averaging 3.0 mm or more in length   2
1 Seeds averaging less than 3.0 mm in length   8

2 Seeds separating into 2 carpels at maturity, surface ribbed and the ribs spiny, *Daucus* sp.
2 Seeds remaining intact at maturity   3

3 Seeds distinctly longitudinally ridged, ribbed, or sulcate   4
3 Seed surface smooth or merely striate   7

4 Seeds with more than 2 ridges, ribs, or sulci   5
4 Seeds with 1 or 2 ridges or sulci   6

5 Seeds coarsely ridged and sulcate, surface with slight if any pubescence, *Thalictrum* spp.
5 Seeds finely 10-ribbed, the sulci strigose, *Erechtites* sp.

6 Seeds with a prominent surface sulcus, rugulose and brown, *Vinca* sp.
6 Seeds with a marginal ridge, surface striate or scalariform, *Celastrus* sp., *Sorbus* spp.

7 Seed axis abruptly curved or hooked at the base, surface dull and faintly striate, *Amelanchier* spp.
7 Axis not curved, surface smooth or striate, *Cirsium arvense*

8 Seeds with a coma   9
8 Coma not present   10

9 Coma apical on the seed, *Epilobium* spp.
9 Coma basal on the seed, *Populus* spp.

10 Surface papillose, *Drosera* spp.
10 Surface not papillose   11

11 Seeds longitudinally ridged, ribbed, or sulcate, transversely rugulose and reticulate, *Euphrasia* sp.
11 Seeds rugulose or reticulate often tending to have membranous ridges, *Ribes* spp.

C / *Seeds elliptic 8–9 in c.s.*

1 Seeds averaging 3.0 mm or more in length   2
1 Seeds averaging less than 3.0 mm in length   11

2 Seeds separating into 2 carpels at maturity, surface ribbed and the ribs spiny, *Daucus* sp.
2 Seeds remaining intact at maturity   3

3 Seeds with a red aril or an obvious caruncle, *Celastrus* sp., *Melampyrum* sp.
3 Seeds not as above   4

4 Seeds longitudinally ridged, ribbed, or sulcate   5
4 Seeds not as above   10

5 Seeds with more than 2 ridges, ribs, or sulci   6
5 Seeds with only 1 or 2 ridges, ribs, or sulci   8

6 Surface coarsely ridged and sulcate, slightly pubescent or glabrous, *Thalictrum* spp.
6 Surface merely ribbed   7

7 Ribs of the seeds transversely rugulose, *Crepis* sp., *Hypochaeris* sp., *Lactuca* spp.
7 Ribs not transversely rugulose, *Cacalia* sp.

8 Ridge marginal and circling the seed, *Anemone quinquefolia*, *Celastrus* sp., *Echinocystis* sp., *Maclura* sp., *Sorbus* spp.
8 Ridge or sulcus on the surface of the seed or only partly encircling it   9

9 Seed with a ridge on only one-half of the margin, *Celastrus* sp., *Sorbus* spp.
9 Seed with surface sulcus, *Cakile* sp., *Shepherdia canadensis*

10 Surface of the seed densely pubescent, *Anemone quinquefolia*
10 Surface glabrous, striate, or reticulate, *Amelanchier* spp., *Cirsium arvense*, *Limonium* sp., *Ptelea* sp.

11 Seed with a coma, or with a remnant style base surrounded by a pappus or pappus rim   12
11 Seeds lacking the above   17

12 Coma basal on the seed, *Populus* spp.
12 Coma or pappus apical   13

13 Coma apical and without a style base or pappus rim, *Epilobium* spp.
13 Style base and pappus or pappus rim present   14

14 Hilum lateral at the base, *Centaurea* spp.
14 Hilum strictly basal   15

15 Seeds with white resinous dots, *Antennaria* spp.
15 Resinous dots not present, surface smooth or longitudinally striate or ribbed   16

16 Surface striate or smooth, *Anaphalis* sp., *Carduus acanthoides*, *Gnaphalium* spp.
16 Surface longitudinally ribbed and transversely rugulose, *Sonchus* spp.

17 Seeds ridged, ribbed, or sulcate, transversely rugulose and reticulate, *Euphrasia* sp.
17 Seeds not as above, surface striate, rugulose, or reticulate, *Campanula rotundifolia*, *Drosera rotundifolia*, *Ribes* spp., *Tofieldia* sp.

D / *Seeds elliptic 10–11 in c.s.*

1 Seeds separating into 2 carpels at maturity, strongly ribbed, *Cryptotaenia* sp.
1 Seeds remaining intact at maturity   2

2 Seeds 15 mm or more in length, margin ridged, surface rugulose and black, *Echinocystis* sp.
2 Seeds less than 10 mm in length, smooth or longitudinally ribbed and transversely rugulose, *Gnaphalium* spp., *Lactuca* spp., *Sonchus* spp.

KEY TO SUB-SERIES 3 (seeds elliptic 3–4 in l.s.)

A Seeds elliptic 6 in c.s., page 14
B Seeds elliptic 7 in c.s., page 15
C Seeds elliptic 8–9 in c.s., page 15
D Seeds elliptic 10–11 in c.s., page 17
E Seeds oblong 18 in c.s., *Myriophyllum* sp., *Sanguisorba minor*
F Seeds oblong 20–21 in c.s., page 18
G Seeds oblong 22–23 in c.s., *Aralia nudicaulis*, *Carum* sp., *Lonicera* spp., *Pedicularis* spp., *Potamogeton* spp.
H Seeds triangular 78–79 in c.s., *Carex* spp., *Cyperus* spp., *Polygonum* spp., *Rumex* spp.
I Seeds triangular 80–81 in c.s., *Scirpus* spp.
J Seeds obtriangular 86, 87–88, or 89–90 in c.s., *Arctostaphylos* sp., *Chamaedaphne* sp., *Cimicifuga* sp., *Cratagus* spp., *Empetrum* sp., *Gaultheria* sp., *Gaylussacia* sp.
K Seeds obtriangular 90–91 in c.s., page 18
L Seeds obtriangular 92–93 in c.s., *Agastache* spp., *Dracocephalum* sp., *Glecoma* sp., *Hedeoma* sp., *Ilex* sp., *Monarda* spp., *Nepeta* sp., *Physostegia* sp., *Prunella* sp., *Pycnanthemum* sp., *Salvia* spp., *Satureja* spp., *Stachys* spp.

14

A / *Seeds elliptic 6 in c.s.*

1 Seeds averaging 3.0 mm or more in length — 2
1 Seeds averaging less than 3.0 mm in length — 11

2 Seeds separating into 2 or more carpels at maturity — 3
2 Seeds remaining intact at maturity — 4

3 Seeds separating into 3 or more carpels, surface glabrous, *Triglochin* spp.
3 Seeds separating into 2 carpels, surface spiny, *Sanicula* spp., *Torilis* sp.

4 Seeds with an aril or caruncle — 5
4 Aril or caruncle not evident — 6

5 Seeds with an aril, *Euonymus* spp., *Taxus* sp.
5 Seeds with a caruncle, *Disporum* sp., *Jeffersonia* sp.

6 Seeds ridged, ribbed, or sulcate — 7
6 Seeds not as above — 10

7 Seeds with a marginal ridge or sulcus, *Euonymus* spp., *Prunus* spp., *Scheuchzeria* sp.
7 Ridges several and not marginal — 8

8 Ridges low and rounded, surface rugulose, *Cornus* spp., *Triosteum* sp.
8 Ridges and sulci coarse and acute — 9

9 Ridges and sulci smooth, glabrous, or slightly pubescent, *Anemonella* sp., *Thalictrum* spp., *Triosteum* sp.
9 Ridges and sulci verrucose and transversely rugulose, *Borago* sp.

10 Surfaces striate or coarsely reticulate or alveolate, *Rubus* spp., *Sanguisorba minor, Zanthoxylum* sp.
10 Surfaces colliculose or smooth, *Brasenia* sp., *Hamamelis* sp., *Lindera* sp., *Scheuchzeria* sp., *Sparganium americanum, Streptopus amplexifolius*

11 Seeds with a basal coma, *Populus* spp.
11 Basal coma not present — 12

12 Seeds with an aril or an obvious caruncle, *Euphorbia* spp., *Juncus* spp., *Luzula* spp., *Nymphaea odorata*
12 Seeds not as above — 13

13 Seeds papillose, muricate, or verrucose — 14
13 Seeds not as above — 15

14 Surface papillose, *Drosera* spp.
14 Surface muricate or verrucose, *Galium* spp., *Heuchera* spp., *Hypoxis* sp., *Penthorum* sp., *Saxifraga* sp.

15 Seeds ridged or sulcate — 16
15 Seeds not ridged or sulcate — 18

16 Surface with several ridges, rugulose and/or reticulate, *Chaenorrhinum* sp., *Odontites* sp.
16 Seeds with a single ridge or sulcus — 17

17 Seeds with a single, marginal ridge, surface puncticulate or reticulate, *Aquilegia* sp., *Rubus* spp.
17 Surface with a single sulcus, areolate or pitted, *Conringia* sp., *Hesperis* sp.

18 Seed surface reticulate, *Epigaea* sp., *Geranium* spp., *Lobelia* spp., *Orobanche* sp., *Pinguicula* spp., *Vaccinium* spp.
18 Surface rugulose, smooth, puncticulate, or areolate. — 19

19 Surface rugulose or only slightly roughened, *Cassiope* sp., *Galium* spp., *Geranium* spp., *Eriocaulon* sp., *Halenia* sp., *Sisyrinchium* spp.
19 Surface smooth, puncticulate, or areolate — 20

20 Surface puncticulate or areolate, *Aquilegia* sp., *Diplotaxis* spp., *Halenia* sp., *Hippuris* sp., *Najas* sp., *Vaccinium* spp.
20 Surface smooth, *Geranium* spp., *Lythrum alatum, Morus* sp., *Myrica gale, Scleranthus* sp., *Trifolium arvense*

B / *Seeds elliptic 7 in c.s.*

1 Seeds averaging 3.0 mm or more in length   2
1 Seeds averaging less than 3.0 mm in length   9

2 Seeds separating into 2 carpels at maturity, surface ridged, smooth, or spiny, *Anethum* sp., *Daucus* sp.
2 Seeds remaining intact at maturity   3

3 Seeds with a red aril, surface smooth or with a marginal ridge, *Celastrus* sp., *Euonymus* spp.
3 Seeds not arillate   4

4 Seeds ridged or sulcate   5
4 Ridges or sulci not present   7

5 Seeds with a single ridge or sulcus, *Celastrus* sp., *Euonymus* spp., *Prunus* spp., *Scheuchzeria* sp., *Sorbus* spp., *Shepherdia argentea*
5 Seeds with several ridges and sulci   6

6 Surface with 4 corky ridges, areolate or rugulose, *Impatiens* spp.
6 Surface with more than 4 ridges, not corky glabrous or slightly pubescent, *Nyssa* sp., *Thalictrum* spp.

7 Seed axis abruptly curved at the base, surface puncticulate, *Amelanchier* spp.
7 Axis not curved, surface smooth, striate, areolate, or reticulate   8

8 Surface striate, areolate, or reticulate, *Cannabis* sp., *Cirsium arvense*, *Myrica asplenifolia*, *Zanthoxylum* sp.
8 Surface smooth and glossy, *Hamamelis* sp., *Scheuchzeria* sp., *Streptopus amplexifolius*

9 Seeds with a persistent style, margin ridged, surface rugulose or areolate, *Ranunculus* spp.
9 Seeds not as above   10

10 Seeds with a basal coma, *Populus* spp.
10 Basal coma not present   11

11 Seeds with an obvious caruncle, *Euphorbia* spp., *Luzula* spp.
11 Caruncle not obvious   12

12 Surface with resinous particles and flower bracts adhering, *Myrica gale*
12 Surface without resinous dots   13

13 Seeds ridged or sulcate   14
13 Seeds not ridged or sulcate   15

14 Surface with several ridges and sulci, transversely rugulose and reticulate, *Euphrasia* sp.
14 Surface with a single ridge or sulcus either marginal or on the surface, *Linnaea* sp., *Raphanus* sp., *Rubus* spp., *Sisymbrium* spp., *Tiarella* sp.

15 Surface papillose, *Drosera* sp.
15 Surface not papillose   16

16 Inner surface with 2 separated or united depressions at the base near the hilum, *Mentha* spp., *Satureja* spp.
16 Basal depressions not present   17

17 Surface rugulose, *Aronia* spp., *Geranium* spp., *Ribes* spp.
17 Surface not rugulose   18

18 Surface reticulate, *Brassica* spp., *Epigaea* sp., *Geranium* spp., *Ribes* spp., *Rubus* spp., *Vaccinium* spp.
18 Surface not reticulate   19

19 Surface areolate or striate, *Aronia* spp., *Brassica* spp., *Campanula* spp., *Coptis* sp., *Diplotaxis* spp., *Erucastrum* sp., *Raphanus* sp., *Thymus* sp., *Vaccinium* spp.
19 Surface smooth, *Geranium* spp., *Melilotus officinalis*, *Salvia* spp., *Trifolium arvense*

C / *Seeds elliptic 8-9 in c.s.*

1 Seeds averaging 3.0 mm or more in length   2
1 Seeds averaging less than 3.0 mm in length   13

2 Seeds separating into 2 carpels at maturity, strongly ribbed, smooth, or spiny, *Anethum* sp., *Daucus* sp.

2 Seeds remaining intact at maturity   3

3 Seeds with a persistent style or flower parts, margins more or less ridged, surface glabrous or pubescent, *Anemone quinquefolia, Maclura* sp., *Myrica asplenifolia.*, *Potamogeton* spp.

3 Seeds not as above   4

4 Axis of seed abruptly curved at the base, hilum more or less lateral, surface areolate, *Amelanchier* spp., *Sorbus* spp.

4 Axis not abruptly curved   5

5 Seeds with one or more distinct longitudinal ridges, ribs, or sulci   6

5 Seeds not as above   10

6 Seeds with a single surface or marginal ridge or sulcus, *Anemone* spp., *Cakile* sp., *Celastrus* sp., *Echinocystis* sp., *Prunus* spp., *Shepherdia* spp.

6 Seeds with 2 or more ridges, ribs, or sulci   7

7 Surface with 4 corky ridges, *Impatiens* spp.

7 Ridges not corky   8

8 Ridges low, rounded, smooth or rugulose, *Myrica asplenifolia, Nyssa* sp., *Viburnum* spp.

8 Ridges or ribs more acute and distinct   9

9 Ridges, ribs, and sulci smooth or areolate, *Cacalia* sp., *Thalictrum* spp.

9 Ribs or surface transversely rugulose, pappus or pappus rim present, *Sonchus* spp.

10 Seeds arillate or with an obvious caruncle, surface smooth, *Celastrus* sp., *Melampyrum* sp., *Ricinus* sp.

10 Seeds without an aril or obvious caruncle   11

11 Surface rugulose or reticulate, *Echinocystis* sp., *Podophyllum* sp., *Ptelea* sp., *Symphoricarpos* spp., *Viburnum* spp.

11 Surface smooth, puncticulate, areolate, or striate   12

12 Surface puncticulate, areolate, or striate, *Lonicera* spp., *Myrica asplenifolia*

12 Surface smooth, *Gleditsia* sp., *Lupinus* sp., *Symphoricarpos* spp.

13 Seeds with a persistent style or remnants of flower parts, or with a remnant style base surrounded by a pappus or pappus rim   14

13 Seeds not as above   16

14 Seeds with a persistent style or remnants of flower parts, *Ranunculus* spp., *Urtica* sp.

14 Seeds with a remnant style base surrounded by a pappus or a pappus rim   15

15 Surface longitudinally ribbed and transversely rugulose, *Sonchus* spp.

15 Surface not ribbed, smooth or striate, *Anaphalis* sp., *Carduus acanthoides*, *Centaurea* spp., *Gnaphalium* spp.

16 Seeds with a basal coma, *Populus* spp.

16 Basal coma lacking   17

17 Seeds with an obvious caruncle, surface areolate or reticulate, *Euphorbia* spp.

17 No obvious caruncle present   18

18 Seeds with 2 separated or fused depressions on the inner surface near the hilum, *Mentha* spp., *Origanum* sp., *Satureja* spp.

18 Seeds without basal depressions   19

19 Seeds longitudinally ridged, ribbed, or sulcate   20

19 Seeds not as above   23

20 Seeds with several ridges, transversely rugulose and reticulate, *Euphrasia* sp.

20 Seeds with a single ridge or sulcus   21

21 Ridge marginal, surface reticulate or alveolate, *Rubus* spp.

21 Ridge or sulcus on the surface and usually indicating the folding of the embryo   22

22 Surface areolate or reticulate, *Arabis* spp., *Camelina* sp., *Capsella* sp., *Draba* sp., *Polanisia* sp., *Raphanus* sp., *Sisymbrium* spp.

22 Surface smooth or faintly roughened, *Barbarea* sp., *Cakile* sp., *Cardamine* spp., *Dentaria* spp., *Linnaea* sp., *Phlox* spp.

23 Surface of the seed papillose, *Drosera* spp.

23 Surface not papillose   24

24 Surface rugulose, *Aralia* spp., *Mitchella* sp., *Ribes* spp., *Sambucus* spp., *Stachys* spp., *Stellaria* spp., *Tofieldia* sp., *Veronica* spp., *Veronicastrum* sp.

24 Surface not rugulose   25

25 Surface reticulate, *Brassica* spp., *Diervilla* sp., *Geranium* spp., *Loiseleuria* sp., *Rubus* spp., *Solanum* spp., *Utricularia* spp., *Vaccinium* spp.

25 Surface areolate, puncticulate, colliculose, striate, or smooth   26

26 Surface areolate, colliculose, puncticulate, or striate, *Aronia* spp., *Brassica* spp., *Campanula* spp., *Diplotaxis* spp., *Erucastrum* sp., *Lonicera* spp., *Polygonum* spp., *Raphanus* sp., *Solanum* spp., *Stachys* spp., *Thymus* sp., *Triodanis* sp., *Vaccinium* spp., *Veronica* spp., *Veronicastrum* sp.

26 Surface smooth, *Andromeda* sp., *Campanula* spp., *Chenopodium* spp., *Lechea* spp., *Melilotus officinalis*, *Salvia* spp., *Trifolium agrarium*, *Triodanis* sp.

D / *Seeds elliptic 10–11 in c.s.*

1 Seeds separating into 2 carpels at maturity, surface ribbed and smooth, *Pastinaca* sp.

1 Seeds remaining intact at maturity   2

2 Seeds averaging 3.0 mm or more in length   3

2 Seeds averaging less than 3.0 mm in length   5

3 Seeds 10 mm or more in length, surface black, dull, and rugulose, *Echinocystis* sp.

3 Seeds less than 10 mm in length   4

4 Surface smooth or areolate, black and glossy, margin light-coloured, *Linum lewisii*

4 Surface rugulose or reticulate, *Symphoricarpos* spp., *Viburnum* spp.

5 Seeds with a persistent style, margins frequently ridged and surface areolate, *Ranunculus* spp.

5 Seeds not as above   6

6 Surface ridged or sulcate   7

6 Surface not ridged or sulcate   8

7 Seeds with a ridge or sulcus showing the folding of the embryo, *Arabis* spp., *Cardamine* spp., *Draba* sp.

7 Ridges and sulci not showing the folding of the embryo, ridges transversely rugulose, pappus or pappus rim and style base evident, *Sonchus* spp.

8 Surface of the seed areolate or reticulate, *Lonicera* spp., *Ribes* spp., *Solanum* spp.

8 Surface rugulose, *Ribes* spp., *Veronica* spp.

F / *Seeds oblong 20–21 in c.s.*

1 Seeds averaging 3.0 mm or more in length	2
1 Seeds averaging less than 3.0 mm in length	3

2 Seeds separating into 2 carpels at maturity, surface ribbed, *Carum* sp., *Taenidia* sp., *Zizia* sp.
2 Seeds remaining intact at maturity, surface rugulose or areolate, *Aralia nudicaulis*, *Lonicera tatarica*, *Panax quinquefolium*

3 Seeds with a persistent style, margin ridged or slightly winged, *Potamogeton* spp.
3 Seeds not as above   4

4 Seeds with a ridge or sulcus indicating the folding of the embryo, surface with fine circular ribs and papillose or rugulose, *Thlaspi* sp.
4 Surface not showing the folding of the embryo, areolate, *Lonicera* spp.

K / *Seeds obtriangular 90–91 in c.s.*

1 Seeds averaging 3.0 mm or more in length	2
1 Seeds averaging less than 3.0 mm in length	3

2 Surface ridged and rugulose, *Crataegus* spp.
2 Surface smooth or only slightly ribbed, *Ilex* sp., *Nemopanthus* sp.

3 Seeds with 2 separated or united depressions on the inner surface near the hilum, *Dracocephalum* sp., *Hedeoma* sp., *Monarda* spp., *Nepeta* sp., *Satureja* spp.
3 Seeds without the basal depressions, or these not evident   4

4 Surface areolate or pubescent, *Agastache* spp., *Prunella* sp.
4 Surface smooth, *Glecoma* sp., *Lechea intermedia*, *Salvia* spp.

KEY TO SUB-SERIES 4 (seeds elliptic 5 in l.s.)

A Seeds elliptic 6 in c.s., page 18
B Seeds elliptic 7 in c.s., page 19
C Seeds elliptic 8–9 in c.s., page 19
D Seeds elliptic 10–11 in c.s., page 20
E Seeds oblong 20–21 or 22–23 in c.s., *Abutilon* sp., *Callitriche* sp., *Lonicera* spp., *Panax quinquefolium*, *Polygonum* spp., *Potamogeton* spp., *Stellaria media*
F Seeds triangular 78–79 in c.s., *Polygonum* spp., *Rumex* spp.
G Seeds obtriangular 86 or 87–88 in c.s., *Cimicifuga* sp., *Gaultheria hispidula*, *Gaylussacia* sp.
H Seeds obtriangular 90–91 or 92–93 in c.s., *Convolvulus sepium*, *Crataegus* spp., *Hedeoma* sp., *Salvia* spp., *Satureja* spp., *Stachys* spp.

A / *Seeds elliptic 6 in c.s.*

1 Seeds averaging 3.0 mm or more in length	2
1 Seeds averaging less than 3.0 mm in length	7

2 Seeds with one or more ridges or sulci   3
2 Seeds not ridged or sulcate   4

3 Surface with several, low, rounded ridges and sulci, smooth or rugulose, *Cornus* spp., *Myrica asplenifolia*, *Triosteum* sp.
3 Surface with a single ridge or sulcus, *Celtis* sp., *Euonymus* spp., *Prunus* spp.

4 Seeds more or less reniform, surface papillose or scurfy, *Hibiscus palustris*
4 Seeds not reniform   5

5 Seeds arillate or with an obvious caruncle, *Euonymus* spp., *Sanguinaria* sp., *Taxus* sp.
5 Neither aril nor obvious caruncle present   6

6 Surface of the seeds areolate, striate, or reticulate, *Smilax* spp., *Taxus* sp., *Zanthoxylum* sp.
6 Surface smooth or colliculose, *Brasenia* sp., *Lindera* sp., *Streptopus amplexifolius*

7 Seeds with an obvious caruncle, *Euphorbia* spp., *Luzula* spp.
7 Seeds lacking an obvious caruncle   8

8 Surface glandular dotted, *Humulus* spp., *Myrica gale*
8 Surface not glandular dotted   9

9 Surface muricate or papillose, *Hypoxis* sp., *Saxifraga* sp.
9 Surface not muricate or papillose   10

10 Surface rugulose, *Eriocaulon* sp., *Galium* spp., *Maianthemum* sp., *Sisyrinchium* spp.
10 Surface not rugulose   11

11 Surface reticulate, *Epigaea* sp., *Vaccinium* spp.
11 Surface smooth or areolate, *Maianthemum* sp., *Scleranthus* sp., *Trifolium arvense*, *Vaccinium* spp.

B / *Seeds elliptic 7 in c.s.*

1 Seeds averaging 3.0 mm or more in length   2
1 Seeds averaging less than 3.0 mm in length   6

2 Seeds with a red aril, inner surface smooth and with a marginal ridge, *Euonymus* spp.
2 Seeds not arillate   3

3 Seeds ridged or sulcate   4
3 Seeds not ridged or sulcate   5

4 Seeds with a single ridge or sulcus, *Celtis* sp., *Prunus* spp., *Euonymus* spp., *Shepherdia argentea*
4 Seeds with 10 low, rounded ridges, *Nyssa* sp.

5 Surface reticulately veined and rugulose, *Cannabis* sp.
5 Surface areolate, striate, or smooth, *Myrica asplenifolia, Smilacina trifolia, Smilax* spp., *Staphylea* sp., *Streptopus amplexifolius, Zanthoxylum* sp.

6 Seeds with a prominent, persistent style, margins usually ridged and surface rugulose or rugose, *Ranunculus* spp.
6 Seeds not as above   7

7 Surface with a single encircling ridge, colliculose and glossy, *Mollugo* sp.
7 Surface not ridged   8

8 Inner surface with 2 separated or united basal depressions near the hilum, *Mentha* spp., *Satureja* spp.
8 Seeds without basal depressions   9

9 Seeds with a distinct caruncle, *Euphorbia* spp., *Mollugo* sp.
9 Seeds lacking an obvious caruncle   10

10 Seed surface papillose, *Stellaria* spp.
10 Surface not papillose   11

11 Surface glandular or with a glandular pubescence, *Chrysosplenium* sp., *Humulus* spp., *Myrica gale*
11 Surface not glandular   12

12 Surface rugulose or reticulate, *Brassica* spp., *Epigaea* sp., *Maianthemum* sp., *Polygonatum* spp., *Rhus copallina, Stellaria* spp., *Vaccinium* spp.
12 Surface smooth or areolate, *Brassica* spp., *Diplotaxis* spp., *Maianthemum* sp., *Polygonatum* spp., *Raphanus* sp., *Salvia* spp., *Stachys* spp., *Thymus* sp., *Trifolium arvense, Vaccinium* spp.

C / *Seeds elliptic 8-9 in c.s.*

1 Seeds averaging 3.0 mm or more in length   2
1 Seeds averaging less than 3.0 mm in length   8

2 Seeds consisting of 2 carpels at maturity, surface ridged and smooth, *Conioselinum* sp., *Conium* sp.

2 Seeds not forming 2 carpels at maturity   3

3 Seeds with a persistent style, marginally ridged or winged and rugulose, *Potamogeton* spp.

3 Seeds not as above   4

4 Surface with one or more ridges or sulci   5

4 Surface not ridged or sulcate   6

5 Surface with a single ridge or sulcus, *Prunus* spp., *Shepherdia argentea*

5 Surface with 10 low, rounded ridges, surface rugulose, *Nyssa* sp.

6 Seeds with an obvious caruncle, surface smooth or areolate, *Ricinus* sp.

6 Seeds lacking an obvious caruncle   7

7 Surface rugulose, *Viburnum* spp.

7 Surface smooth, areolate, or striate, *Gleditsia* sp., *Lonicera* spp., *Lupinus* sp., *Myrica asplenifolia, Rhamnus frangula, Smilacina trifolia, Staphylea* sp.

8 Seeds separating into 2 carpels at maturity, surface ridged and smooth, *Cicuta bulbifera, Sium* sp.

8 Seeds remaining intact at maturity   9

9 Seeds with a persistent style or a remnant style base surrounded by a pappus or pappus rim   10

9 Seeds not as above   11

10 Seeds with a persistent style, margins usually ridged and the surface rugulose or areolate, *Potamogeton* spp., *Ranunculus* spp.

10 Seeds with a persistent style base surrounded by a pappus or pappus rim, hilum in a lateral notch at the base, *Centaurea* spp.

11 Seeds with an obvious caruncle, *Euphorbia* spp., *Mollugo* sp., *Portulaca* sp.

11 Caruncle absent or not obvious   12

12 Seeds ridged or sulcate   13

12 Surfaces not ridged or sulcate   14

13 Ridge or sulcus indicating the folding of the embryo, *Arabis* spp., *Camelina* sp., *Cardamine* spp., *Dentaria* spp., *Nasturtium* sp., *Polanisia* sp.

13 Ridge encircling the seed, surface colliculose and glossy, *Mollugo* sp.

14 Inner surface of the seed with 2 separated or united depressions at the base near the hilum, *Mentha* spp., *Origanum* sp., *Satureja* spp., *Thymus* sp.

14 Seeds without the basal depressions   15

15 Surface papillose, *Portulaca* sp., *Stellaria* spp.

15 Surface not papillose   16

16 Surface rugulose or reticulate, *Aralia nudicaulis, Brassica* spp., *Diervilla* sp., *Menyanthes* sp., *Mitchella* sp., *Rhus copallina, Ribes* spp., *Solanum* spp., *Stachys* spp., *Vaccinium* spp., *Veronica* spp.

16 Surface areolate, striate, or smooth   17

17 Surface areolate or striate, *Barbarea* sp., *Brassica* spp., *Diplotaxis* spp., *Houstonia caerulea, Lonicera* spp., *Raphanus* sp., *Solanum* spp., *Stachys* spp., *Thymus* sp., *Vaccinium* spp., *Veronica* spp.

17 Surface smooth, *Campanula* spp., *Chenopodium* spp., *Lechea* spp., *Melilotus alba, Menyanthes* sp., *Polygonum* spp., *Rhus copallina, Salvia* spp., *Trifolium agrarium, Triodanis* sp.

D / *Seeds elliptic 10–11 in c.s.*

1 Seeds averaging 3.0 mm or more in length   2

1 Seeds averaging less than 3.0 mm in length   4

2 Seeds separating into 2 carpels at maturity, surface ridged, *Pastinaca* sp.

2 Seeds remaining intact at maturity   3

3 Seeds with a persistent style, margin ridged and surface areolate, *Ranunculus* spp.

3 Seeds without a persistent style, surface with longitudinal ridges and nerves, rugulose, *Viburnum* spp.

4 Seeds with a persistent style, margin ridged and surface rugulose or areolate, *Ranunculus* spp.

4 Seeds not as above   5

5 Seeds with a ridge or sulcus indicating the folding of the embryo, *Arabis* spp., *Cardamine* spp., *Nasturtium* sp.

5 Seeds not showing the folding of the embryo   6

6 Surface rugulose or reticulate, *Ribes* spp., *Veronica* spp.

6 Surface smooth, areolate, or striate, *Lonicera villosa*, *Solanum* spp., *Veronica* spp.

KEY TO SUB-SERIES X (seeds elliptic 6 in l.s.)

A / *Seeds elliptic 6 in c.s.*

1 Seeds averaging 3.0 mm or more in length   2

1 Seeds averaging less than 3.0 mm in length   11

2 Seeds pubescent or with uncinate bristles or hairs, *Galium* spp., *Sanicula* spp., *Tilia* sp.

2 Seeds not as above   3

3 Seeds ridged or sulcate   4

3 Seeds not ridged or sulcate   5

4 Seeds with a ridge or sulcus partially or completely encircling the surface, *Celtis* sp., *Hydrophyllum* spp., *Prunus* spp., *Sassafras* sp.

4 Seeds with several ridges or sulci, *Cornus* spp., *Comandra* spp.

5 Seeds more or less reniform, surface papillose or scurfy, *Hibiscus palustris*

5 Seeds not reniform   6

6 Seeds with a caruncle, surface smooth, *Cubelium* sp.

6 Seeds without an obvious caruncle   7

7 Seeds with a blue outer coating with a whitish bloom, *Caulophyllum* sp.

7 Seed coating not blue   8

8 Surface muricate and covered with white, waxy particles, *Myrica pensylvanica*

8 Surface not muricate   9

9 Outer seed coating irregularly or reticulately veined, *Comandra* sp., *Humulus* spp.

9 Outer seed covering not irregularly veined   10

10 Surface rugulose or reticulate, *Arisaema dracontium*, *Hydrophyllum* spp., *Medeola* sp., *Smilacina* spp., *Symplocarpus* sp.

10 Surface smooth, areolate, or striate, *Lathyrus* spp., *Medeola* sp., *Sassafras* sp., *Smilacina* spp., *Smilax* spp., *Vicia* spp.

11 Seeds ridged or sulcate   12
11 Seeds not ridged or sulcate   13

12 Seed with a ridge encircling the surface, *Allium tricoccum*
12 Seeds with several ridges and sulci, *Cymbalaria* sp.

13 Seeds with an obvious caruncle, surface areolate, *Luzula* spp.
13 Seeds without an obvious caruncle   14

14 Surface verrucose or coarsely papillose, *Saururus* sp., *Scleria* sp.,
    Scutellaria spp.
14 Surface not as above   15

15 Surface rugulose or reticulate, *Brassica* spp., *Comandra* spp.,
    Galium spp., *Hydrophyllum* spp., *Maianthemum* sp., *Sisyrinchium*
    spp., *Vaccinium* spp.
15 Surface smooth or areolate, *Brassica* spp., *Collinsonia* sp.,
    Maianthemum sp., *Vaccinium* spp., *Vicia* spp.

B / *Seeds elliptic 7 in c.s.*

1 Seeds averaging 3.0 mm or more in length   2
1 Seeds averaging less than 3.0 mm in length   4

2 Surface smooth or areolate, *Smilax* spp., *Staphylea* sp.
2 Surface with one or more ridges or sulci   3

3 Seeds with a single ridge or sulcus partially or completely
    encircling the surface, *Celtis* sp., *Prunus* spp.
3 Surface with several ridges or sulci, *Cornus* spp.

4 Seeds with a persistent style, margins usually ridged and surface
    rugulose or areolate, *Ranunculus* spp.
4 Seeds not as above   5

5 Surface with an encircling ridge, black and glossy, *Amaranthus* spp.
5 Surface not ridged   6

6 Seeds with 2 separated or united basal depressions on the inner
    surface near the hilum, *Mentha* spp.
6 Seeds without basal depressions   7

7 Seeds glandular or glandular pubescent, *Chrysosplenium* sp.,
    Humulus spp.
7 Surface not glandular   8

8 Surface verrucose or papillose, *Lychnis* sp., *Scutellaria* spp.,
    Stellaria spp., *Vaccaria* sp.
8 Surface not as above   9

9 Surface alveolate, style persistent, pods indehiscent, *Neslia* sp.
9 Surface not alveolate   10

10 Surface rugulose or reticulate, *Fumaria* sp., *Maianthemum* sp.,
    Polygonatum spp., *Stellaria* spp., *Vaccinium* spp.
10 Surface smooth, areolate, or puncticulate, *Chenopodium* spp.,
    Maianthemum sp., *Polygonatum* spp., *Rhus copallina*, *Stachys*
    spp., *Vaccinium* spp.

C / *Seeds elliptic 8–9 in c.s.*

1 Seeds averaging 3.0 mm or more in length   2
1 Seeds averaging less than 3.0 mm in length   4

2 Seeds 2-carpelled at maturity, surface ribbed and smooth,
    Conioselinum sp., *Conium* sp.
2 Seeds not 2-carpelled at maturity   3

3 Seeds with one or more ribs or sulci, *Cornus* spp., *Rhamnus*
    frangula
3 Seeds not ribbed, surface smooth, *Gymnocladus* sp., *Staphylea* sp.

4 Seeds separating into 2 carpels at maturity, surface ribbed and
    smooth, *Cicuta bulbifera*, *Sium* sp.
4 Seeds remaining intact at maturity   5

5 Seeds with a persistent style   6
5 Seed with style lacking   7

6 Margin of the achenes ridged or winged, surface rugulose or areo-late, *Potamogeton* spp., *Ranunculus* spp.
6 Surface of the fruits alveolate, pods indehiscent, *Neslia* sp.

7 Seeds with an obvious caruncle, *Claytonia* spp., *Corydalis* spp., *Dicentra* spp., *Portulaca* sp.
7 Seeds without an obvious caruncle   8

8 Surface with a ridge or sulcus   9
8 Surface not ridged or sulcate   10

9 Ridge marginal, surface black and glossy, *Amaranthus* spp.
9 Ridge or sulcus on the surface and indicating the folding of the embryo, *Berteroa* sp., *Nasturtium* sp.

10 Surface papillose, *Arenaria* spp., *Lychnis* sp., *Portulaca* sp., *Saponaria* sp., *Spergula* sp., *Stellaria* spp.
10 Surface not papillose   11

11 Surface rugulose or reticulate, *Fumaria* sp., *Menyanthes* sp., *Physalis* spp., *Ribes* spp., *Stellaria* spp., *Vaccinium* spp.
11 Surface areolate or smooth, *Adlumia* sp., *Amaranthus* spp., *Atriplex* sp., *Berteroa* sp., *Chenopodium* spp., *Cycloloma* sp., *Houstonia caerulea*, *Menyanthes* sp., *Physalis* spp., *Phytolacca* sp., *Potentilla palustris*, *Rhus copallina*, *Stachys* spp., *Suaeda* sp., *Vaccinium* spp.

D / *Seeds elliptic 10-11 in c.s.*

1 Seeds with a persistent style, margin ridged and surface rugulose, *Laportea* sp., *Ranunculus* spp.
1 Seeds lacking a style   2

2 Seeds with an obvious caruncle, *Claytonia* spp., *Corydalis* spp.
2 Seeds without an obvious caruncle   3

3 Surface with a ridge or sulcus indicating the folding of the embryo, *Nasturtium* sp.
3 Surface without a ridge or sulcus   4

4 Surface papillose, *Linaria vulgaris*, *Saponaria* sp.
4 Surface not papillose   5

5 Surface rugulose or reticulate, *Physalis* spp., *Solanum* spp., *Vaccinium* spp.
5 Surface areolate or smooth, *Berteroa* sp., *Cycloloma* sp., *Physalis* spp., *Solanum* spp., *Vaccinium* spp.

F / *Seeds oblong 20-21 in c.s.*

1 Seeds averaging 3.0 mm or more in length   2
1 Seeds averaging less than 3.0 mm in length   3

2 Seeds tending to be reniform, hilar area deeply notched, raphe conspicuous, surface scaly, setose, and areolate, *Abutilon* sp.
2 Seeds with the embryo forming a marginal ring and this trans-versely rugulose or ridged, *Menispermum* sp.

3 Surface papillose, *Arenaria stricta*, *Silene* spp., *Stellaria media*
3 Surface not papillose   4

4 Surface areolate, slightly pubescent particularly near the hilum, raphe evident, *Malva* spp.
4 Surface smooth and glossy, *Polygonum* spp.

KEY TO SUB-SERIES 5 (seeds elliptic 7 in l.s.)

A Seeds elliptic 6 in c.s., *Arisaema triphyllum*, *Cornus* spp. *Scutellaria* spp.
B Seeds elliptic 7 in c.s., page 24
C Seeds elliptic 8-9 in c.s., page 24
D Seeds elliptic 10-11 in c.s., page 24
E Seeds oblong 18 in c.s., *Myriophyllum* spp.

F Seeds oblong 20–21 in c.s., *Arenaria serpyllifolia*, *Datura* sp., *Hibiscus trionum*, *Silene* spp., *Solanum* spp.

G Seeds oblong 22–23 in c.s., *Astragalus* spp., *Datura* sp., *Lycium* sp., *Nicandra* sp., *Solanum* spp.

H Seeds ovate 38–39 in c.s., *Lappula* sp.

I Seeds ovate 47–48 or 49–50 in c.s., *Hydrastis* sp., *Malva* spp.

B / *Seeds elliptic 7 in c.s.*

1 Seeds averaging 3.0 mm or more in length 2
1 Seeds averaging less than 3.0 mm in length 3

2 Surface of the seed smooth, *Lathyrus tuberosus*
2 Surface of the seed rugulose, *Uvularia* sp.

3 Seed a fruit with a persistent style, surface alveolate or rugulose, *Fumaria* sp., *Neslia* sp.
3 Seeds without a persistent style 4

4 Seed with an obvious caruncle, longitudinally rugose and transversely rugulose, *Trillium* spp.
4 Seeds without an obvious caruncle 5

5 Surface verrucose or papillose, *Lychnis* sp., *Scutellaria* spp., *Vaccaria* sp.
5 Surface not verrucose or papillose 6

6 Surface rugulose or reticulate, *Asparagus* sp.
6 Surface smooth, *Glycyrrhiza* sp.

C / *Seeds elliptic 8–9 in c.s.*

1 Seeds averaging 3.0 mm or more in length 2
1 Seeds averaging less than 3.0 mm in length 3

2 Surface of the seeds rugulose or with undulating ridges, *Rhus radicans*, *Uvularia* sp.
2 Surface smooth, *Amphicarpa* sp., *Lathyrus tuberosus*, *Robinia* sp.

3 Seeds separating into 2 carpels at maturity, surface ridged, *Erigenia* sp.
3 Seeds remaining intact at maturity 4

4 Seed an indehiscent fruit with a persistent style, surface alveolate or rugulose, *Fumaria* sp., *Neslia* sp.
4 Seeds not as above 5

5 Seeds with an obvious caruncle, surface smooth, rugulose, or reticulate, *Arenaria* spp., *Corydalis* spp., *Dicentra* spp., *Papaver* sp., *Trillium* spp.
5 Seeds without an obvious caruncle 6

6 Surface papillose, *Arenaria* spp., *Lychnis* sp., *Saponaria* sp.
6 Surface not papillose 7

7 Surface reticulate, *Papaver* sp., *Physalis* spp.
7 Surface not reticulate 8

8 Surface areolate, puncticulate, or colliculose, *Arenaria* spp., *Adlumia* sp., *Baptisia* sp., *Physalis* spp., *Phytolacca* sp.
8 Surface smooth, *Glycyrrhiza* sp., *Lespedeza* spp., *Lotus* sp., *Medicago* sp., *Rhus* spp., *Suaeda* sp., *Tephrosia* sp.

D / *Seeds elliptic 10–11 in c.s.*

1 Seeds averaging 3.0 mm or more in length, surface smooth, *Amphicarpa* sp., *Robinia* sp.
1 Seeds averaging less than 3.0 mm in length 2

2 Seeds separating into 2 carpels at maturity, surface ribbed and smooth, *Erigenia* sp.
2 Seeds remaining intact at maturity 3

3 Seeds with an obvious caruncle, surface papillose or colliculose, *Corydalis* spp.
3 Seeds without an obvious caruncle 4

4 Surface papillose, *Saponaria* sp.
4 Surface not papillose — 5

5 Surface reticulate, *Physalis* spp., *Solanum* spp.
5 Surface areolate, colliculose, or smooth, *Arenaria* spp., *Lespedeza* spp., *Physalis* spp., *Solanum* spp., *Tephrosia* sp.

KEY TO SUB-SERIES 6 (seeds elliptic 8–9 in l.s.)

A Seeds elliptic 6 in c.s., *Arisaema triphyllum, Cornus* spp., *Scutellaria* spp.
B Seeds elliptic 7 in c.s., *Asparagus* sp., *Glycyrrhiza* sp., *Lathyrus tuberosus, Scutellaria* spp., *Trillium* spp., *Uvularia* sp.
C Seeds elliptic 8–9 in c.s., page 25
D Seeds elliptic 10–11 in c.s., page 25
E Seeds oblong 20–21 or 22–23 in c.s., *Arenaria serpyllifolia, Astragalus* spp., *Datura* sp., *Hibiscus trionum, Solanum* spp.
F Seeds ovate 38–39 or 40–41 in c.s., *Hackelia* sp.
G Seeds obovate 47–48 or 49–50 in c.s., *Hydrastis* sp.

C / *Seeds elliptic 8–9 in c.s.*

1 Seeds averaging 3.0 mm or more in length — 2
1 Seeds averaging less than 3.0 mm in length — 3

2 Surface rugulose or with undulating ridges, *Rhus radicans, Uvularia* sp.
2 Surface smooth, *Amphicarpa* sp., *Lathyrus tuberosus, Robinia* sp.

3 Seeds separating into 2 carpels at maturity, surface ribbed and smooth, *Erigenia* sp.
3 Seeds remaining intact at maturity — 4

4 Seeds with an obvious caruncle, *Chelidonium* sp., *Dicentra* spp., *Papaver* sp., *Trillium* spp.
4 Seeds without an obvious caruncle — 5

25

5 Surface verrucose or papillose, *Arenaria* spp., *Scutellaria* spp.
5 Surface not verrucose or papillose — 6

6 Surface rugulose or reticulate, *Papaver* sp., *Plantago* spp., *Utricularia* spp.
6 Surface not rugulose or reticulate — 7

7 Surface areolate, puncticulate, or colliculose, *Arenaria* spp., *Baptisia* sp., *Utricularia* spp.
7 Surface smooth, *Glycyrrhiza* sp., *Lespedeza* spp., *Lotus* sp., *Medicago* sp., *Rhus* spp., *Tephrosia* sp.

D / *Seeds elliptic 10–11 in c.s.*

1 Seeds averaging 3.0 mm or more in length, surface smooth, *Amphicarpa* sp., *Robinia* sp.
1 Seeds averaging less than 3.0 mm in length — 2

2 Seeds separating into 2 carpels at maturity, surface ridged and smooth, *Erigenia* sp.
2 Seeds remaining intact at maturity — 3

3 Seeds peltate, smooth or rugulose, *Plantago* spp.
3 Seeds not peltate — 4

4 Surface areolate or reticulate, *Physalis* spp.
4 Surface smooth or colliculose, *Arenaria lateriflora, Lespedeza* spp., *Tephrosia* sp.

KEY TO SUB-SERIES 7 (seeds elliptic 10–11 in l.s.)

A Seeds elliptic 8–9 or 10–11 in c.s., *Plantago* spp.
B Seeds ovate 38–39 or 40–41 in c.s., *Hackelia* sp.

Series B (seeds oblong in l.s.)

KEY TO SUB-SERIES 1 (seeds oblong 12 in l.s.)

A Seeds elliptic 8–9 in c.s., *Apocynum cannabinum*
B Seeds oblong 18 in c.s., *Eupatorium* spp., *Inula* sp.
C Seeds oblong 20–21 in c.s., *Leontodon* sp., *Osmorhiza* spp.

KEY TO SUB-SERIES 2 (seeds oblong 13–14 in l.s.)

A Seeds elliptic 6 in c.s., page 26
B Seeds elliptic 7 in c.s., *Cirsium arvense, Prenanthes* spp., *Salix* spp., *Senecio* spp., *Solidago* spp.
C Seeds elliptic 8–9 in c.s., page 26
D Seeds elliptic 10–11 in c.s., *Arabis* spp., *Aster* spp., *Coreopsis* sp., *Erigeron pulchellus, Gnaphalium* spp., *Zannichellia* sp.
E Seeds oblong 18 in c.s., *Anthemis* sp., *Dipsacus* sp., *Eupatorium* spp., *Hieracium* spp., *Rudbeckia hirta, Senecio* spp.
F Seeds oblong 19, 20–21, or 22–23 in c.s., *Chamaelirium* sp., *Leontodon* sp., *Osmorhiza* spp., *Vernonia* sp.
G Seeds rhombic 32–33 in c.s., *Anthemis tinctoria, Helianthus tuberosus, Rudbeckia laciniata*
H Seeds triangular 80–81 in c.s., *Dulichium* sp.
I Seeds obtriangular 90–91 or 92–93 in c.s., *Spiraea* spp., *Verbena* spp.

A / *Seeds elliptic 6 in c.s.*

1 Seeds averaging 3.0 mm or more in length   2
1 Seeds averaging less than 3.0 mm in length   3

2 Seeds separating into 2 carpels at maturity, coarsely longitudinally ridged, *Ligusticum* sp.
2 Seeds not separating into 2 carpels, surface longitudinally ribbed, *Alliaria* sp.

3 Seeds with a remnant style base surrounded by a pappus or a pappus rim, *Anthemis* spp., *Krigia* sp., *Petasites* spp., *Senecio* spp., *Solidago* spp.
3 Seeds not as above   4

4 Surface ridged or striate   5
4 Surface not ridged or striate, coarsely reticulate, somewhat irridescent from the thick mucilaginous coating, *Vallisneria* sp.

5 Ridges or striae papillose, *Mimulus* sp.
5 Ridges not papillose; striate, reticulate, or rugulose, *Hypericum* spp., *Ludwigia* sp., *Triadenum* sp.

C / *Seeds elliptic 8–9 in c.s.*

1 Seeds averaging 3.0 mm or more in length   2
1 Seeds averaging less than 3.0 mm in length   4

2 Seeds with a remnant of a style base surrounded by a pappus or pappus rim, surface ribbed or striate, *Cacalia* sp., *Cirsium arvense, Prenanthes* spp., *Sonchus* spp., *Tussilago* sp., *Rudbeckia laciniata*
2 Seeds not as above   3

3 Seeds with an obvious caruncle, surface areolate, *Melampyrum* sp.
3 Seeds without a caruncle, terminal coma present, surface smooth, *Apocynum cannabinum*

4 Seeds with the remnant of a style base surrounded by a pappus or pappus rim, *Aster* spp., *Conyza* sp., *Erigeron* spp., *Gnaphalium* spp., *Solidago* spp.
4 Seeds not as above   5

5 Surface smooth, basal coma present, *Salix* spp.
5 Surface ridged or sulcate   6

6 Ridges marginal, outer ridge toothed, *Zannichellia* sp.
6 Ridges or sulci on the surface   7

7 Ridge or sulcus showing the folding of the embryo, margin often winged, *Arabis* spp.
7 Surface ridges several and transversely rugulose, *Sonchus* spp.

KEY TO SUB-SERIES 3 (seeds oblong 15-16 in l.s.)

A Seeds elliptic 6 in c.s., page 27
B Seeds elliptic 7 in c.s., page 27
C Seeds elliptic 8-9 in c.s., page 27
D Seeds elliptic 10-11 in c.s., *Arabis* spp., *Coreopsis* sp., *Erigeron pulchellus*, *Gnaphalium* spp., *Sonchus* spp., *Zannichellia* sp.
E Seeds oblong 18 in c.s., *Anthemis* sp., *Euphorbia* spp., *Linaria canadensis*, *Sisymbrium* spp.
F Seeds oblong 20-21 or 22-23 in c.s., *Chamaelirium* sp.
G Seeds rhombic 32 in c.s., *Anthemis tinctoria*
H Seeds obtriangular 90-91 or 92-93 in c.s., *Echium* sp., *Pycnanthemum* sp.

A / *Seeds elliptic 6 in c.s.*

1 Seeds with a remnant style base surrounded by a pappus or a pappus rim, *Krigia* sp., *Solidago* spp.
1 Seeds not as above   2

2 Seeds with an obvious caruncle, *Euphorbia* spp.
2 Caruncle not obvious, surface ridged or sulcate   3

3 Ridge or sulcus showing the folding of the embryo, *Hesperis* sp.
3 Ridge or sulcus not showing the folding of the embryo   4

4 Surface reticulate or transversely rugulose, *Hypericum* spp., *Ludwigia* sp., *Triadenum* sp., *Verbascum* spp.
4 Surface ridges papillose, *Mimulus* sp.

B / *Seeds elliptic 7 in c.s.*

1 Seeds averaging 3.0 mm or more in length   2
1 Seeds averaging less than 3.0 mm in length   3

2 Seeds separating into 2 carpels at maturity, surface strongly ribbed, *Angelica* sp.
2 Seeds with only one apparent carpel, style base present and surrounded by a pappus rim, surface striate, *Cirsium arvense*

3 Seeds with a remnant style base surrounded by a pappus or pappus rim, *Senecio* spp., *Solidago* spp.
3 Seeds not as above   4

4 Seeds with a basal coma, *Salix* spp.
4 Coma not present, surface ridged or sulcate   5

5 Surface with a single ridge or sulcus, surface areolate or reticulate, *Sisymbrium* spp., *Pedicularis canadensis*
5 Lower half of the seed with several low, rounded ridges, upper half with numerous pits, *Calla* sp.

C / *Seeds elliptic 8-9 in c.s.*

1 Seeds averaging 3.0 mm or more in length   2
1 Seeds averaging less than 3.0 mm in length   4

2 Seeds separating into 2 carpels at maturity, surface strongly ribbed, *Angelica* sp.
2 Seeds remaining intact at maturity   3

28

3 Seeds with a remnant of a style base surrounded by a pappus or pappus rim, surface ribbed or striate, *Cacalia* sp., *Cirsium arvense*
3 Seeds with a remnant of a style base surrounded by a pappus or pappus rim, surface ribbed or striate, *Cacalia* sp., *Cirsium arvense*
3 Seeds not as above, base with an obvious caruncle, surface areolate, *Melampyrum* sp.

4 Seeds with a remnant of a style base surrounded by a pappus or pappus rim, *Erigeron* spp., *Gnaphalium* spp., *Solidago* spp., *Sonchus* spp.
4 Seeds not as above   5

5 Seeds with a basal coma, *Salix* spp.
5 No coma present, surface ridged or sulcate   6

6 Ridges marginal, outer ridge toothed, *Zannichellia* sp.
6 Ridges or sulci not marginal   7

7 Surface with a single ridge or sulcus showing the folding of the embryo, *Arabis* spp., *Cardamine* spp., *Sisymbrium* spp.
7 Surfaces with several ridges and these transversely rugulose, *Butomus* sp.

KEY TO SUB-SERIES 4 (seeds oblong 17 in l.s.)

A Seeds elliptic 6 in c.s., *Verbascum* spp.
B Seeds elliptic 7 or 8–9 in c.s., *Angelica* sp.
C Seeds oblong 18 in c.s., *Euphorbia* spp., *Linaria canadensis*
D Seeds obtriangular 90–91 in c.s., *Echium* sp.

KEY TO SUB-SERIES X (seeds oblong 18 in l.s.)

A Seeds elliptic 6 in c.s., *Agrimonia* spp.

KEY TO SUB-SERIES 5 (seeds oblong 19 in l.s.)

A Seeds elliptic 7 or 8–9 in c.s., *Lathyrus latifolius*
B Seeds oblong 15–16, 17, or 19 in c.s., *Apios* sp., *Lathyrus latifolius*
C Seeds ovate 40–41 in c.s., *Cynoglossum* spp.

KEY TO SUB-SERIES 6 (seeds oblong 20–21 in l.s.)

A Seeds oblong 15–16 or 17 in c.s., *Apios* sp.
B Seeds oblong 19 or 20–21 in c.s., *Lathyrus latifolius*, *Strophostyles* sp.
C Seeds ovate 38–39 or 40–41 in c.s., *Cynoglossum* spp., *Dianthus* sp.

KEY TO SUB-SERIES 7 (seeds oblong 22–23 in l.s.)

A Seeds elliptic 7 or 8–9 in c.s., *Houstonia canadensis*
B Seeds ovate 38–39 or 40–41 in c.s., *Cynoglossum* spp., *Dianthus* sp.

Series D (seeds ovate in l.s.)

Sub-series 2: Seeds ovate 36–37 in l.s., page 28
Sub-series 3: Seeds ovate 38–39 in l.s., page 28
Sub-series 4 & X: Seeds ovate 40–41 in l.s., page 31

KEY TO SUB-SERIES 2 (seeds ovate 36–37 in l.s.)

A Seeds elliptic 5 in c.s., *Erythronium* spp.
B Seeds elliptic 6 in c.s., *Hepatica* spp., *Streptopus roseus*
C Seeds elliptic 7 in c.s., *Apocynum androsaemifolium*, *Hepatica* spp., *Phryma* sp., *Rosa* spp., *Streptopus roseus*
D Seeds elliptic 8–9 in c.s., *Apocynum androsaemifolium*, *Erysimum* spp., *Phryma* sp., *Rosa* spp.
E Seeds obovate 49–50 or 50–51 in c.s., *Dalibarda* sp., *Lycopsis* sp.
F Seeds triangular 78–79 in c.s., *Rumex* spp.
G Seeds obtriangular 87–88 or 89–90 in c.s., *Rosa* spp.

KEY TO SUB-SERIES 3 (seeds ovate 38–39 in l.s.)

A Seeds elliptic 5 in c.s., *Asarum* sp., *Erythronium* spp., *Rosa* sp.
B Seeds elliptic 6 in c.s., page 29
C Seeds elliptic 7 in c.s., page 29

D Seeds elliptic 8–9 in c.s., page 30
E Seeds elliptic 10–11 in c.s., *Pilea* sp., *Polygonum* spp.
F Seeds obovate 47–48 in c.s., *Fragaria* spp.
G Seeds obovate 49–50, 50–51, or 52–53 in c.s., *Dalibarda* sp., *Fragaria* spp., *Lithospermum* spp., *Mertensia* spp.
H Seeds triangular 78–79 or 80–81 in c.s., *Polygonum* spp., *Proserpinaca* sp., *Rumex* spp.
I Seeds obtriangular 87–88, 89–90, or 90–91 in c.s., *Rosa* spp., *Sida* sp., *Symphytum* sp.

B / *Seeds elliptic 6 in c.s.*

1	Seeds averaging 3.0 mm or more in length	2
1	Seeds averaging less than 3.0 mm in length	11
2	Seeds covered with long, hooked spines, *Xanthium* sp.	
2	Seeds not spiny	3
3	Seeds with persistent styles or adhering flower parts	4
3	Seeds not as above	5
4	Seeds with a style only, marginally ridged, surface pubescent, *Hepatica* spp.	
4	Seeds with flower parts adhering, surface ridged and rugulose, *Pontederia* sp.	
5	Seeds with a prominent caruncle, surface longitudinally rugose and transversely rugulose, *Erythronium* spp.	
5	Seeds without an obvious caruncle	6
6	Surface ridged, striate, or sulcate	7
6	Surface not ridged or sulcate	10
7	Surface with a single ridge or sulcus	8
7	Surface with several ridges or sulci	9

8	Seeds with a ridge on the inner surface, basal scar large, vascular scars obvious, dorsal surface smooth or with occasional pits, *Lithospermum* spp., *Onosmodium* sp.	
8	Surface glabrous or with pubescence at the base or apex, *Rosa* sp.	
9	Surface faintly ridged or striate and rugulose, basal scar large, *Cladium* sp.	
9	Ridges obvious, low and rounded, surface rugulose, *Cornus* spp.	
10	Surface of the seed rugose, *Dirca* sp.	
10	Surface reticulately veined, *Rubus chamaemorus*	
11	Surface longitudinally ridged	12
11	Surface not ridged	13
12	Seeds ridged on the inner surface, basal scar large, vascular scars evident, surface smooth or verrucose, *Lithospermum* spp.	
12	Seeds with several ridges and transversely rugulose, *Streptopus roseus*	
13	Surface papillose and areolate, *Hudsonia* sp.	
13	Surface not papillose	14
14	Surface striate and areolate, frequently caudate, *Juncus* spp.	
14	Surface smooth, *Morus* sp.	

C / *Seeds elliptic 7 in c.s.*

1	Seeds averaging 3.0 mm or more in length	2
1	Seeds averaging less than 3.0 mm in length	8
2	Seeds separating into 2 carpels at maturity, surface ridged, *Cicuta maculata*	
2	Seeds remaining intact at maturity	3
3	Surface ridged or sulcate	4
3	Surface not ridged or sulcate	5

4 Ridge marginal, style persistent, surface pubescent, *Hepatica* spp.

4 Surface with a ridge or sulcus, glabrous or with a basal or apical tuft of hairs, *Rosa* spp.

5 Surface smooth or merely undulate, *Asimina* sp.

5 Surface not smooth   6

6 Surface reticulately veined, *Rubus chamaemorus*

6 Surface not reticulately veined   7

7 Surface longitudinally rugose and transversely rugulose, base with an obvious caruncle, *Erythronium* spp.

7 Surface rugulose and with prominent resin glands, no caruncle present. *Juniperus* spp.

8 Surface of the seeds hidden by the dense pubescence, *Potentilla fruticosa*

8 Surface not densely pubescent   9

9 Seeds with a persistent style and/or adhering flower parts   10

9 Seeds not as above, ridged or sulcate   11

10 Seeds with a spongy calyx adhering, surface dotted, areolate, and puberulent, *Boehmeria* sp.

10 Seeds without adhering flower parts, style prominent, somewhat asymmetrical, surface smooth and dark-dotted, *Ruppia* sp.

11 Ridge or sulcus showing the folding of the embryo, *Descurania* sp., *Erysimum* spp., *Sisymbrium* spp.

11 Ridge or sulcus not showing the folding of the embryo   12

12 Ridge marginal, surface smooth or reticulate or alveolate, *Rubus* spp.

12 Ridges not strictly marginal   13

13 Ridges on both the margin and surface, very irregular, smooth or areolate, *Potentilla* spp.

13 Ridges and sulci regular, transversely rugulose, *Streptopus roseus*

D / *Seeds elliptic 8-9 in c.s.*

1 Seeds averaging 3.0 mm or more in length   2

1 Seeds averaging less than 3.0 mm in length   7

2 Seeds separating into 2 carpels at maturity, surface ridged, *Cicuta maculata*

2 Seeds remaining intact at maturity   3

3 Surface with resinous glands, *Juniperus* spp.

3 Surface lacking resinous glands   4

4 Seeds often with 2 long, hooked styles, surface glossy, *Polygonum virginianum*

4 Styles not evident   5

5 Surface with a longitudinal ridge and sulcus, glabrous, or with an apical or basal tuft of hairs, *Rosa* spp.

5 Surface not longitudinally ridged   6

6 Surface smooth, undulate, or transversely rugulose, *Asarum* sp., *Asimina* sp.

6 Surface smooth and with white longitudinal nerves, *Ostrya* sp.

7 Seeds with a persistent style and/or adhering flower parts   8

7 Seeds not as above   9

8 Seeds with spongy calyx lobes adhering or with a loose, rugulose, or areolate ovary wall, *Boehmeria* sp., *Pilea* sp., *Urtica* sp.

8 Persistent style evident, ovary wall closely investing the seed, surface smooth and dotted, *Ruppia* sp.

9 Seeds ridged or sulcate   10

9 Seeds not ridged or sulcate, surface reticulate, alveolate, areolate, or puncticulate, *Rubus* spp.

10 Ridges marginal, surface smooth or reticulate, *Myosotis* spp., *Potentilla* spp., *Rubus* spp.

10 Ridges not entirely marginal   11

11 Surface with a single ridge or sulcus showing the folding of the embryo, *Erysimum* spp., *Sisymbrium* spp.
11 Surface with several ridges, marginal or on the surface, surface ridges irregular, *Potentilla* spp.

KEY TO SUB-SERIES 4 & X (seeds ovate 40–41 in l.s.)

A Seeds elliptic 6 in c.s., *Lithospermum* spp., *Onosmodium* sp. *Rosa* spp., *Rubus chamaemorus*
B Seeds elliptic 7 in c.s., page 31
C Seeds elliptic 8–9 in c.s., page 31
D Seeds elliptic 10–11 in c.s., *Carex* spp., *Pilea* sp., *Polygonum* spp.
E Seeds oblong 20–21 or 22–23 in c.s., *Polygonum* spp.
F Seeds ovate 38–39 or 40–41 in c.s., *Lycopsis* sp.
G Seeds obovate 47–48 or 49–50–51 in c.s., *Fragaria* spp., *Lithospermum* spp., *Panax trifolium*
H Seeds triangular 78–79 or 80–81 in c.s., *Polygonum* spp., *Proserpinaca* sp., *Rumex* spp.
I Seeds obtriangular 90–91 in c.s., *Sida* sp.

B / *Seeds elliptic 7 in c.s.*

1 Seeds averaging 3.0 mm or more in length   2
1 Seeds averaging less than 3.0 mm in length   5

2 Seeds separating into 2 carpels at maturity, surface ridged, *Cicuta maculata*
2 Seeds remaining intact at maturity   3

3 Surface with resinous glands, *Juniperus* spp.
3 Surface lacking resinous glands   4

4 Surface longitudinally ridged, slightly granular, *Carpinus* sp.
4 Surface reticulately veined, *Rubus chamaemorus*

5 Seeds with persistent styles and/or persistent flower parts   6
5 Seeds not as above   7

6 Seeds with a persistent, spongy calyx, surface dotted, areolate, and puberulent, *Boehmeria* sp.
6 Calyx absent, style persistent, somewhat asymmetrical in form, surface smooth and dotted, *Ruppia* sp.

7 Seeds with a marginal ridge, surface irregularly ridged, *Potentilla* spp.
7 Margin ridged, surface areolate or reticulate, *Rubus* spp.

C / *Seeds elliptic 8–9 in c.s.*

1 Seeds averaging 3.0 mm or more in length   2
1 Seeds averaging less than 3.0 mm in length   6

2 Seeds forming 2 carpels at maturity, longitudinally ribbed, *Cicuta maculata*, *Conium* sp.
2 Seeds not forming 2 carpels at maturity   3

3 Surface with resinous glands, *Juniperus* spp.
3 Surface lacking resinous glands   4

4 Surface longitudinally ridged, and somewhat granular, *Carpinus* sp.
4 Surface not longitudinally ridged   5

5 Seeds with a red aril, inner surface smooth or slightly rugulose, *Magnolia* sp.
5 Seeds not arillate, apex with two hooked, somewhat persistent styles, surface smooth and glossy, *Polygonum virginianum*

6 Seeds with persistent styles or adhering flower parts   7
6 Seeds not as above   8

7 Seeds with a spongy calyx adhering, or a loose, rugulose ovary wall, *Boehmeria* sp., *Pilea* sp.
7 Style persistent, ovary wall closely investing the seed, somewhat asymmetrical in form, surface dotted and smooth, *Ruppia* sp.

8   Surface ridged or sulcate   9
8   Surface not ridged or sulcate but areolate or alveolate, *Carex* spp.,
    Polygonum spp., *Potentilla* spp.

9   Ridges both marginal and on the surface, surface ridges irregular,
    Potentilla spp.
9   Ridges marginal only   10

10 Surface reticulate or areolate, *Rubus* spp.
10 Surface smooth and glossy, *Myosotis* spp.

Series E (seeds obovate in l.s.)

Sub-series 2: Seeds obovate 45–46 in l.s., **page 32**
Sub-series 3: Seeds obovate 47–48 in l.s., **page 34**
Sub-series 4 & X: Seeds obovate 49–50 in l.s., **page 38**
Sub-series 6: Seeds obovate 52–53 in l.s., **page 40**

KEY TO SUB-SERIES 2 (seeds obovate 45–46 in l.s.)

A   Seeds elliptic 6 in c.s., page 32
B   Seeds elliptic 7 in c.s., page 32
C   Seeds elliptic 8–9 in c.s., page 33
D   Seeds elliptic 10–11 in c.s., *Achillea* spp., *Aster umbellatus*, *Bellis*
    sp., *Clematis* spp., *Geum* spp., *Grindelia* sp.
E   Seeds oblong 20–21 in c.s., *Salicornia* sp., *Chrysanthemum
    balsamita*
F   Seeds oblong 22–23 or 24 in c.s., *Bidens* spp.
G   Seeds rhombic 32–33 or 34–35 in c.s., *Bidens coronata*
H   Seeds triangular 78–79 or 80–81 in c.s., *Cyperus* spp.,
    Eriophorum spp., *Scirpus* spp.
I   Seeds obtriangular 85–86 or 87–88 in c.s., *Grindelia* sp.
J   Seeds obtriangular 90–91 or 92–93 in c.s., *Hyssopus* sp.

A / *Seeds elliptic 6 in c.s.*

1   Seeds averaging 3.0 mm or more in length   2
1   Seeds averaging less than 3.0 mm in length   4

2   Seeds with a long, twisted awn, surface pubescent, *Erodium* sp.
2   Seeds not awned   3

3   Seeds separating into 3 carpels at maturity, *Triglochin palustris*
3   Seeds remaining intact, surface longitudinally ribbed, muricate,
    and strigose, *Oxybaphus* sp.

4   Surface covered with uncinate bristles, *Circaea alpina*
4   Surface not bristly   5

5   Seeds with an apical coma or a remnant style base surrounded
    by a pappus or a pappus rim   6
5   Seeds not as above   7

6   Seeds with an apical coma only, surface glabrous, *Epilobium* spp.
6   Seeds with a style base and a pappus or pappus rim, *Solidago
    graminifolia*

7   Seeds with an obvious white caruncle, surface pubescent and rugu-
    lose or areolate, *Polygala* spp.
7   Seeds lacking an obvious caruncle   8

8   Surface with a single ridge, areolate and glossy, *Mitella* spp.
8   Surface not ridged but reticulate or rugulose, *Caltha* sp., *Gratiola*
    sp., *Lobelia* spp.

B / *Seeds elliptic 7 in c.s.*

1   Seeds averaging 3.0 mm or more in length   2
1   Seeds averaging less than 3.0 mm in length   4

2 Seeds with a long style hooked at the apex, surface pubescent particularly at the apex, *Geum* spp.

2 Seeds with a remnant of a style base surrounded by a pappus or pappus rim   3

3 Seeds with a long beak on which is the pappus, apex of the achene muricate, *Taraxacum* spp.

3 Seeds lacking a beak, pappus at the apex of the achene, *Aster* spp.

4 Seeds with a terminal coma with no trace of a style base, *Epilobium* spp.

4 Seeds with a style base surrounded by a pappus or a pappus rim   5

5 Hilum lateral at the base in a crescent-shaped indentation, surface ribbed or striate, pappus absent or of scales, *Centaurea* spp.

5 Hilum strictly basal   6

6 Surface with 10 white longitudinal ribs, glandular dotted and papillose, pappus none, *Chrysanthemum leucanthemum*

6 Surface not 10-ribbed   7

7 Surface longitudinally ribbed and with a pappus of capillary bristles, *Aster* spp.

7 Surface not ribbed but striate, glabrous, and lacking a pappus, *Artemisia vulgaris*

C / *Seeds elliptic 8-9 in c.s.*

1 Seeds averaging 3.0 mm or more in length   2

1 Seeds averaging less than 3.0 mm in length   11

2 Seeds with a persistent style or a style base surrounded by a pappus or pappus rim   3

2 Seeds without a style, surface rugulose and slightly glossy, *Berberis* sp.

3 Seeds with a style only   4

3 Seeds with the style base and a pappus or a pappus rim   5

4 Style long, flexuous, and plumose, surface pubescent, *Clematis* spp.

4 Style shorter, stiff, hooked at the apex, surface pubescent particularly at the apex, *Geum* spp.

5 Seeds short- or long-beaked, pappus on the apex of the beak, surface longitudinally ribbed and transversely rugulose or muricate, *Lactuca serriola, Taraxacum* spp.

5 Seeds not beaked, pappus on the apex of the achene   6

6 Surface striate   7

6 Surface definitely ribbed   8

7 Surface with transverse markings, pappus of barbed bristles, *Carduus nutans*

7 Surface not transversely marked, pappus of plumose bristles, *Cirsium* spp.

8 Achenes with numerous ribs   9

8 Achenes with few ribs   10

9 Surface smooth, pappus none, *Lapsana* sp.

9 Surface finely transversely rugulose, pappus of bristles, *Sonchus* spp.

10 Surface rugulose, mottled, pappus of deciduous scales, *Arctium* spp.

10 Surface smooth, glabrous or pubescent, pappus of bristles, *Aster* spp.

11 Seeds with a remnant style base surrounded by a pappus or a pappus rim   12

11 Seeds not as above   14

12 Seeds lacking a pappus or the pappus of deciduous scales, surface ridged or striate, *Artemisia* spp., *Chrysanthemum leucanthemum*, *Centaurea* spp., *Grindelia* sp., *Matricaria matricarioides*
12 Seeds with a distinct pappus   13

13 Pappus of stiff or bristle-tipped scales, *Cichorium* sp., *Helenium* spp.
13 Pappus of capillary bristles, *Aster* spp., *Sonchus* spp.

14 Seeds with a coma, surface striate, *Epilobium* spp.
14 Seeds lacking a coma, surface rugulose and with a longitudinal sulcus, *Collomia* sp.

KEY TO SUB-SERIES 3 (seeds obovate 47–48 in l.s.)

A / *Seeds elliptic 6 in c.s.*

1 Seeds averaging 3.0 mm or more in length   2
1 Seeds averaging more than 3.0 mm in length   7

2 Seeds with apical, broad-based spines, *Ambrosia* spp.
2 Seeds without apical spines   3

3 Seeds with an aril or caruncle, inner surface areolate, *Nymphaea tuberosa*
3 Seeds without an aril or obvious caruncle   4

4 Surfaces ridged or sulcate   5
4 Surfaces not ridged or sulcate   6

5 Surface rugulose, *Crataegus* spp., *Peltandra* sp.
5 Surface muricate and strigose, *Oxybaphus* sp.

6 Surface papillose, *Convolvulus arvensis*
6 Surface smooth, *Nuphar variegatum*, *Peltandra* sp. (fresh specimens)

7 Seeds with a remnant style base surrounded by a pappus or a pappus rim, *Anthemis cotula*, *Solidago graminifolia*
7 Seeds not as above   8

8 Surfaces covered with uncinate bristles, *Circaea alpina*
8 Surfaces not bristly   9

9 Seeds with an obvious caruncle, *Acalypha* sp., *Euphorbia* spp., *Polygala* spp., *Viola* sp.
9 Seeds without an obvious caruncle   10

10 Surfaces ridged or sulcate   11
10 Surfaces not ridged or sulcate   13

11 Ridge or sulcus showing the folding of the embryo, *Lepidium* spp.
11 Ridge or sulcus not showing the folding of the embryo   12

12 Surface with a single marginal ridge and areolate, *Isopyrum* sp., *Mitella* spp.
12 Surface with several ridges and areolate or alveolate, *Sanguisorba canadensis*, *Sedum* spp.

13 Surface pubescent, *Waldsteinia* sp.
13 Surface reticulate or rugulose, *Caltha* sp., *Lobelia* spp., *Sisyrinchium* spp.

B / *Seeds elliptic 7 in c.s.*

1 Seeds averaging 3.0 mm or more in length   2
1 Seeds averaging less than 3.0 mm in length   7

2 Surface covered with uncinate bristles, *Circaea quadri sulcata*
2 Surface without uncinate bristles   3

3 Surface strongly muricate, *Floerkea* sp.
3 Surface not muricate   4

4 Seed with a remnant style base surrounded by a pappus or a pappus rim, *Aster* spp., *Helianthus divaricatus*, *Polymnia* sp.
4 Seeds not as above   5

5 Seeds with a persistent style, margin ridged or winged, surface rugulose, *Potamogeton* spp.
5 Seeds not as above   6

6 Surface papillose, *Convolvulus* spp.
6 Surface smooth, inner face with a large elliptical depression at the base, *Galeopsis* sp.

7 Seeds with a remnant style base surrounded by a pappus or a pappus rim, *Artemisia* spp., *Aster* spp., *Centaurea* spp., *Chrysanthemum leucanthemum*
7 Seeds not as above   8

8 Seeds with a persistent style or more or less conical style base or tubercle   9
8 Seeds without a style   10

9 Margins ridged or winged, surface rugulose or areolate, *Potamogeton* spp., *Ranunculus* spp.
9 Margins not ridged, achenes subtended by a series of barbed bristles, *Eleocharis* spp.

10 Seeds with an obvious caruncle, surface pubescent, *Polygala* spp.
10 Seeds without an obvious caruncle   11

11 Seeds with a terminal coma, surface striate, *Epilobium* spp.
11 Seeds lacking a coma   12

12 Seeds with a marginal ridge, surface rugulose or areolate, *Isopyrum* sp.
12 Marginal ridge not present   13

13 Surface coarsely reticulate or alveolate, *Trichostema* sp.
13 Surface areolate or smooth, *Ceanothus* spp., *Coptis* sp., *Stachys* spp.

C / *Seeds elliptic 8–9 in c.s.*

1 Seeds averaging 3.0 mm or more in length   2
1 Seeds averaging less than 3.0 mm in length   12

2 Surface covered with uncinate bristles, *Circaea quadrisulcata*
2 Surface not bristly   3

3 Seeds with a prominent, persistent style   4
3 Styles not prominent   7

4 Styles as long as or longer than the achenes, slender   5
4 Styles shorter than the achenes, usually broad-based   6

5 Styles long, flexuous, and plumose, surface pubescent, *Clematis* spp.

5 Styles shorter, rigid, and hooked at the apex, surface hirsute particularly at the apex, *Geum* spp.

6 Surface covered with a dense, woolly pubescence, *Anemone* spp.

6 Surface not woolly pubescent, margins ridged or winged, surface areolate or rugulose, *Potamogeton* spp., *Ranunculus* spp.

7 Seeds with a remnant style base surrounded by a pappus or a pappus rim   8

7 Seeds not as above   9

8 Pappus of capillary bristles, plumose or barbellate, *Carduus nutans*, *Cirsium* spp., *Onopordum* sp.

8 Pappus of deciduous scales or lacking, *Arctium* spp., *Helianthus divaricatus*, *Polymnia* sp.

9 Seeds with a ridge or sulcus, surface smooth, *Rhamnus* spp.

9 Surface not ridged or sulcate   10

10 Surface strongly muricate, *Floerkea* sp.

10 Surface not muricate   11

11 Surface rugulose, *Berberis* sp., *Podophyllum* sp.

11 Surface areolate or smooth, *Cercis* sp., *Galeopsis* sp., *Pyrus* sp., *Sicyos* sp.

12 Seeds with a remnant of a style base surrounded by a pappus or a pappus rim   13

12 Seeds not as above   17

13 Pappus of capillary bristles, *Aster* spp.

13 Pappus not of capillary bristles   14

14 Pappus of awns or scales, surface ribbed or striate, glabrous or strigose, glandular or eglandular, *Cichorium* sp., *Helenium* spp.

14 Seeds without a pappus   15

15 Hilum lateral in a basal, crescent-shaped cavity, *Centaurea* spp.

15 Hilum basal   16

16 Surface with 10 longitudinal, white ribs, papillose, *Chrysanthemum leucanthemum*

16 Surface not strongly ribbed, striate, *Artemisia* spp., *Grindelia* sp., *Iva* sp., *Ratibida* sp.

17 Seeds ridged or sulcate   18

17 Seeds not ridged or sulcate   23

18 Seeds with a persistent style, margins ridged or winged, surface rugulose or areolate, *Potamogeton* spp., *Ranunculus* spp.

18 Seeds not as above   19

19 Surface transversely ridged, outer coating aril-like and transparent, *Oxalis stricta*

19 Surface longitudinally ridged or sulcate   20

20 Surface with several ridges or sulci, nodulose and areolate, *Oxalis acetosella*

20 Surface with a single ridge or sulcus   21

21 Ridge or sulcus showing the folding of the embryo, *Cardaria* sp., *Lepidium* spp.

21 Ridge or sulcus not showing the folding of the embryo   22

22 Surface rugulose, *Collomia* sp., *Phlox* spp.

22 Surface smooth or almost so, *Ceanothus* spp.

23 Seeds with a caruncle, surface pubescent and areolate, *Polygala* spp.

23 Seeds without an obvious caruncle   24

24 Seeds subtended by retrorsely barbed bristles   25

24 Seeds lacking the barbed bristles   26

25 Seeds with a prominent style base or tubercle, *Eleocharis* spp., *Rhynchospora* spp.

25 Seeds without a prominent tubercle, merely a peg-like style base, *Scirpus* spp.

26 Seeds with a coma   27
26 Seeds lacking a coma   28

27 Coma apical on the seed, *Epilobium* spp.
27 Coma basal, *Populus* spp.

28 Surface papillose, *Iva* sp., *Spergularia* sp.
28 Surface not papillose   29

29 Surface reticulate or alveolate, *Trichostema* sp., *Vaccinium* spp.
29 Surface not reticulate   30

30 Surface rugulose, *Sambucus* spp., *Stachys* spp., *Veronica* spp.
30 Surface areolate or smooth, *Iva* sp., *Nuphar microphyllum*,
   Polygonum spp., *Stachys* spp., *Vaccinium* spp., *Veronica* spp.

D / *Seeds elliptic 10–11 in c.s.*

1 Seeds averaging 3.0 mm or more in length   2
1 Seeds averaging less than 3.0 mm in length   6

2 Seeds separating into 2 carpels at maturity, margin ridged or
  winged, surface ridged, *Heracleum* sp.
2 Seeds remaining intact at maturity   3

3 Seeds with persistent styles   4
3 Seeds without persistent styles, surface rugulose, undulate, or
  ridged, *Viburnum opulus*

4 Surface densely woolly pubescent, *Anemone* spp.
4 Surface not woolly pubescent   5

5 Style long, flexuous, and plumose, or rigid and hooked at the
  apex, *Clematis* spp., *Geum* spp.
5 Style short, thick-based, surface rugulose and ridged,
  Ranunculus spp.

6 Seeds with a remnant style base surrounded by a pappus or a
  pappus rim, *Achillea* spp., *Aster* spp., *Bellis* sp., *Grindelia* sp.
6 Seeds not as above, surface ridged or sulcate   7

7 Surface transversely ridged, covered with a transparent aril-like
  coating, *Oxalis stricta*
7 Surface longitudinally ridged or sulcate   8

8 Surface with several ridges and sulci, nodulose and areolate, *Oxalis
  acetosella*
8 Surface not as above   9

9 Surface with a single ridge or sulcus showing the folding of the
  embryo, *Lepidium* spp.
9 Ridges marginal, style persistent, surface rugulose or areolate,
  Ranunculus spp.

G / *Seeds oblong 20–21 in c.s.*

1 Seeds averaging 3.0 mm or more in length   2
1 Seeds averaging less than 3.0 mm in length   4

2 Seeds with a persistent style, dorsal margin ridged or winged,
  surface rugulose, *Potamogeton* spp.
2 Seeds not as above   3

3 Surface coarsely nodulose and alveolate, *Proboscidia* sp.
3 Surface smooth or rugulose, *Rhamnus* spp.

4 Seeds with a persistent style, dorsal margin ridged, sulcate, or
  winged, *Potamogeton* spp., *Alisma* spp.
4 Seeds not as above   5

5 Seeds with a ridge or sulcus showing the folding of the embryo,
  surface with more or less circular, papillose ridges, *Thlaspi* sp.
5 Seeds not ridged or sulcate   6

6 Surface papillose, *Cerastium arvense*
6 Surface glabrous or pubescent, *Kochia* sp., *Salicornia* sp.

P / *Seeds obtriangular 90–91 in c.s.*

1	Seeds averaging 3.0 mm or more in length	2
1	Seeds averaging less than 3.0 mm in length	4

2 Surface papillose and rugulose, *Convolvulus* spp.

2 Surface not papillose   3

3 Surface areolate and rugulose, angles prominently ridged, *Actaea* spp.

3 Surface smooth, angles rounded, large elliptical depression at the base on the inner surface, *Galeopsis* sp.

4 Surface coarsely reticulate or alveolate, *Trichostema* sp.

4 Surface areolate or smooth   5

5 Inner face with 2 depressions near the hilum, *Hyssopus* sp., *Lamium* spp., *Satureja* spp.

5 Inner face without evident depressions, *Marrubium* sp., *Melissa* sp., *Prunella* sp., *Stachys* spp.

Q / *Seeds obtriangular 92–93 in c.s.*

1 Seeds averaging 3.0 mm or more in length, surface smooth, inner surface with an elliptical depression near the hilum, *Galeopsis* sp.

1 Seeds averaging less than 3.0 mm in length   2

2 Seeds with a remnant style base and a toothed pappus or pappus rim, angles with whitish margins, *Matricaria maritima*

2 Seeds not as above   3

3 Surface reticulate or alveolate, *Isanthus* sp., *Trichostema* sp.

3 Surface smooth or areolate   4

4 Inner surface with 2 depressions near the hilum, *Hyssopus* sp., *Lamium* spp., *Satureja* spp.

4 Inner surface without obvious depressions, *Lycopus* spp., *Marrubium* sp., *Melissa* sp., *Prunella* sp., *Stachys* spp.

KEY TO SUB-SERIES 4 & X (seeds obovate 49–50 in l.s.)

A Seeds elliptic 6 in c.s., page 38

B Seeds elliptic 7 in c.s., page 39

C Seeds elliptic 8–9 in c.s., page 39

D Seeds elliptic 10–11 in c.s., page 40

E Seeds oblong 18 in c.s., *Euphorbia* spp.

F Seeds oblong 19 in c.s., *Potamogeton* spp.

G Seeds oblong 20–21 or 22–23 in c.s., *Alisma* spp., *Cerastium* spp., *Kochia* sp., *Potamogeton* spp., *Rhamnus* spp.

H Seeds obovate 47–48 or 49–50 in c.s., *Potamogeton* spp.

I Seeds obovate 50–51 or 52–53 in c.s., *Vitis* sp.

J Seeds triangular 78–79 in c.s., *Bulbostylis* sp., *Carex* spp., *Fimbristylis* sp.

K Seeds obtriangular 87–88 or 89–90 in c.s., *Parthenocissus* sp.

L Seeds obtriangular 90–91 in c.s., page 40

M Seeds obtriangular 92–93 in c.s., page 40

A / *Seeds elliptic 6 in c.s.*

1	Seeds averaging 3.0 mm or more in length	2
1	Seeds averaging less than 3.0 mm in length	5

2 Seeds with an aril or a caruncle, *Nymphaea tuberosa*

2 Seeds without an obvious aril or caruncle   3

3 Seeds more or less reniform, surface papillose or scurfy, *Hibiscus palustris*

3 Seeds not reniform   4

4 Surface with a ridge or sulcus, rugulose, *Crataegus* spp.

4 Surface smooth, *Convolvulus spithamaeus*, *Nuphar variegatum*

5 Seeds with an obvious caruncle, *Acalypha* sp., *Euphorbia* spp., *Viola* spp.

5 Seeds without an obvious caruncle, surface smooth or slightly roughened, *Brassica* spp., *Cuscuta* sp., *Sisyrinchium* spp.

B / *Seeds elliptic 7 in c.s.*

1 Seeds averaging 3.0 mm or more in length   2
1 Seeds averaging less than 3.0 mm in length   6

2 Seeds with a persistent style, margin ridged or slightly winged, surface rugulose, *Potamogeton* spp.
2 Seeds not as above   3

3 Surface muricate or papillose   4
3 Surface not muricate or papillose   5

4 Surface coarsely muricate, *Floerkea* sp.
4 Surface papillose, *Agrostemma* sp.

5 Surface rugulose, ridged, or sulcate, *Viburnum* spp.
5 Surface smooth, *Convolvulus* spp., *Galeopsis* sp., *Staphylea* sp.

6 Seeds with a persistent style or somewhat conical style base (tubercle)   7
6 Seeds not as above   8

7 Margin of the achene ridged or winged, surface rugulose, *Potamogeton* spp.
7 Margin not ridged, achene with a number of barbed bristles at the base, *Eleocharis* spp.

8 Seeds with an obvious caruncle, surface pubescent and areolate or reticulate, *Polygala* spp.
8 Seeds without an obvious caruncle   9

9 Surface coarsely reticulate or alveolate, *Teucrium* sp., *Trichostema* sp.
9 Surface smooth or areolate, *Ceanothus* spp., *Cuscuta* sp., *Stachys* spp.

C / *Seeds elliptic 8-9 in c.s.*

1 Seeds averaging 3.0 mm or more in length   2
1 Seeds averaging less than 3.0 mm in length   6

2 Seeds with a persistent style, margin ridged or slightly winged, *Potamogeton* spp.
2 Seeds not as above   3

3 Surface muricate or papillose   4
3 Surface not muricate or papillose   5

4 Surface coarsely muricate, *Floerkea* sp.
4 Surface papillose, *Agrostemma* sp.

5 Surface rugulose, ridged, or sulcate, *Rhamnus* spp.
5 Surface smooth or areolate, *Cercis* sp., *Galeopsis* sp., *Rhamnus* spp., *Sicyos* sp., *Staphylea* sp.

6 Seeds with a persistent style or style base (tubercle)   7
6 Seeds not as above   10

7 Seeds with a marginal ridge or wing, surface rugulose, smooth, or areolate, *Potamogeton* spp., *Ranunculus* spp.
7 Seeds not marginally ridged   8

8 Style base or tubercle large and more or less conical, base of the achene with barbed bristles, *Eleocharis* spp.
8 Style base very small or peg-like   9

9 Achenes subtended by barbed bristles, *Scirpus* spp.
9 Bristles not present at the base of the achene, *Iva* sp.

10 Surface striate, ridged, or sulcate   11
10 Surface not as above   14

11 Surface with a single ridge or sulcus   12
11 Surface with several ridges or sulci   13

12 Ridge or sulcus showing the folding of the embryo, *Cardaria* sp., *Lepidium* spp.
12 Ridge marginal and encircling the seed, *Amaranthus* spp.

13 Ridges on the surface transverse, outer covering aril-like and transparent, *Oxalis stricta*
13 Ridges longitudinal, surface nodulose and areolate, *Oxalis acetosella*

14 Seeds with an obvious caruncle, surface pubescent and areolate or reticulate, *Polygala* spp.
14 Seeds without an obvious caruncle   15

15 Surface papillose, *Spergularia* sp.
15 Surface not papillose   16

16 Surface rugulose, reticulate, or alveolate, *Teucrium* sp., *Trichostema* sp., *Vaccinium* spp., *Veronica* spp.
16 Surface areolate or smooth, *Ceanothus* spp., *Chenopodium album*, *Nuphar microphyllum*, *Polygonum* spp., *Stachys* spp., *Vaccinium* spp., *Veronica* spp.

D / *Seeds elliptic 10–11 in c.s.*

1 Seeds averaging 3.0 mm or more in length   2
1 Seeds averaging less than 3.0 mm in length   3

2 Seeds separating into 2 carpels at maturity, margins ridged or winged, surface ridged, *Heracleum* sp.
2 Seeds remaining intact at maturity, surface rugulose or areolate, *Ranunculus* spp.

3 Seeds with a persistent style, margin ridged, surface smooth or areolate, *Ranunculus* spp.
3 Seeds not as above, surface ridged or sulcate   4

4 Surface with a single ridge or sulcus showing the folding of the embryo, *Lepidium* spp.
4 Surface with several ridges or sulci   5

5 Ridges transverse, surface with a thin, transparent aril-like coating, *Oxalis stricta*
5 Ridges longitudinal, nodulose, and areolate, *Oxalis acetosella*

L / *Seeds obtriangular 90–91 in c.s.*

1 Seeds coarsely reticulate or alveolate, *Trichostema* sp.
1 Surface not reticulate or alveolate   2

2 Surface papillose, *Convolvulus* spp.
2 Surface not papillose, smooth or areolate, *Decodon* sp., *Galeopsis* sp., *Stachys* spp.

M / *Seeds obtriangular 92–93 in c.s.*

1 Surface coarsely reticulate or alveolate, *Isanthus* sp., *Trichostema* sp.
1 Surface areolate or smooth, *Decodon* sp., *Galeopsis* sp., *Lycopus* spp., *Stachys* spp.

KEY TO SUB-SERIES 6 (seeds obovate 52–53 in l.s.)

A Seeds elliptic 6 or 7 in c.s., *Trientalis* sp.

Series I (seeds obtriangular in l.s.)

Sub-series 1 & 2: Seeds obtriangular 84–85–86 in l.s., page 41
Sub-series 3: Seeds obtriangular 87–88 in l.s., page 41
Sub-series 4 & X: Seeds obtriangular 89–90 in l.s., page 41
Sub-series X & 5 or 6: Seeds obtriangular 90–91 or 92–93 in l.s., page 41

KEY TO SUB-SERIES 1 & 2 (seeds obtriangular 84–86 in l.s.)

A Seeds elliptic 6 in c.s., *Anthemis* spp., *Erodium* sp.
B Seeds elliptic 7 in c.s., *Galinsoga* sp., *Lythrum salicaria*
C Seeds elliptic 8–9 or 10–11 in c.s., *Bidens cernua, Lythrum salicaria, Platanus* sp.
D Seeds oblong 18 in c.s., *Anthemis arvensis, Chrysanthemum parthenium*
E Seeds oblong 20–21 in c.s., *Chrysanthemum balsamita*
F Seeds rhombic 30 in c.s., *Tanacetum* sp.
G Seeds rhombic 31 or 32–33 in c.s., *Anthemis tinctoria, Cephalanthus* sp., *Liatris* spp.
H Seeds obtriangular 90–91 in c.s., *Galinsoga* sp.

KEY TO SUB-SERIES 3 (seeds obtriangular 87–88 in l.s.)

A Seeds elliptic 6 or 7 in c.s., *Anthemis* spp., *Galinsoga* sp., *Lythrum salicaria*
B Seeds elliptic 8–9 or 10–11 in c.s., *Bidens cernua, Lythrum salicaria*
C Seeds oblong 18 in c.s., *Anthemis arvensis*
D Seeds rhombic 32–33 in c.s., *Anthemis tinctoria*
E Seeds obtriangular 90–91 or 92–93 in c.s., *Galinsoga* sp., *Leonurus* sp.

KEY TO SUB-SERIES 4 & X (seeds obtriangular 89–90 in l.s.)

A Seeds elliptic 6 in c.s., *Agrimonia* spp.
B Seeds obtriangular 90–91 or 92–93 in c.s., *Decodon* sp.

KEY TO SUB-SERIES X & 5 OR 6 (seeds obtriangular 90–91 or 92–93 in l.s.)

A Seeds elliptic 6 in c.s., *Salsola* sp.

Key to Series X (seeds obviously winged)

1 Ovaries 2 or more, or occasionally one by abortion 2
1 Ovaries one or apparently so 3

2 Seeds splitting longitudinally into 2 carpels at maturity, style and stylopodium usually persistent, surface longitudinally ribbed, *Anethum* sp., *Heracleum* sp., *Pastinaca* sp.
2 Seeds separating transversely into 2 or more carpels or occasionally remaining intact, wings originating from the ovary or from modified bracts, *Acer* spp., *Corylus* spp.

3 Wings obviously terminal or sub-terminal on the seed 4
3 Wings not as above 7

4 Wings lobed 5
4 Wings not lobed 6

5 Wing a 3-lobed bract more or less subtending the ovary, *Carpinus* sp.
5 Wing 2-lobed and strictly terminal, membranous or coriaceous, surface strigose, *Axyris* sp.

6 Wing coriaceous and more or less linear, *Fraxinus* spp., *Liriodendron* sp.
6 Wing membranous, short, more or less oval or elliptical, *Abies* sp., *Larix* sp., *Liquidambar* sp., *Picea* spp., *Pinus* spp., *Tsuga* sp.

7 Wing lateral on the ovary or ovule 8
7 Wing circling the ovary or ovule 11

8 Wing on only one side of the ovary or ovule 9
8 Wing on two sides of the ovary or ovule 10

9 Seeds with a persistent style, surface areolate, *Sagittaria* spp.
9 Seeds without a style, surface areolate or reticulate, *Chamaelirium* sp., *Pedicularis lanceolata, Rhinanthus* sp.

10 Seeds with 2 persistent styles, wings usually membranous or chartaceous, *Alnus* spp., *Betula* spp.

10 Seeds with narrow wings, coriaceous or spongy, style base persistent and usually a pappus rim, *Achillea* spp., *Bellis* sp., *Coreopsis* sp., *Silphium* sp.

11 Seeds with a persistent style or some flower parts   12
11 Seeds without persistent styles or flower parts   13

12 Wings derived from the calyx lobes, *Cycloloma* sp., *Salsola* sp.
12 Wing derived from the ovary wall, style usually persistent, *Anemone canadensis*, *Corispermum* sp., *Oxyria* sp., *Potamogeton* spp., *Ptelea* sp., *Ulmus* spp.

13 Seeds averaging 3.0 mm or more in length   14
13 Seeds averaging less than 3.0 mm in length   17

14 Seeds irregular in l.s. and c.s., irregularly ridged, *Linaria dalmatica*, *Veratrum* sp.
14 Seeds more regular in form   15

15 Seeds elliptic in l.s., *Chelone* sp., *Dioscorea* sp., *Swertia* sp., *Thuja* sp.
15 Seeds obovate in l.s.   16

16 Wings narrow and thin, *Asclepias* spp., *Cynanchum* sp.
16 Wings broad and spongy, *Dioscorea* sp., *Lilium* spp.

17 Wing very narrow, more or less ridge-like, not membranous or spongy, *Alyssum* sp., *Callitriche* sp., *Campanula* spp., *Cimicifuga* sp., *Lepidium virginicum*, *Spergula* sp., *Triodanis* sp.
17 Wing comparatively broad, membranous or spongy   18

18 Surface with resinous glands, *Thuja* sp.
18 No resinous glands present   19

19 Wing broadest at the seed apex, surface with a sulcus, *Arabis* spp.
19 Seeds not as above   20

20 Seed body papillose, *Linaria vulgaris*
20 Seed surface not papillose   21

21 Wing lacerate, *Rhododendron* sp.
21 Wing entire, surface striate or reticulate, *Gentiana andrewsii*, *Parnassia* spp.

Key to Series Y (the Gramineae)

1 Spikelets strongly flattened   2
1 Spikelets not strongly flattened   7

2 Spikelets almost circular in outline, transversely rugose, *Beckmannia* sp.
2 Spikelets not circular   3

3 Spikelets without a palea, margins of the glumes fused at the base and the keels ciliate, lemmas awned, *Alopecurus* spp.
3 Spikelets with a palea   4

4 Keels of the lemmas glabrous, apex bifid, *Spartina* sp.
4 Keels of the lemmas ciliate or hispid   5

5 Lemmas sharp-pointed or short-awned, *Dactylis* sp.
5 Lemmas not sharp-pointed   6

6 Keel of the lemma hispid, palea small and narrow, glumes absent, *Leersia* spp.
6 Keel of the lemma glabrous, lemma and palea hyaline, keels of the glumes stiff-ciliate and the apex awned, *Phleum* sp.

7 Inflorescence or spikelets with unisexual flowers, or both perfect and staminate or neuter florets   8
7 Florets of an inflorescence or spikelet all bisexual; sterile florets present as awns in some species   16

8 Inflorescence and spikelets either staminate or pistillate, glumes absent, fertile lemma with 3 prominent ribs, *Zizania* sp.
8 Inflorescence and spikelets not as above   9

9 Spikelets in pairs with one perfect spikelet and one sessile or pedicelled staminate spikelet, or merely the pedicel, *Andropogon* spp., *Sorghastrum* sp.
9 Spikelets each containing one perfect floret and one or more staminate or sterile florets   10

10 Spikelets each with a perfect floret and one or two staminate florets, *Anthoxanthum* sp., *Arrhenatherum* sp., *Hierochloe* sp.
10 Spikelets containing a perfect floret and one or more neuter or sterile florets   11

11 Sterile florets 2 and represented by hairy bracts at the base of the fertile floret, *Phalaris* sp.
11 Sterile florets represented by a sterile lemma, fertile lemma and palea usually indurate   12

12 Surface with long, stout, sharp spines, *Cenchrus* sp.
12 Surface not spiny   13

13 Surface of the glumes and sterile lemma with short, stiff hairs often with pustulate bases, *Echinochloa* spp.
13 Surface without stiff pubescence   14

14 Fertile lemmas transversely rugulose or reticulate, *Setaria* spp.
14 Lemmas not rugulose   15

15 Fertile lemma coriaceous, margins of the sterile lemmas hyaline or translucent, glabrous or pubescent, *Digitaria* spp.
15 Fertile lemma and palea chartaceous or indurate with the lemma overlapping the palea, *Panicum* spp.

16 Grains remaining enclosed in the persistent lemmas or palea or both   17
16 Grains not permanently enclosed in the lemmas and paleas   21

17 Awns long, twisted, and geniculate, *Avena* sp., *Stipa* sp.
17 Awns straight or merely flexuous   18

18 Awns originating just below the bifid apex of the lemmas, *Bromus* spp., *Schizachne* sp.
18 Awns terminating the lemmas   19

19 Awns long and slender, almost capillary, usually accompanied by the awns of sterile spikelets, *Hordeum* sp.
19 Awns stouter and more rigid   20

20 Lemmas glabrous, soft-pubescent or hispid, *Agropyron* spp., *Brachyelytrum* sp., *Elymus* spp., *Hystrix* sp.
20 Lemmas soft-pubescent, surface usually glossy, base with a well-developed callus which is pubescent or glabrous, *Oryzopsis* spp.

21 Lemmas awned   22
21 Lemmas not awned   26

22 Awn arising from near the middle or the base of the lemma   23
22 Awn terminal or sub-terminal on the lemma   24

23 Awn hair-like or capillary, scarcely distinguishable from the basal hairs and about as long as the lemmas, *Calamagrostis* spp.
23 Awns stouter, as long as or longer than the lemma, straight or geniculate, *Deschampsia* spp.

24 Awn terminal on the lemma, keel of the lemma hispid, *Dactylis* sp.
24 Awn originating from between or just behind the bifid apex of the lemma   25

25 Awn very short, about 1.0 mm long, lemma about 4.0 mm long, *Cinna* spp.
25 Awn long, flexuous or geniculate, *Danthonia* sp., *Schizachne* sp.

26 Lemma with the nerves parallel and not arching towards the apex, *Glyceria* spp.
26 Lemma nerves not parallel, sometimes very faint or absent   27

27 Lemma more or less keeled or acute on the back   28
27 Lemmas rounded or obtuse on the back   30

28 Keels ciliate or hispid particularly at the apex, also sometimes glandular, *Eragrostis* spp.
28 Keels not ciliate or hispid   29

29 Lemma 3–5-nerved and the nerves close together forming a keel, *Eleusine* sp.
29 Lemma 1-nerved, keel scabrous or hispid, *Sporobolus* sp., *Koeleria* sp.

30 Lemma nerveless, rather coriaceous, margins inrolled and clasping the palea, *Milium* sp.
30 Lemma nerved or obscurely nerved   31

31 Lemma distinctly nerved and with a long, tapering apex; long, silky-pubescent on the short remnant of the rachis attached to the floret, *Phragmites* sp.
31 Lemma nerves indistinct   32

32 Callus at the base of the floret with a tuft of hairs, *Ammophila* sp., *Calamovilfa* sp., *Muhlenbergia* sp.
32 Base of the floret without a tuft of hairs   33

33 Surface of the lemma inconspicuously but densely papillose, dorsal nerve short-hispid, *Sphenopholis* sp.
33 Surface of the lemma inconspicuously hispid, dorsal nerve scabrous, *Koeleria* sp.

Key to Series Z (seeds very irregular in form)

1 Seeds nut-like, i.e., consisting of a fruit with a hard outer wall and partially or completely enclosed in an involucre or husk, *Carya* spp., *Corylus* spp., *Fagus* sp., *Juglans* spp., *Quercus* spp.
1 Seeds not as above   2

2 Seeds with a spongy or net-like reticulum surrounding or extending from the embryo, *Aureolaria* sp., *Buchnera* sp., *Castilleja* sp., *Chimaphila* spp., *Drosera rotundifolia*, *Gerardia* spp., *Ledum* sp., *Moneses* sp., *Monotropa* spp., *Parnassia* spp., *Pterospora* sp., *Pyrola* spp., and species of the family Orchidaceae
2 Seeds not as above   3

3 Seeds averaging 3.0 mm or more in length   4
3 Seeds averaging less than 3.0 mm in length   12

4 Seeds with a number of stout apical spines surrounding a short, stout style, *Ambrosia* spp.
4 Apex of the seeds not spiny   5

5 Seeds with a persistent style and ovary wall   6
5 Seeds lacking a persistent style   9

6 Seeds obtriangular in l.s. with a persistent style, longitudinally ridged and glossy, *Sparganium eurycarpum*
6 Seeds not obtriangular in l.s.   7

7 Seeds ovate in l.s., 3-carpelled and 3-angled in c.s., surface reticulate or alveolate, *Panax trifolium*
7 Seeds either flattened or more or less cylindrical in form   8

8 Seed segments flattened (loments), surface with uncinate, glandular pubescence or glabrous, reticulately veined and indehiscent, *Desmodium* spp.
8 Seeds cylindrical pods, longitudinally ridged and with one or more transverse constrictions, *Cakile* sp., *Raphanus* sp.

9 Seeds more or less linear in l.s., longitudinally ridged and sulcate, *Clintonia* sp., *Zygadenus* sp.
9 Seeds more or less rounded in l.s., rounded or flattened in c.s.   10

10 Seeds flattened in c.s., *Actaea* spp., *Iris* spp.
10 Seeds more or less rounded in c.s.   11

11 Seeds papillose, *Agrostemma* sp.
11 Seeds longitudinally ridged or finely ribbed, *Alliaria* sp.,
Erythronium spp., *Peltandra* sp.

12 Seeds peltate, the hilum on the more or less flattened base,
Anagallis sp., *Glaux* sp., *Lysimachia* spp., *Plantago* spp., *Primula*
sp., *Samolus* sp., *Trientalis* sp.
12 Seeds not peltate   13

13 Surface of the seeds papillose or verrucose, *Gentiana procera*,
Helianthemum spp., *Hypoxis* spp., *Saxifraga* sp., *Stellaria* spp.,
Lechea spp., *Rhexia* sp.
13 Seeds not papillose   14

14 Seeds with a longitudinal ridge or sulcus showing the folding of
the embryo, *Coronopus* spp., *Descurania* sp., *Erysimum* spp.,
Nasturtium sp., *Raphanus* sp., *Rorippa* spp., *Sisymbrium* spp.
14 Seeds not as above   15

15 Seeds more or less linear in c.s., longitudinally ridged and
reticulate, *Allium schoenoprasum*, *Linaria dalmatica*, *Spiraea* spp.
15 Seeds elliptic or variously angled in c.s.   16

16 Surface of the seeds reticulate, *Conopholis* sp., *Epifagus* sp.,
Houstonia spp., *Liquidambar* sp. (sterile seeds), *Phyllodoce* sp.,
Oenothera spp., *Orobanche* sp., *Penstemon* spp., *Scrophularia*
lanceolata, *Utricularia* spp.
16 Surface rugulose or areolate, *Aquilegia* sp., *Arctostaphylos* sp.,
Chamaedaphne sp., *Cimicifuga* sp., *Gaultheria procumbens*,
Oenothera spp., *Ribes* spp., *Typha* spp.

Description of seeds and fruits

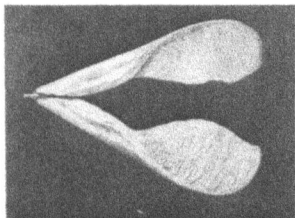
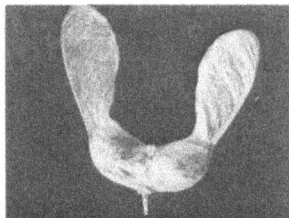
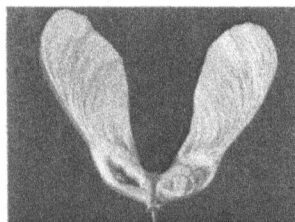

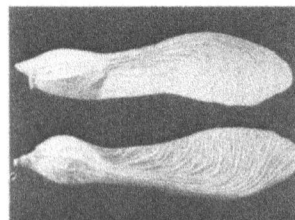
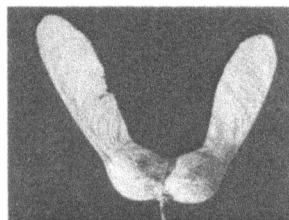

The fruit of the Aceraceae is derived from a single, 2-locular ovary which at maturity forms 2 separate carpels, but by abortion, only one carpel may reach maturity. Each carpel develops a dorsal wing and forms a samara.

1 **Acer negundo** L. (x1) Box Elder. Samaras 37.0 x 10.0 mm, wings at an angle of about 150° with the pedicel, body ridged or sulcate and the wings reticulately veined.

2 **Acer nigrum** Michx. f. (x0.8) Black Maple. Samaras 25.0 x 7.0 mm, wings almost vertically parallel and at an angle of about 180° with the pedicel, reticulately veined, body plump and reticulately veined.

3 **Acer pensylvanicum** L. (x1) Striped Maple. Samaras 27.0 x 11.0 mm, wings at an angle of about 145° with the pedicel, conspicuously veined.

4 **Acer rubrum** L. (x1) Red Maple. Samaras 24.0 x 8.0 mm, maturing about the time that the leaves are fully expanded, wings at an angle of about 160° with the pedicel, body, and wings reticulately veined.

5 **Acer saccharinum** L. (x0.8) Silver Maple. Samaras 43.0 x 10.0 mm, maturing in the early summer, wings at an angle of about 150° with the pedicel, one samara of the pair often aborted, fruit body plump and wings reticulately veined.

6 **Acer saccharum** Marsh. (x0.8) Sugar Maple. Samaras 30.0 x 8.0 mm, wings almost vertically parallel and at an angle of about 180° with the pedicel, reticulately veined, body plump.

7 **Acer spicatum** Lam. (x1.1) Mountain Maple. Samaras 22.0 x 8.0 mm, wings at an angle of about 140° with the pedicel, fruit body and wings reticulately veined.

1 **Mollugo verticillata** L. (x30) Carpet Weed. Seeds 0.6 x 0.5 x 0.4 mm, elliptic 5 in l.s., elliptic 7-8 in c.s.; embryo curved and seeds more or less reniform, remains of the funiculus often persistent; margin with a distinct rib
or ridge, surface colliculose and with fine concentric lines or ridges, glossy
brown.

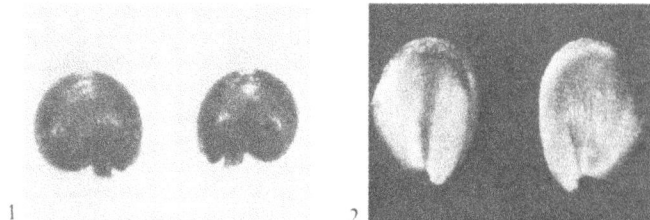

ALISMATACEAE

2 **Alisma plantago-aquatica** L. var. **americana** Roem. & Schult. (x8) Water
Plantain. Achenes 2.8 x 2.1 x 0.8 mm, obliquely obovate 48-49 in l.s.,
oblong 21-22 in c.s.; the embryo is curved and this is indicated by the
lateral sulcus, hilar end oblique; narrow dorsal surface sulcate, base of the
style persistent.

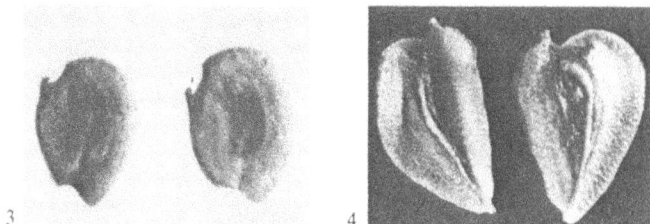

3 **Alisma subcordatum** Raf. (x11) Achenes 1.8 x 1.1 x 0.4 mm, obliquely
obovate 47-48 in l.s., hilar end asymmetrical, oblong 21-22 in c.s.; embryo
curved and indicated by a lateral sulcus, dorsal surface sulcate, style base
persistent.

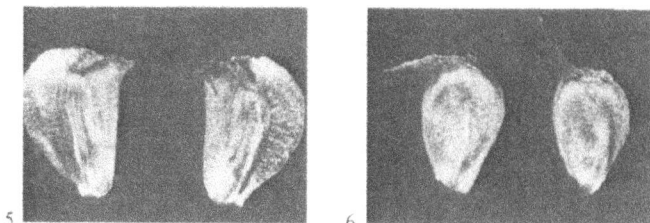

4 **Sagittaria cuneata** Sheldon (x10) Wapato. Achenes 2.5 x 2.0 x 0.5 mm,
obliquely obovate 48-49 in l.s., oblong 22-23 in c.s.; margin with a broad,
thick, spongy wing and finely areolate, style short and erect, lateral sulcus
showing the curving of the embryo.

5 **Sagittaria latifolia** Willd. (x7) Duck-potato. Achenes 2.8 x 2.0 x 0.4 mm,
obliquely obovate 48-49 in l.s., oblong 22-23 in c.s.; margin broadly
winged especially at the apex; beak almost horizontal and broad-based;
seed body grayish, wing whitish, and surface faintly areolate, embryo
folded and lateral sulcus present.

6 **Sagittaria rigida** Pursh (x7) Achenes 2.5 x 1.6 x 0.7 mm, obliquely obovate
47-48 in l.s., oblong 21-22 in c.s.; margin narrowly winged, beak stout and
ascending or sometimes horizontal; surface faintly reticulate or slightly
roughened, embryo folded, and sulcus present.

50

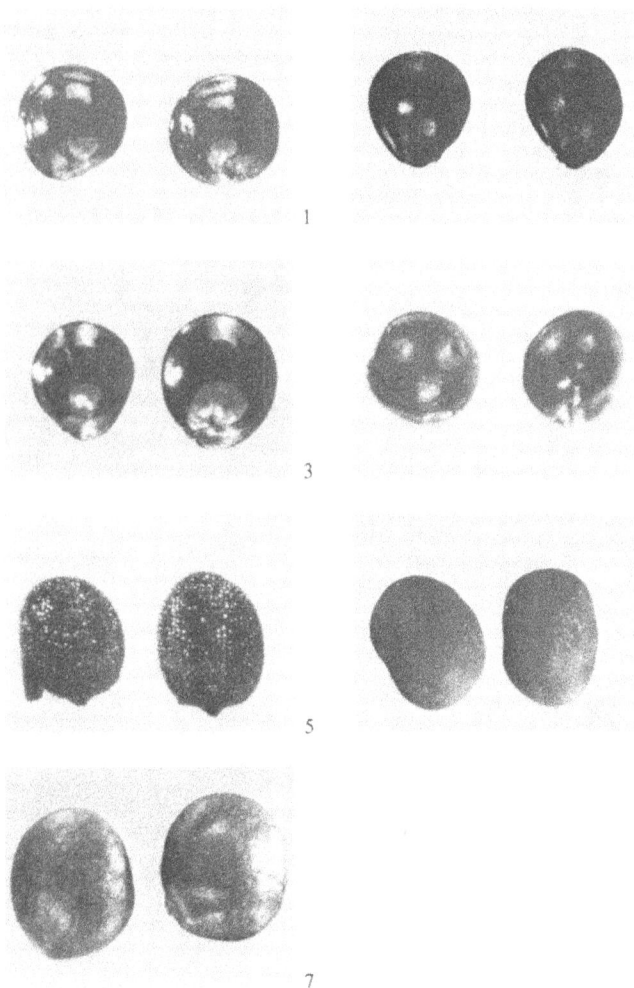

AMARANTHACEAE

1 **Amaranthus albus** L. (x17) Tumbleweed. Seeds 0.9 x 0.9 x 0.5 mm, elliptic 6 in l.s., elliptic 8–9 in c.s., margin ridged; surface glossy and black.

2 **Amaranthus graecizans** L. (x10) Prostrate Pigweed. Seeds 1.4 x 1.4 x 0.8 mm, elliptic 6 in l.s., elliptic 8–9 in c.s., margin with a narrow ridge; surface glossy, black and minutely colliculose.

3 **Amaranthus retroflexus** L. (x15) Redroot Pigweed. Seeds 1.0 x 0.9 x 0.5 mm, elliptic 6 to obovate 49–50 in l.s., elliptic 8–9 in c.s., margin slightly ridged; surface black, glossy and smooth, or inconspicuously colliculose.

4 **Amaranthus tuberculatus** (Moq.) Sauer (x25) Water Hemp. Seeds 0.6 x 0.6 x 0.4 mm, elliptic 6 in l.s., elliptic 8 in c.s., slightly reniform, margin slightly ridged; surface glossy, black, and faintly colliculose.

AMARYLLIDACEAE

5 **Hypoxis hirsuta** (L.) Cov. (x15) Stargrass. Seeds 1.2 x 0.9 x 0.9 mm, irregular in form or elliptic 4–5 in l.s., elliptic 6 in c.s.; surface black, lustrous and muricate or papillate with the processes in concentric rows; ovules anatropous and raphe prominent at the hilar end.

ANACARDIACEAE

6 **Rhus aromatica** Ait. (x4) Fragrant Sumac. Seeds covered by a thin, red, hairy fruit coat, 3.4 x 4.3 x 2.3 mm, elliptic 7–8 in l.s., elliptic 8–9 in c.s.; surface smooth, dull, and brown.

7 **Rhus copallina** L. (x6.5) Shining Sumac. Seeds with a red, hairy, fruit coat, 2.9 x 2.5 x 1.8 mm, elliptic 5–6 in l.s., elliptic 7–8 in c.s.; surface smooth or slightly roughened, dull, and brown.

ANACARDIACEAE

1 **Rhus radicans** L. (x5.5) Poison Ivy. Seeds 3.1 x 3.9 x 2.0 mm, elliptic 7–8 in l.s., elliptic 8–9 in c.s.; surface smooth, dull, undulate, and grayish.

2 **Rhus typhina** L. (x6.5) Staghorn Sumac. Seeds with a thin, red, hairy fruit coat, 2.5 x 3.0 x 1.6 mm, elliptic 7 in l.s., slightly enlarged at one end, elliptic 8–9 in c.s.; surface smooth and dull, gray or brown.

3 **Rhus vernix** L. (x5.5) Poison Sumac. Seeds 2.6 x 4.4. x 2.5 mm, elliptic 8–9 in l.s., elliptic 8–9 in c.s.; ovule on a curved stalk and hilum depressed; surface smooth but undulate and brownish.

ANNONACEAE

4 **Asimina triloba** (L.) Dunal (x1.5) Pawpaw. Seeds 23.0 x 13.0 x 10.0 mm, ovate 38–39 in l.s., elliptic 7–8 in c.s.; surface undulate, smooth, and brown.

APOCYNACEAE

5 **Apocynum androsaemifolium** L. (x8) Spreading Dogbane. Seeds 1.9 x 0.5 x 0.4 mm, ovate 36–37 in l.s., elliptic 7–8 in c.s.; surface slightly longitudinally grooved and brown, seeds terminating with a coma.

6 **Apocynum cannabinum** L. (x5) Indian Hemp. Seeds 5.8 x 0.7 x 0.4 mm, oblong 12–13 in l.s., elliptic 8–9 in c.s.; surface longitudinally ridged and brown, seeds with a terminal coma.

7 **Vinca minor** L. (x4) Periwinkle. Seeds 7.2 x 2.2 x 2.0 mm, elliptic 1–2 or oblong 13–14 in l.s., elliptic 6–7 in c.s.; surface with a deep longitudinal groove on one face, rugose or verrucose, and brown.

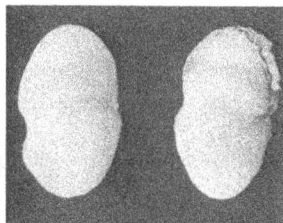
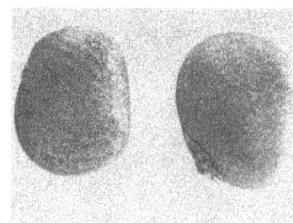
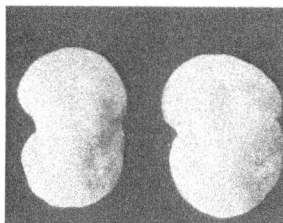
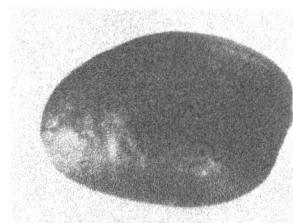
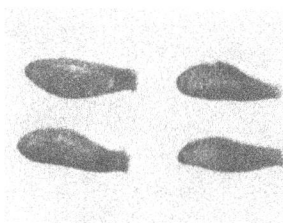
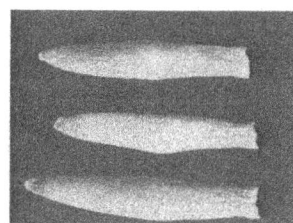

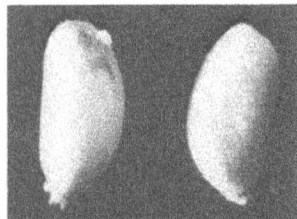
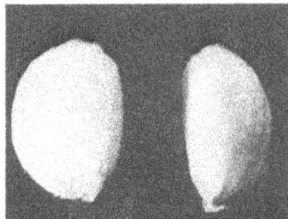
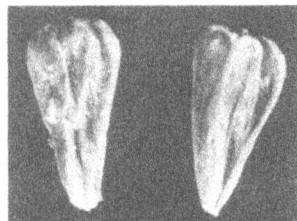
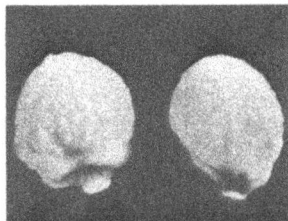
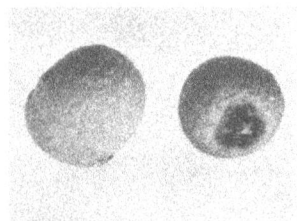
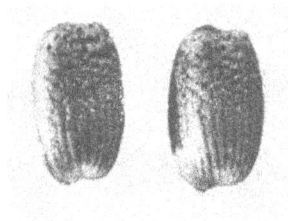
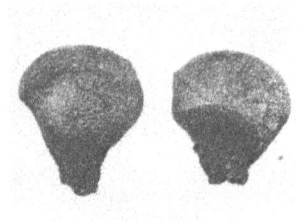

AQUIFOLIACEAE

1 **Ilex verticillata** (L.) Gray (x7) Winterberry. Seeds 3.6 x 1.7 x 1.3 mm, obliquely elliptic 2–3 in l.s., obtriangular 91–92 in c.s. with the dorsal surface rounded; surface smooth or longitudinally ridged and whitish.

2 **Nemopanthus mucronatus** (L.) Trel. (x5.5) Mountain Holly. Seeds 4.0 x 2.6 x 2.5 mm, obliquely elliptic 3–4 in l.s., obtriangular 90–91 in c.s., rounded on the dorsal surface; surface smooth but a raphe ridge present, dull and whitish.

ARACEAE

3 **Acorus calamus** L. (x5) Sweetflag. Fruits 5.4 x 2.7 x 1.8 mm, obovate 47 in l.s., obliquely rhombic 32 in c.s.; surfaces irregularly ridged or grooved and wrinkled, slightly glossy, and light brown.

4 **Arisaema dracontium** (L.) Schott (x5) Green Dragon. Seeds 3.5 x 3.5 x 3.5 mm, elliptic 6 in l.s. and in c.s.; hilum in a small concavity and with a prominent remains of the funiculus; surface rugulose and brown-dotted.

5 **Arisaema triphyllum** (L.) Schott (x5) Jack-in-the-pulpit. Seeds 3.0 x 3.7 x 3.7 mm, elliptic 7–8 in l.s., with a small concavity at the base, elliptic 6 in c.s.; surface faintly rugulose, brown and with darker brown dots.

6 **Calla palustris** L. (x7.5) Water Arum. Seeds 2.9 x 1.8 x 1.5 mm, oblong 15–16 in l.s., with the sides slightly curved, truncate at both ends (barrel-shaped), elliptic 7 in c.s. with a distinct longitudinal ridge; upper half of the seed with shallow pits and the lower half with low, rounded, longitudinal ridges, brown in colour.

7 **Peltandra virginica** (L.) Kunth (x2.2) Arrow Arum. Fruits 10.0 x 6.5 x 6.5 mm, very irregular when dry, when fresh obliquely obovate 47–48 in l.s., elliptic 6 in c.s.; surface with a gelatinous coating, smooth, green, or brown.

1 **Symplocarpus foetidus** (L.) Nutt. (x1.5) Skunk Cabbage. Fruits 10.0 x 10.0 x 10.0 mm, obliquely elliptic 6 in l.s. with a concavity at the base, elliptic 6 in c.s.; surface rugulose and black.

ARALIACEAE

2 **Aralia hispida** Vent. (x8) Bristly Sarsaparilla. Seeds 2.8 x 2.0 x 1.2 mm, obliquely elliptic 4–5 in l.s., elliptic 8–9 in c.s.; surface slightly roughened, dull, and brown.

3 **Aralia nudicaulis** L. (x6) Wild Sarsaparilla. Seeds 3.4 x 2.2 x 0.8 mm, obliquely elliptic 3–4 in l.s., oblong 21–22 in c.s.; surface rugulose and grayish or purplish.

4 **Aralia racemosa** L. (x10) Spikenard. Seeds 2.1 x 1.4 x 0.7 mm, elliptic 4 in l.s., elliptic 9 in c.s.; surface slightly and unevenly wrinkled or sulcate, light gray to brown.

5 **Panax quinquefolium** L. (x3.5) Wild Ginseng. Seeds 6.3 x 4.7 x 2.7 mm, obliquely elliptic 4–5 in l.s., oblong 20–21 in c.s.; surface roughened and gray.

6 **Panax trifolium** L. (x7) Dwarf Ginseng. Fruits 3.6 x 3.8 x 3.8 mm, ovate 41–42 in l.s., 3-carpelled and 3-angled in c.s. with the angles rounded; surface coarsely reticulate or rugose, green to gray. All fruits collected lacked seeds.

ARISTOLOCHIACEAE

7 **Asarum canadense** L. (x6) Wild Ginger. Seeds 4.0 x 2.5 x 1.5 mm, ovate 38–39 in l.s., elliptic 5 in c.s.; ovules anatropous and the hollow inner surface with a large, conspicuous, rather spongy raphe, outer surface transversely ridged, somewhat glossy, and green to brown.

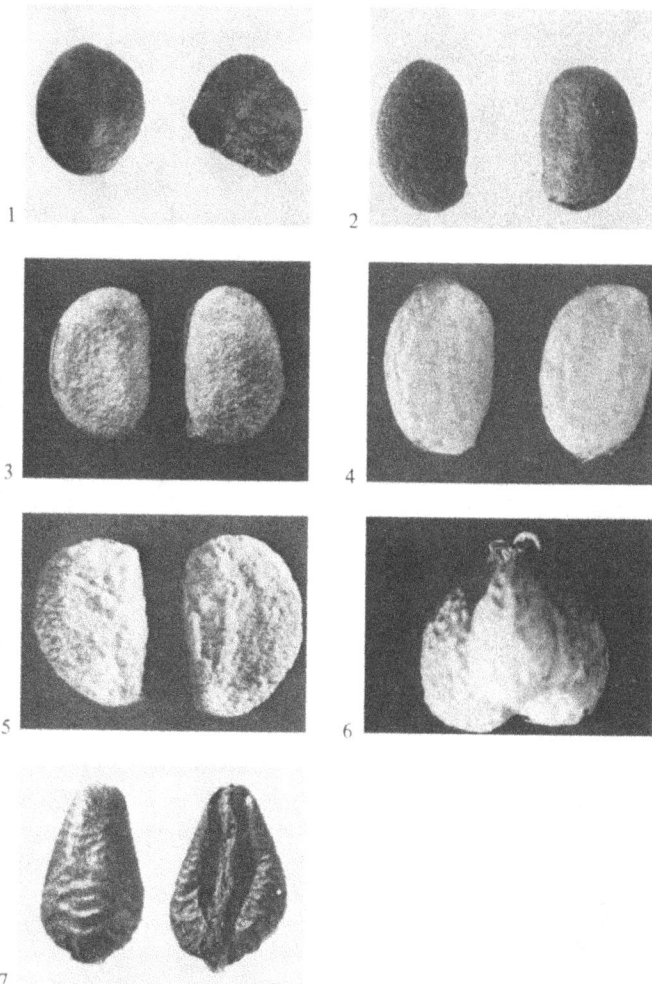

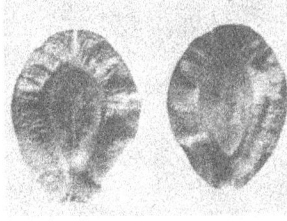
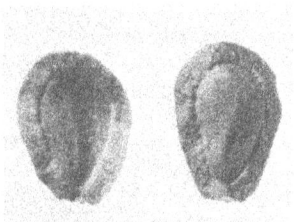
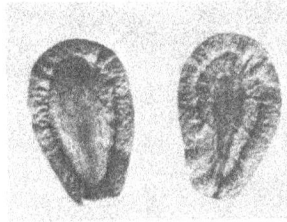
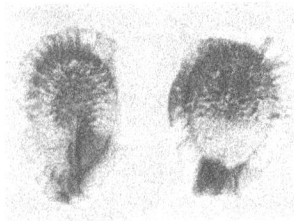
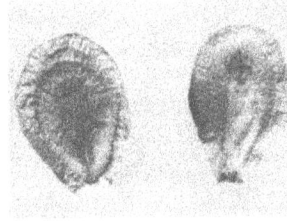

The seeds of the genus *Asclepias* have a distinct, rather spongy marginal wing and a silky coma at the hilum end which in the pod is actually the upper end. One surface is slightly convex and the other surface slightly concave or convex and shows the position of the raphe by a series of striations. The position of the radicle may be indicated by a low ridge.

1 **Asclepias exaltata** L. (x3) Poke Milkweed. Seeds 7.2 x 4.7 x 1.0 mm, obovate 47–48 in l.s., elliptic 10–11 in c.s.; convex surface reticulately veined or rugulose and the other surface showing the raphe and slightly reticulate.

2 **Asclepias incarnata** L. (x2.5) Swamp Milkweed. Seeds 9.5 x 7.2 x 0.8 mm, obovate 48–49 in l.s., elliptic 11 in c.s.; surfaces rather undulate and finely veined or reticulate.

3 **Asclepias sullivantii** Engelm. (x2.8) Seeds 7.1 x 4.8 x 0.8 mm, obovate 48–49 in l.s., elliptic 11 in c.s.; surfaces reticulately veined.

4 **Asclepias syriaca** L. (x3.1) Common Milkweed. Seeds 7.3 x 4.8 x 0.8 mm, obovate 48–49 in l.s., elliptic 11 in c.s.; wings broad and rather thickened, both surfaces reticulately veined.

5 **Asclepias tuberosa** L. (x4) Butterfly Weed. Seeds 6.0 x 4.4 x 0.8 mm, obovate 48–49 in l.s., elliptic 10–11 in c.s.; broadly winged and both surfaces prominently veined.

6 **Asclepias verticillata** L. (x4) Whorled Milkweed. Seeds 5.3 x 3.4 x 0.5 mm, obovate 47–48 in l.s., elliptic 10–11 in c.s.; margin narrowly winged, surfaces rugulose.

7 **Asclepias viridiflora** Raf. (x3) Seeds 7.4 x 5.0 x 0.8 mm, obovate 48–49 in l.s., elliptic 11 in c.s. and broadly winged; surface smooth or slightly roughened.

ASCLEPIADACEAE

1 **Cynanchum medium** R. Br. (x3.7) Dog-strangling Vine. Seeds 7.0 x 4.1 x 1.2 mm, obovate 47–48 in l.s., elliptic 10–11 in c.s., marginal wing very narrow; raphe ridge evident, surfaces smooth and brown.

BALSAMINACEAE

2 **Impatiens biflora** Walt. (x6) Touch-me-not. Seeds 4.1 x 2.5 x 1.8 mm, elliptic 3–4 in l.s., elliptic 7–8 in c.s. with 4 corky, longitudinal ridges; surface areolate or sometimes papillose, mottled green to brown in colour.

3 **Impatiens pallida** Nutt. (x4.2) Seeds 4.7 x 3.0 x 1.7 mm, elliptic 3–4 in l.s., elliptic 8–9 in c.s. with 4 corky or chartaceous, longitudinal ridges; surface smooth to rugulose and reticulate, green to brown.

BERBERIDACEAE

4 **Berberis vulgaris** L. (x4) Common Barberry. Seeds 5.7 x 2.3 x 1.5 mm, obovate 46–47 in l.s., elliptic 8–9 in c.s.; surface finely rugulose and slightly glossy, brown.

5 **Caulophyllum thalictroides** (L.) Michx. (x2.4) Blue Cohosh. Seeds 6.7 x 6.7 x 6.7 mm, elliptic 6 in l.s. and in c.s.; seed with a thin blue, outer seed coat giving it a fruitlike appearance, inner seed coat dark and rugulose. This is a rare example of a naked seed in the dicotyledons.

6 **Jeffersonia diphylla** (L.) Pers. (x5) Twinleaf. Seeds 5.5 x 2.4 x 2.4 mm, elliptic 2–3 in l.s., elliptic 6 in c.s.; hilum with a prominent, lacerated caruncle; surface rugulose and glossy, dark brown, margin with a prominent raphe ridge.

7 **Podophyllum peltatum** L. (x5) May Apple. Seeds 5.7 x 3.4 x 1.9 mm, obliquely elliptic 3–4 to obovate 47–48 in l.s., elliptic 8–9 in c.s.; surface rugulose and light brown.

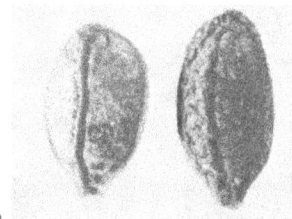

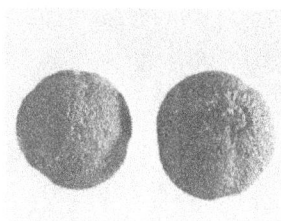
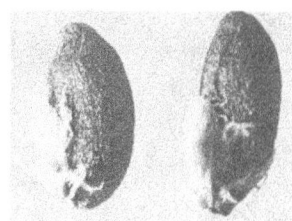
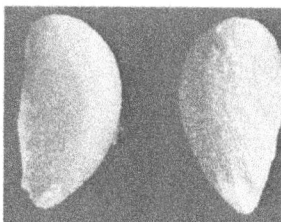

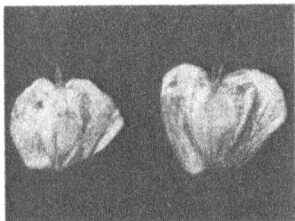
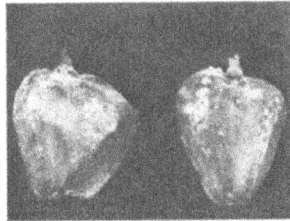
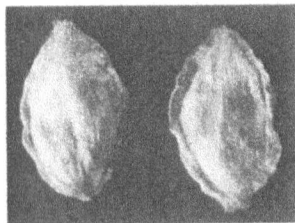
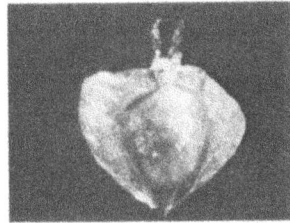

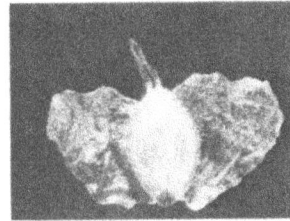

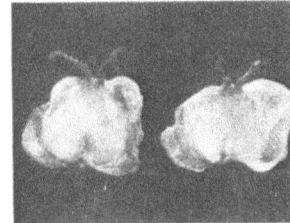

1 **Alnus crispa** (Ait.) Pursh (x6) Green Alder. Samaras 2.5 x 3.0 x 0.5 mm, elliptic 7 to obovate 51 in l.s., more or less cordiform, elliptic 11 in c.s.; wings broad, thin and membranous; the two stigmas persistent and prominent in the apical notch.

2 **Alnus rugosa** (DuRoi) Spreng. (x7) Speckled Alder. Samaras 2.7 x 2.7 x 0.5 mm, obovate 50 in l.s., elliptic 10–11 in c.s.; wings narrow and somewhat corky, surface smooth or wrinkled, styles persistent.

3 **Betula glandulosa** Michx. (x13) Samaras 1.7 x 1.3 x 0.5 mm, elliptic 3–4 in l.s., elliptic 9–10 in c.s.; wing narrow, surface smooth or slightly puberulent.

4 **Betula lenta** L. (x7.5) Black Birch. Samaras 2.7 x 2.9 x 0.4 mm, obliquely obtriangular 90–91 or obovate 50–51 in l.s., elliptic 11 in c.s.; wing about 0.6 mm at the broadest point and chartaceous, surface smooth, styles persistent.

5 **Betula lutea** Michx. f. (x6) Yellow Birch. Samaras 4.3 x 4.1 x 0.7 mm, obovate 49–50 in l.s., elliptic 11 in c.s.; wings thin, chartaceous and about 1.0 mm at the broadest point, body slightly pubescent.

6 **Betula papyrifera** Marsh. (x8.5) White Birch. Samaras 2.0 x 3.7 x 0.4 mm, obovate 52–53 in l.s., elliptic 11 in c.s.; wings membranous and translucent, each about 1.4 mm at the broadest point, surface slightly pubescent.

7 **Betula populifolia** Marsh. (x10) Old-field Birch. Samaras 2.3 x 2.7 x 0.3 mm, obovate 50–51 in l.s., elliptic 11 in c.s.; wing membranous and cordiform, about 1.0 mm at the broadest point.

8 **Betula pumila** L. (x8) Swamp Birch. Samaras 1.7 x 2.1 x 0.3 mm, obovate 51–52 in l.s., elliptic 11 in c.s.; wings membranous, narrow, about 0.4 mm wide.

1 **Carpinus caroliniana** Walt. (x3.7) Blue Beech. Fruit 4.8 x 4.1 x 3.0 mm,
nutlet attached to a single, usually 3-lobed, reticulately veined bract about
2.3 cm long; nutlets ovate 40–41 in l.s., acute at the apex, elliptic 7–8 in
c.s.; surface sharply longitudinally ribbed, slightly granular particularly at
the apex, green to brown.

2 **Corylus americana** Walt. (x0.8) American Hazel. Fruit a nut or a cluster of
nuts surrounded by a flat, laciniate, persistent, pubescent involucre; nuts
about 15 x 12 x 10 mm, ovate 39–40 in l.s., elliptic 7 in c.s.; apex abruptly
pointed and base truncate, surface smooth or slightly ridged.

3 **Corylus cornuta** Marsh. (x0.8) Beaked Hazel. Fruit a nut, either single or
frequently in pairs, or subtended by a number of aborted ovaries, involucre
long, beaked and laciniate or lacerate at the apex; nut about 13 x 12 x 10
mm, ovate 40–41 in l.s., elliptic 6–7 in c.s.; surface similar to the previous
species.

4 **Ostrya virginiana** (Mill.) K. Koch (x3.1) Ironwood. Fruit a nutlet enclosed
in an inflated, membranous bract; nutlet 8.0 x 4.3 x 2.2 mm, ovate 38–39
in l.s., elliptic 8–9 in c.s.; surface smooth, somewhat glossy, dark green or
olive colour with whitish longitudinal nerves.

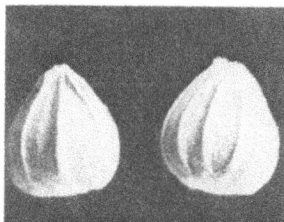

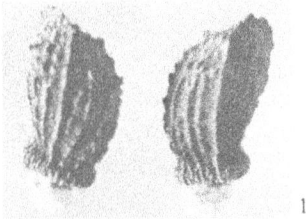

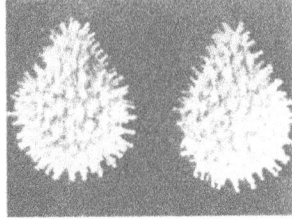

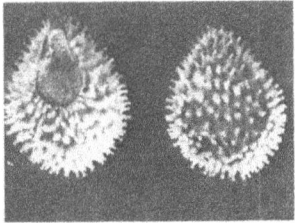

The seed (nutlet) of this family is developed from a single, 4-lobed or 4-parted ovary which at maturity forms 4 separate nutlets. The hilum may appear sub-terminal, lateral, sub-basal, or basal. Because of the lateral contact with other nutlets, each one often appears obtriangular or obovate in cross section if one considers the dorsal surface the apex of the cross section.

1 **Borago officinalis** L. (x5) Borage. Nutlet 4.5 x 2.8 x 2.8 mm, obliquely elliptic 3–4 in l.s. with an acute apex and a broad, concave base with an enlarged, prominently ridged collar, elliptic 6 in c.s.; surface longitudinally ridged and finely verrucose, brown to black, remains of the funiculus often obvious.

2 **Cynoglossum boreale** Fern. (x6) Northern Wild Comfrey. Nutlet 2.8 x 4.0 x 4.5 mm, oblong 19-20 in l.s., ovate 40–41 in c.s.; large hilar scar near the apex of the inner surface; surfaces covered with spines terminating in star-like processes, stalks of the spines glandular, base of the spines and the body surface minutely verrucose, gray to light brown.

3 **Cynoglossum officinale** L. (x3.3) Hound's-tongue. Nutlets 2.5 x 5.0 x 6.0 mm, oblong 21 in l.s., ovate 40 in c.s.; style beaklike, large hilar scar near the apical end; surface covered with spines terminating in star-like processes, stalks of the spines glandular, body surface smooth and gray.

4 **Echium vulgare** L. (x8.5) Blueweed. Nutlets 2.4 x 1.6 x 1.5 mm, obliquely oblong 16–17 in l.s., rounded and pointed above the middle, obliquely obtriangular 90–91 in c.s.; outer rounded surface rugose and inconspicuously verrucose, base broad and with a corneous collar and prominent vascular scars.

BORAGINACEAE

1 **Hackelia virginiana** (L.) Johnst. (x5.5) Stickseed. Nutlets 1.0 x 2.2 x 3.5 mm, obliquely elliptic 9–10 in l.s., ovate 38–39 in c.s.; hilum central on the inner face; rounded dorsal surface covered with spines terminating in star-shaped processes, surface minutely verrucose and dark brown.

2 **Lappula echinata** Gilib. (x8.5) Stickseed. Nutlets 1.0 x 1.2 x 2.3 mm, obliquely elliptic 7 in l.s., ovate 38–39 in c.s.; rounded dorsal surface verrucose, margin with a double row of slender bristles terminating in star-shaped, spiny processes.

3 **Lithospermum arvense** L. (x8) Corn Gromwell. Nutlets 2.8 x 1.6 x 1.6 mm, obliquely ovate 38–39 in l.s., rather pointed at the apex, base truncate and with two prominent vascular bundle scars, elliptic 6 to obovate 50 in c.s.; surface rugose, coarsely pitted and verrucose, gray in colour.

4 **Lithospermum canescens** (Michx.) Lehm. (x10) Puccoon. Nutlets 2.0 x 1.4 x 1.4 mm, obliquely ovate 39–40 in l.s., pointed at the apex, truncate at the base, elliptic 6 to obovate 50 in c.s.; strongly ridged on the inner surface, outer surface glossy and gray.

5 **Lithospermum caroliniense** (Walt.) McMill. (x6) Nutlets 3.6 x 2.6 x 2.6 mm, obliquely ovate 39–40 in l.s., pointed at the apex, elliptic 6 to obovate 50 in c.s., inner surface with a distinct ridge and dorsal surface glossy and white.

6 **Lithospermum latifolium** Michx. (x5.5) Broad-leaved Gromwell. Nutlets 4.0 x 3.2 x 3.2 mm, obliquely ovate 39–40 in l.s., pointed at the apex and truncate at the base, elliptic 6 to obovate 50 in c.s., with a slight ridge on the inner face; outer surface smooth, basal scar large and vascular bundle scars prominent, inner surface slightly pitted, outer surface glossy and white.

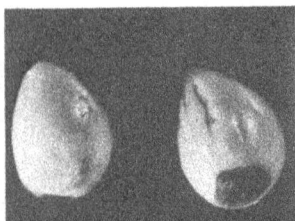
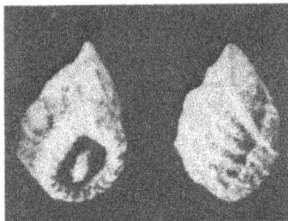

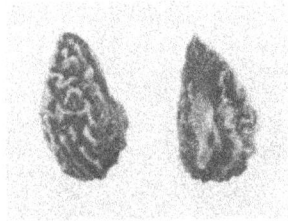
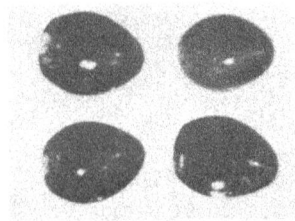
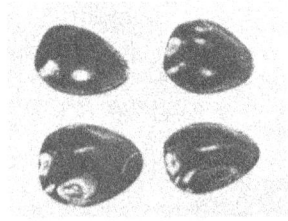
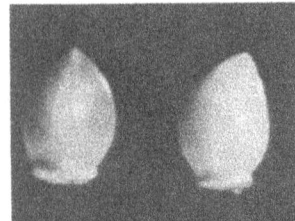

1 **Lithospermum officinale** L. (x7) European Gromwell. Nutlets 2.6 x 2.0 x 2.0 mm, obliquely ovate 39–40 in l.s., rather pointed at the apex, elliptic 6 to obovate 50 in c.s., slightly ridged on the inner surface and sulcate or pitted on each side of the ridge; basal scar and vascular bundle scars prominent; surface smooth, glossy, and white.

2 **Lycopsis arvensis** L. (x5.5) Small Bugloss. Nutlets 3.5 x 2.4 x 2.2 mm, obliquely ovate 39–40 in l.s., obovate 50–51 in c.s.; surface strongly ridged and finely verrucose, hilum in the oblique, concave base which has a prominent, ridged collar, remains of the funiculus prominent, gray to brown in colour.

3 **Mertensia maritima** (L.) S.F. Gray (x4.5) Nutlets 4.2 x 3.2 x 2.3 mm, obliquely ovate 38–39 in l.s., pointed at the apex, obovate 51–52 in c.s.; surface irregularly wrinkled and colliculose.

4 **Mertensia paniculata** (Ait.) G. Don (x5) Lungwort. Nutlets 4.1 x 2.8 x 2.4 mm, obliquely ovate 38–39 in l.s., obovate 50–51 in c.s.; surface rugose and minutely verrucose and papillose, ridges white and the intervals black or brown.

5 **Myosotis laxa** Lehm. (x12) Forget-me-not. Nutlets 1.2 x 0.9 x 0.6 mm, ovate 39–40 in l.s., elliptic 8 in c.s., margin narrowly ridged or winged; surface smooth, reticulate or scalariform, occasionally papillate, glossy, and gray to black.

6 **Myosotis scorpioides** L. (x11) Nutlets 1.3 x 1.1 x 0.7 mm, ovate 40–41 in l.s., elliptic 8–9 in c.s., margin narrowly winged or ridged; surface smooth, reticulate, scalariform or occasionally papillose, glossy, brown or black.

7 **Onosmodium molle** Michx. (x5) False Gromwell. Nutlets 3.5 x 2.6 x 2.6 mm, ovate 39–40 in l.s., with the apex pointed, elliptic 6 in c.s., with a slight ridge on the inner face; basal scar large, surface with scattered pits, gray and glossy.

1 **Symphytum officinale** L. (x5) Common Comfrey. Nutlets 4.5 x 2.5 x 3.2 mm, obliquely ovate 38–39 in l.s., obliquely obtriangular 88–89 in c.s., with a prominent ridge on the inner surface; oblique at the base and concave with a prominent, serrated collar; outer surface undulate and puncticulate, shiny, and black.

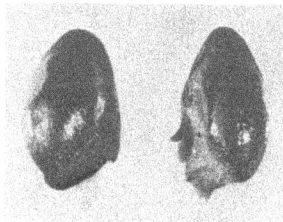

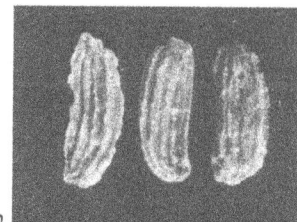

BUTOMACEAE

2 **Butomus umbellatus** L (x14) Flowering Rush. Seeds 1.4 x 0.5 x 0.3 mm, oblong 14–15 in l.s., elliptic 8–9 in c.s.; surface strongly ribbed, the ribs slightly winged and finely transversely rugulose.

CALLITRICHACEAE

3 **Callitriche verna** L. (x10) Water Starwort. Fruits 1.2 x 1.0 x 0.4 mm, elliptic 5 in l.s., oblong 21–22 in c.s., with rounded, scarious-winged margins; surface scalariform, strongly grooved; fruit eventually separates into 4 semicircular, flattened nutlets.

CAMPANULACEAE

4 **Campanula americana** L. (x11) Tall Bellflower. Seeds 1.3 x 0.9 x 0.5 mm, elliptic 4–5 in l.s., elliptic 8–9 in c.s., margin with a narrow, light-coloured wing; surface smooth or inconspicuously colliculose, glossy, and dark brown.

5 **Campanula aparinoides** Pursh (x15) Marsh Bellflower. Seeds 0.8 x 0.5 x 0.4 mm, elliptic 3–4 in l.s., elliptic 7–8 in c.s.; surface smooth or inconspicuously colliculose, semiglossy, and yellowish or brown.

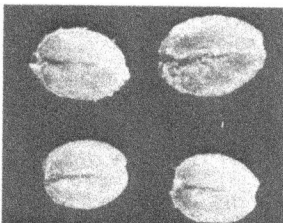

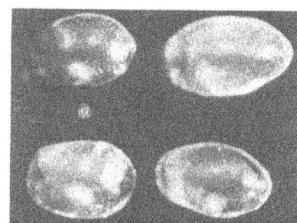

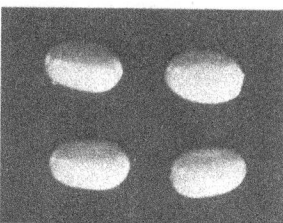

62

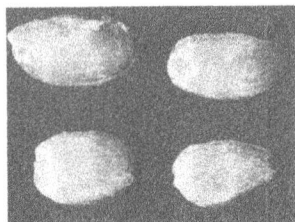
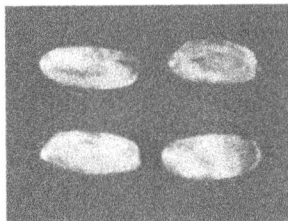
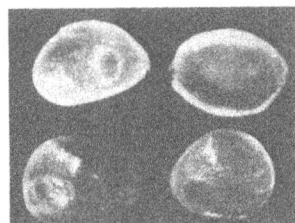
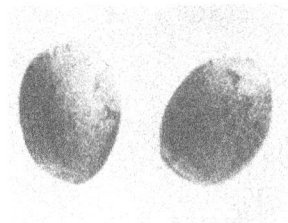
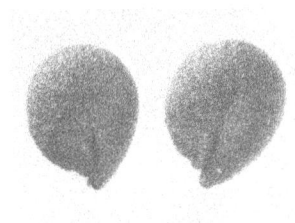
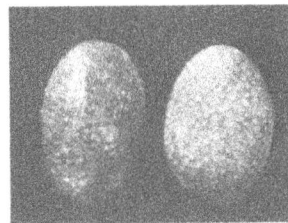

CAMPANULACEAE

1 **Campanula rapunculoides** L. (x10) Bellflower. Seeds 1.5 x 0.8 x 0.5 mm, elliptic 3–4 in l.s., elliptic 8–9 in c.s., slightly winged on the margin; surface finely striate, glossy, and light brown.

2 **Campanula rotundifolia** L. (x16) Bluebell. Seeds 0.9 x 0.4 x 0.2 mm, elliptic 2–3 in l.s., elliptic 9 in c.s., margin slightly winged; surface finely striate and brown.

3 **Triodanis perfoliata** (L.) Nieuwl. (x30) Venus's Looking-glass. Seeds 0.5 x 0.4 x 0.2 mm, elliptic 4–5 in l.s., elliptic 9 in c.s., margin winged and the wing deflexed; surface smooth, slightly glossy, and brownish.

CANNABINACEAE

4 **Cannabis sativa** L. (x5.5) Hemp or Marijuana. Achenes 3.4 x 2.4 x 2.0 mm, elliptic 4–5 in l.s., elliptic 7 in c.s., fruit coatings membranous, reticulately veined, green, brown, black, or mottled.

CAPPARIDACEAE

5 **Polanisia dodecandra** (L.) DC. (x9.5) Clammyweed. Seeds 2.2 x 1.8 x 1.2 mm, elliptic 4–5 in l.s., asymmetrical at the base, embryo folded, indicated by the lateral sulcus and somewhat reniform shape, elliptic 8 in c.s.; surface areolate and brown.

CAPRIFOLIACEAE

6 **Diervilla lonicera** Mill. (x22) Bush Honeysuckle. Seeds 1.0 x 0.8 x 0.5 mm, obliquely elliptic 4–5 in l.s., elliptic 8–9 in c.s.; surface reticulate and brown.

1 **Linnaea borealis** L. (x11) Twinflower. Seeds 2.1 x 1.1 x 0.8 mm, elliptic 3–4 in l.s., elliptic 7–8 in c.s. with a single lateral, longitudinal sulcus; surface green to brown and smooth.

2 **Lonicera canadensis** Marsh. (x6.3) Fly Honeysuckle. Seeds 3.4 x 2.3 x 1.3 mm, elliptic 4–5 in l.s., elliptic 8–9 in c.s.; surface finely areolate and light brown, slight ridge or sulcus may be evident indicating the anatropous ovule.

3 **Lonicera dioica** L. (x6) Trumpet Honeysuckle. Seeds 3.4 x 2.6 x 1.4 mm, elliptic 4–5 in l.s., elliptic 8–9 in c.s.; surface areolate and light brown, slight sulcus may be evident.

4 **Lonicera hirsuta** Eat. (x6.5) Hairy Honeysuckle. Seeds 2.8 x 2.2 x 1.4 mm, elliptic 4–5 in l.s., elliptic 8–9 in c.s.; surface areolate and light brown.

5 **Lonicera involucrata** (Richards.) Banks (x8). Seeds 2.7 x 1.8 x 0.7 mm, elliptic 4 in l.s., obliquely oblong 21–22 in c.s.; surface areolate, slightly shiny and black.

6 **Lonicera oblongifolia** (Goldie) Hook. (x9.5) Swamp Honeysuckle. Seeds 2.7 x 1.7 x 0.7 mm, elliptic 3–4 in l.s., oblong 21–22 in c.s.; surface smooth or puncticulate and gray to brown, slight sulcus may show the inverted ovule.

7 **Lonicera tatarica** L. (x6) Tartarian Honeysuckle. Seeds 3.0 x 2.3 x 1.1 mm, elliptic 4–5 in l.s., oblong 21–22 in c.s.; surface irregularly ridged or grooved, areolate, and brown.

8 **Lonicera villosa** (Michx.) R.&S. (x11) Mountain Fly Honeysuckle. Seeds 1.7 x 1.3 x 0.6 mm, slightly asymmetrical at the hilar end, elliptic 4–5 in l.s., elliptic 9–10 in c.s.; surface areolate, glossy, and brown.

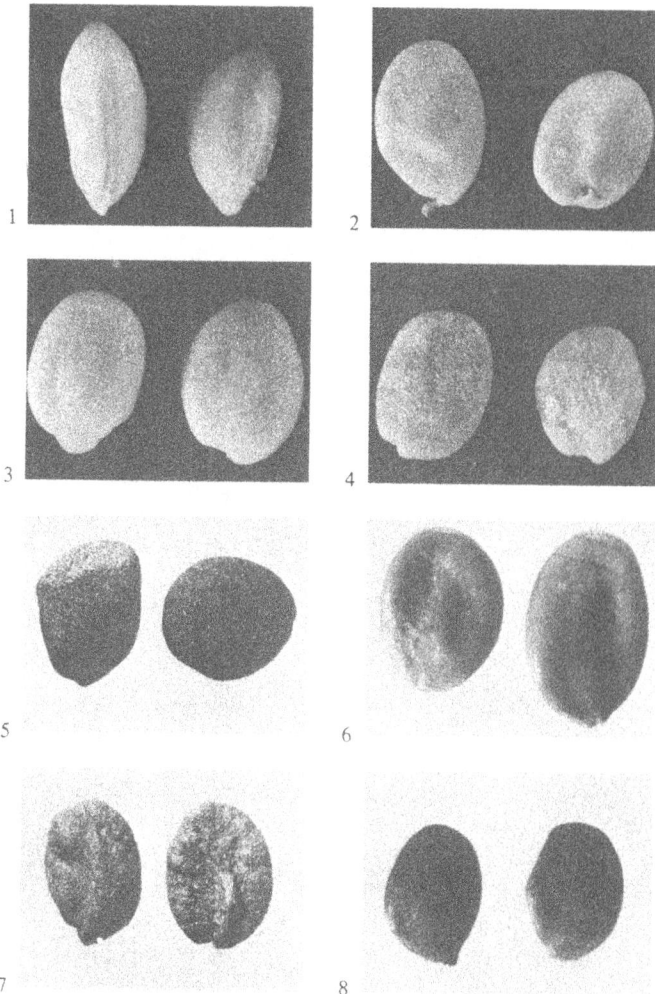

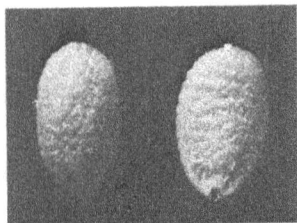

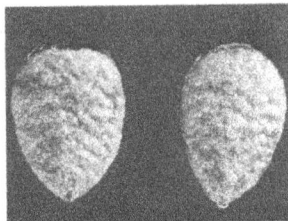

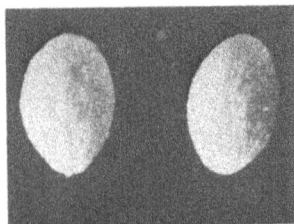

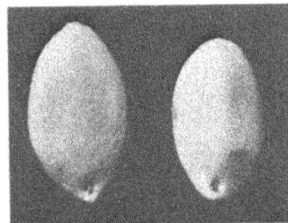

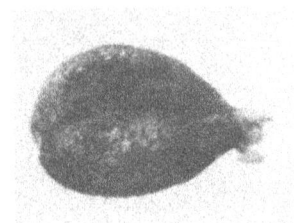

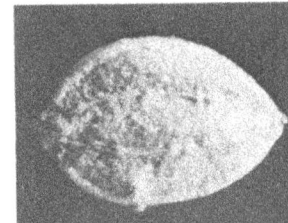

1 **Sambucus canadensis** L. (x7) Common Elderberry. Seeds 2.8 x 1.5 x 1.0 mm, obovate 47–48 l.s., elliptic 8 in c.s., somewhat plano-convex; surface rugulose and light brown.

2 **Sambucus pubens** Michx. (x7.5) Red-berried Elder. Seeds 2.6 x 1.6 x 1.0 mm, elliptic 3–4 to obovate 47–48 in l.s., elliptic 8–9 in c.s., tending to be plano-convex; surface rugulose and brown.

3 **Symphoricarpos albus** (L.) Blake (x5.5) Snowberry. Seeds 3.6 x 2.4 x 1.4 mm, elliptic 4 in l.s., elliptic 8–9 in c.s.; surface rugulose and white to yellowish.

4 **Symphoricarpos occidentalis** Hook (x5.5). Seeds 4.5 x 2.5 x 1.2 mm, elliptic 3–4 in l.s., elliptic 9–10 in c.s.; surface smooth and grayish.

5 **Triosteum perfoliatum** L. (x3) Horse Gentian. Stone 7.0 x 5.8 x 5.8 mm, elliptic 4–5 in l.s., elliptic 6 in c.s.; surface roughened, ridged, and sulcate; eventually the stone separates into three rather triangular sections or nutlets.

6 **Viburnum acerifolium** L. (x2.7) Maple-leaved Viburnum. Stone 6.3 x 5.1 x 2.4 mm, elliptic 4–5 in l.s., elliptic 9–10 in c.s.; surfaces slightly roughened and with two ridges on one surface and one on the other, sulci nerved.

7 **Viburnum alnifolium** Marsh. (x3.5) Hobblebush. Stone 7.9 x 5.2 x 2.5 mm, ovate 39 in l.s., elliptic 9–10 in c.s.; surface roughened and with two ridges on one surface and one on the other.

8. **Viburnum cassinoides** L. (x4.3) Wild Raisin, Stone 6.1 x 4.4 x 1.8 mm, elliptic 4–5 to ovate 39–40 in l.s., elliptic 9–10 in c.s.; surface slightly roughened and undulate.

1 **Viburnum dentatum** L. (x3.8) Southern Arrow Wood. Stone 4.7 x 4.2 x 3.0 mm, elliptic 5–6 in l.s., elliptic 7–8 in c.s.; deeply grooved on one surface and slightly roughened.

2 **Viburnum edule** (Michx.) Raf. (x3.5) Squashberry. Stone 4.9 x 5.2 x 2.0 mm, ovate 40–41 in l.s., elliptic 9–10 in c.s., more or less plano-convex; surface rugulose.

3 **Viburnum lentago** L. (x4) Nannyberry. Stone 7.5 x 6.5 x 2.5 mm, ovate 40–41 in l.s., elliptic 9–10 in c.s.; surfaces roughened and one surface with an obscure ridge.

4 **Viburnum opulus** L. var. **americanum** Ait. (x4) Highbush Cranberry. Stone 9.0 x 7.0 x 1.8 mm, ovate 39–40 in l.s., elliptic 10–11 in c.s.; surface roughened, not ridged.

5 **Viburnum rafinesquianum** Schult. (x4.5) Downy Arrow Wood. Stone 5.5 x 4.5 x 2.4 mm, elliptic 4–5 in l.s., obliquely elliptic 8–9 in c.s., more or less plano-convex; flattened surface with distinct ridges and sulci, convex surface inconspicuously grooved, slightly roughened.

CARYOPHYLLACEAE

Most of the seeds of this family have a curved embryo coiled around a central mass of albumen. This often gives a rather reniform appearance to the seed although the overall shape is most frequently elliptic. The surface in the majority of species is finely or coarsely papillose or rugulose.

6 **Agrostemma githago** L. (x5) Purple Cockle. Seeds 3.0 x 2.8 x 2.2 mm, irregular in form or obliquely obovate 49–50 in l.s., tapering from the apex to the base in radial section, obliquely elliptic 7–8 in c.s.; surface with curved rows of spiny papillae, dull and black.

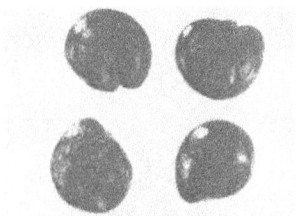

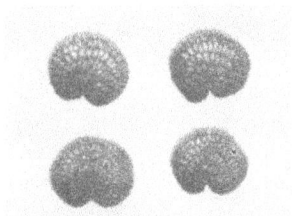

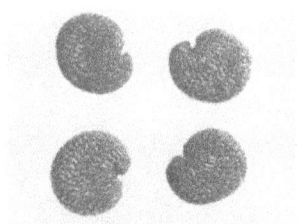

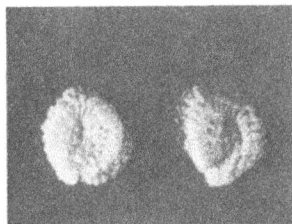

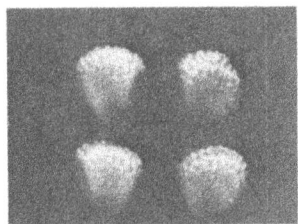

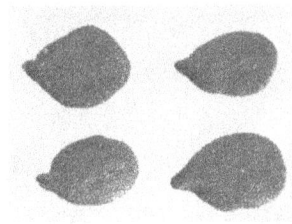

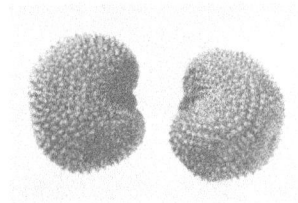

1 **Arenaria lateriflora** L. (x10) Grove Sandwort. Seeds 0.9 x 1.1 x 0.5 mm, elliptic 7–8 in l.s., slightly reniform, elliptic 9–10 in c.s., margin faintly ridged; strophiole or caruncle prominent; surface faintly colliculose black, and shiny.

2 **Arenaria serpyllifolia** L. (x18) Thyme-leaved Sandwort. Seeds 0.4 x 0.5 x 0.3 mm, elliptic 7–8 in l.s., slightly reniform, oblong 20–21 to elliptic 8–9 in c.s., margins rounded; surface papillose with the papillae bases somewhat elongated and in curved rows, dull and black.

3 **Arenaria stricta** Michx. (x14) Rock Sandwort. Seeds 0.7 x 0.7 x 0.4 mm, elliptic 6 in l.s., slightly reniform, oblong 20–21 in c.s.; papillae elongate and more or less irregularly distributed over the surface, dull and black.

4 **Cerastium arvense** L. (x12) Field Chickweed. Seeds 1.0 x 0.8 x 0.5 mm, obovate 48–49 in l.s., tapering from apex to base in radial section, oblong 20–21 in c.s. with the faces somewhat concave and showing the coiling of the embryo; papillae small and in concentric rows, brown.

5 **Cerastium vulgatum** L. (x13) Mouse-ear Chickweed. Seeds 0.65 x 0.6 x 0.4 mm, obovate 49–50 in l.s., tapering from apex to base in radial section, oblong 20 in c.s.; papillae in rather irregular, curved rows, brown in colour.

6 **Dianthus armeria** L. (x13) Deptford Pink. Seeds 0.3 x 0.7 x 1.1 mm, oblong 21–22 in l.s., concave-convex and peltate, ovate 40–41 in c.s. with an abruptly pointed anterior end; surface finely papillose with the papillae in irregular curved rows, dull, and black.

7 **Lychnis alba** Mill. (x13) White Cockle. Seeds 1.1 x 1.3 x 0.9 mm, elliptic 6–7 in l.s., slightly reniform, tapering from apex to base in radial section, elliptic 7–8 in c.s.; papillae in concentric rows and black-tipped, elongate or elliptical at the base; hilum with a wide orifice and a prominent pad on each side, gray.

1 **Saponaria officinalis** L. (x9.5) Bouncing Bet. Seeds 1.6 x 1.9 x 0.8 mm, elliptic 6–7 in l.s., reniform, elliptic 9–10 in c.s.; surface papillose with the papillae in concentric rows, blunt-tipped and more or less elongate, dull, and black.

2 **Scleranthus annuus** L. (x12) Knawel. Seeds enclosed in a hard, tough, persistent, 10-ribbed calyx tube, calyx lobes longer than the tube; seeds 1.0 x 0.7 x 0.7 mm, elliptic 4–5 in l.s., apiculate or beaked at the apex, elliptic 6 in c.s.; surface smooth, white to brown.

3 **Silene antirrhina** L. (x25) Sticky Catchfly. Seeds 0.6 x 0.7 x 0.4 mm, elliptic 6–7 in l.s., slightly reniform, oblong 20–21 in c.s.; surface papillose with the papillae in concentric rows, blunt and black-tipped, orifice open and the pads conspicuous; surface almost black.

4 **Silene cucubalus** Wibel (x10) Bladder Campion. Seeds 1.2 x 1.4 x 0.9 mm, elliptic 6–7 in l.s., slightly reniform, oblong 20–21 in c.s.; orifice with prominent pads; surface papillose with the papillae in more or less concentric rows on the dorsal surface, irregular and rather small on the sides, black-tipped, surface gray.

5 **Silene noctiflora** L. (x11) Night-flowering Catchfly. Seeds 1.2 x 1.4 x 0.9 mm, elliptic 6–7 in l.s., slightly reniform, oblong 20–21 in c.s.; surface gray, papillose, the papillae small and more or less rounded at the base or slightly elliptic, black-tipped, not consistently in concentric rows, particularly on the sides, orifice not large and pads small.

6 **Spergula arvensis** L. (x18) Spurrey. Seeds 1.1 x 1.1 x 0.7 mm, elliptic 6 in l.s., elliptic 8–9 in c.s., margin with a light-coloured wing; surface dull, black with scattered white rather clavate hairs or spines.

7 **Spergularia rubra** (L.) J.&C. Presl (x25) Sand Spurrey. Seeds 0.6 x 0.45 x 0.2 mm, obovate 48–49 in l.s., elliptic 9 in c.s.; surface faintly papillose, roughened, dark brown.

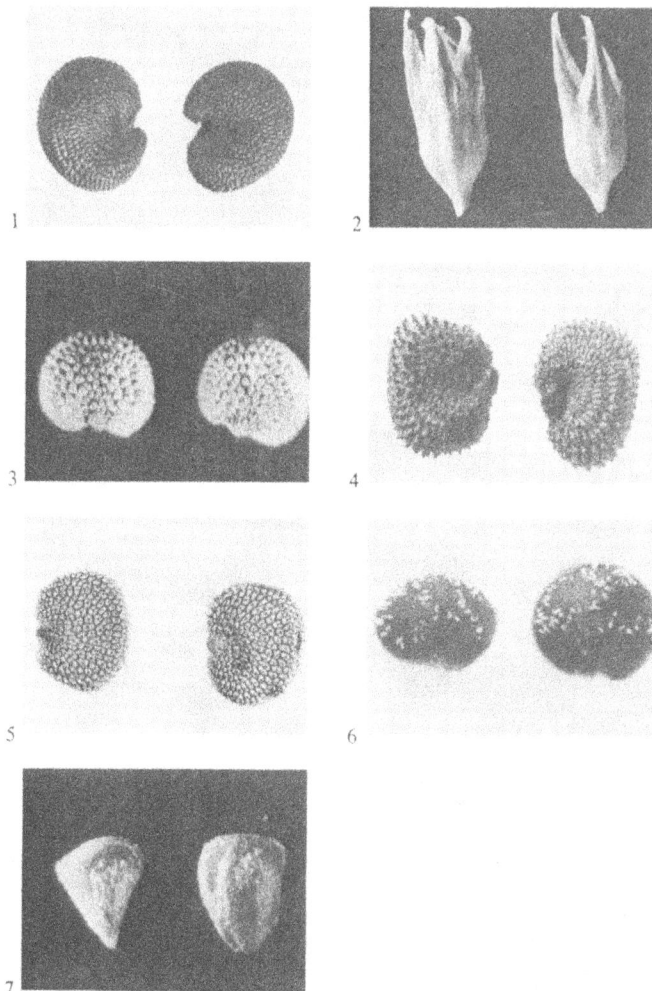

CARYOPHYLLACEAE

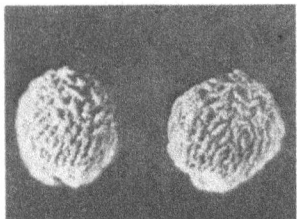
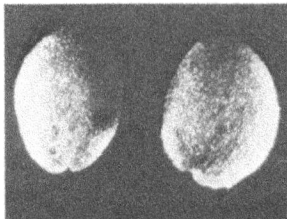
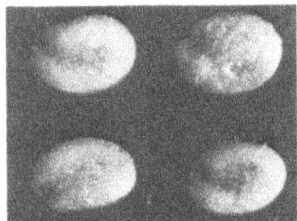
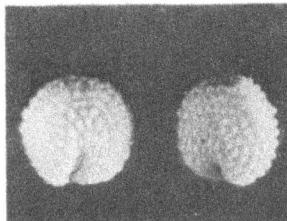
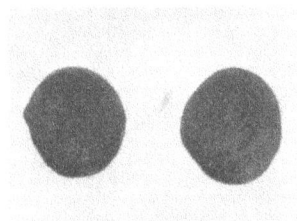

1 **Stellaria graminea** L. (x15) Grass-leaved Stitchwort. Seeds 1.0 x 0.9 x 0.5 mm, irregular or elliptic 5–6 in l.s., elliptic 8–9 in c.s.; surface papillose, the papillae elongate and irregular in arrangement, brown to almost black.

2 **Stellaria longifolia** Muhl. (x16) Seeds 1.1 x 0.9 x 0.6 mm, irregular or elliptic 4–5 in l.s., elliptic 8 in c.s.; surface slightly roughened, colliculose or undulate, dull, and brown.

3 **Stellaria longipes** Goldie (x18) Stitchwort. Seeds 0.8 x 0.7 x 0.5 mm, irregular or elliptic 5–6 in l.s., elliptic 7–8 in c.s.; surface faintly roughened, dull, and brown.

4 **Stellaria media** (L.) Cyrill. (x12) Common Chickweed. Seeds 1.1 x 1.0 x 0.6 mm, irregular in form or elliptic 5–6 in l.s., slightly reniform, oblong 20–21 in c.s.; papillae low and rounded.

5 **Vaccaria segetalis** (Neck.) Garcke (x6.5) Cow Cockle. Seeds 2.3 x 2.4 x 2.0 mm, elliptic 6–7 in l.s., elliptic 7 in c.s.; surface faintly papillose or colliculose, dull, and black.

CELASTRACEAE

6 **Celastrus scandens** L. (x4.8) Climbing Bittersweet. Seeds 5.2 x 2.3 x 1.9 mm, obliquely elliptic 2–3 in l.s., elliptic 7–8 in c.s.; seeds covered with a fleshy, red aril, surface with one prominent lateral longitudinal ridge at the position of the raphe and a number of fine ridges, areolate and brown or gray.

7 **Euonymus atropurpureus** Jacq. (x4.6) Wahoo. Seeds 5.4 x 2.8 x 2.8 mm, elliptic 3–4 in l.s., elliptic 6 in c.s.; seed covered with a fleshy, red aril, surface with one longitudinal ridge, otherwise smooth, dull, white or light brown.

1 **Euonymus obovatus** Nutt. (x4.8) Strawberry Bush. Seeds 4.6 x 3.4 x 3.0 mm, elliptic 4–5 in l.s., elliptic 6–7 in c.s.; seed covered with a fleshy red aril, surface with one prominent ridge, rugulose and light brown.

CHENOPODIACEAE

The embryo is folded or coiled in the seed and this is often evident in the form or the external markings.

2 **Atriplex patula** L. (x6) Spreading Orache. Fruits enclosed by 2 triangular or rhombic veiny bracts often covered with farinaceous particles; seeds when removed from the thin, gray fruit coat 2.4 x 2.4 x 1.2 mm, elliptic 6 in l.s., elliptic 9 in c.s.; surface shiny, black, and faintly areolate.

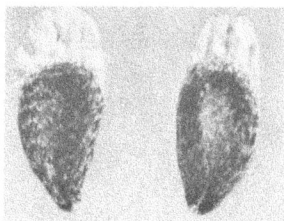

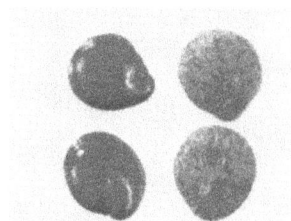

3 **Axyris amaranthoides** L. (x9) Russian Pigweed. Fruits 2.8 x 1.3 x 1.0 mm, obovate 46–47 in l.s., with a retuse terminal wing, elliptic 7–8 in c.s.; surface black or brown with stiff, whitish, appressed hairs; a small sulcus at the hilar end indicates the folding of the embryo.

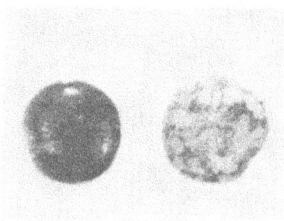

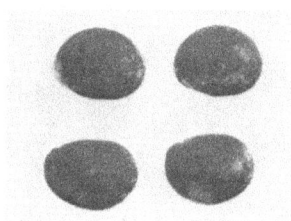

4 **Chenopodium album** L. (x10) Lamb's-quarters. Fruit coat a dull, membranous covering with a small circular apical opening; seeds 1.2 x 1.0 x 0.5 mm, elliptic 5 to obovate 49 in l.s. with the hilar end asymmetrical, elliptic 9 in c.s.; surface glossy, smooth or obscurely striate, and black.

5 **Chenopodium botrys** L. (x18) Jerusalem Oak. Seeds enclosed in grayish, slightly mottled fruit coats; seeds 0.7 x 0.7 x 0.5 mm, elliptic 6 in l.s., elliptic 7–8 in c.s., margin with a narrow groove; surface shiny, black, or dark brown.

6 **Chenopodium capitatum** (L.) Aschers. (x10) Strawberry Blite. Fruits enclosed in a bright red fleshy calyx; fruit coat present but thin and not conspicuous; seeds 1.0 x 0.8 x 0.5 mm, elliptic 4–5 in l.s., elliptic 8–9 in c.s., margin slightly ridged; surface black and shiny.

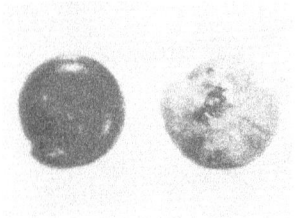

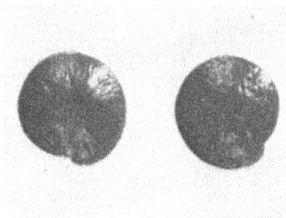

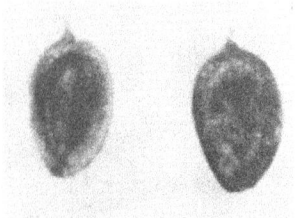

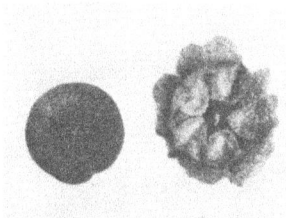

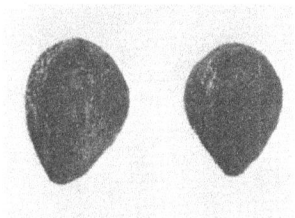

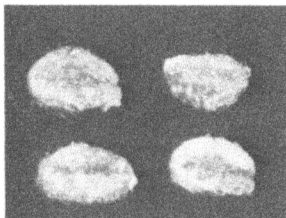

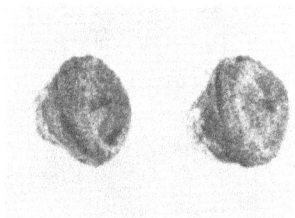

1 **Chenopodium glaucum** L. (x15) Oak-leaved Goosefoot. Fruit coat dull, gray and membranous; seeds 0.7 x 0.7 x 0.4 mm, elliptic 6 in l.s., elliptic 8–9 in c.s., margin with a conspicuous ridge; surface smooth or collicu-lose, shiny, and black.

2 **Chenopodium hybridum** L. (x7.5) Maple-leaved Goosefoot. Fruit coat dull, gray, and membranous; seeds 1.5 x 1.5 x 0.8 mm, elliptic 6 in l.s., elliptic 8–9 in c.s.; surface faintly and irregularly roughened or undulate, very faintly areolate, black, and glossy.

3 **Corispermum hyssopifolium** L. (x5.5) Bugseed. Seeds enclosed in a per-sistent ovary wall, fruit 3.2 x 2.0 x 0.5 mm, elliptic 3–4 to obovate 47–48 in l.s., oblong 22–23 in c.s., winged margins concave-convex, greenish, glossy, and translucent, faintly areolate.

4 **Cycloloma atriplicifolium** (Spreng.) Coult. (x7) Winged Pigweed. Fruits surrounded by the horizontally winged calyx; seed enclosed in a membra-nous pericarp which has a whitish pubescence and 5 nerves radiating from the hilum; seeds 1.8 x 1.8 x 0.7 mm, elliptic 6 in l.s., elliptic 9–10 in c.s.; surface dull and black.

5 **Kochia scoparia** (L.) Schrader (x12) Summer Cypress. Backs of the sepals of the fruiting calyx winged and hardened; seeds enclosed in a membranous coating, 1.5 x 1.2 x 0.8 mm, obovate 48–49 in l.s. and somewhat reniform, oblong 20 in c.s. with the faces concave, surface black, dull, and granular.

6 **Salicornia europaea** L. (x10) Glasswort. Fruits 1.7 x 0.8 x 0.5 mm, obovate 46–47 in l.s., oblong 20–21 in c.s.; seeds finely white pubescent with short, curved hairs, body brown.

7 **Salsola kali** L. var. **tenuifolia** G.W.F. Meyer (x8) Russian Thistle. Fruits en-closed in the horizontally winged calyx, 1.4 x 1.7 x 1.7 mm, obliquely obtriangular 91–92 in l.s., elliptic 6 in c.s.; surface with a spiral groove in-dicating the inner coiled embryo.

1 Suaeda maritima (L.) Dumort. (x9) Sea-blite. Seeds 1.6 x 1.8 x 0.9 mm, elliptic 6–7 in l.s., elliptic 9 in c.s.; asymmetrical or reniform at the base; surface smooth, glossy, and black.

CISTACEAE

2 Helianthemum bicknellii Fern. (x10) Open flower seeds irregularly angular to rounded in form, surface spotted with brown and indistinctly reticulate; cleistogamous flower seeds 1.6 x 1.0 x 0.5 mm; surface obscurely reticulate, reddish brown.

3 Helianthemum canadense (L.) Michx. (x10) Frostweed. Seeds of open and cleistogamous flowers similar and irregularly angular or dome-shaped, 1.1 x 1.1 x 0.1 mm, surface papillose, black, or brown.

4 Hudsonia tomentosa Nutt. (x11) Beach Heath. Seeds 1.3 x 0.8 x 0.8 mm, ovate 38–39 in l.s., elliptic 6 in c.s.; young seeds with a white coating having longitudinal lines of elliptically based papillae, older seeds without the coating, indistinctly areolate and brown.

5 Lechea intermedia Leggett (x10) Pinweed. Seeds 1.1 x 0.5 x 0.5 mm, irregular in shape or elliptic 2–3 in l.s., obtriangular 90 in c.s., rounded on the dorsal surface; surface smooth and brown, often with a wrinkled exterior coating.

6 Lechea stricta Leggett (x13) Seeds 1.1 x 0.9 x 0.6 mm, irregular in form or elliptic 4–5 in l.s., obliquely elliptic 8 in c.s. tending to be plano-convex; surface smooth and brown.

7 Lechea villosa Ell. (x13) Seeds 0.9 x 0.5 x 0.4 mm, irregular or elliptic 3–4 in l.s., obtriangular 91–92 to elliptic 7–8 in c.s., tending to be concave-convex; dorsal surface smooth and brown.

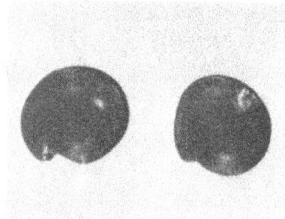
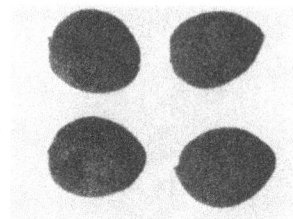

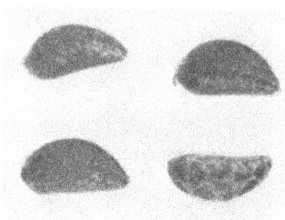
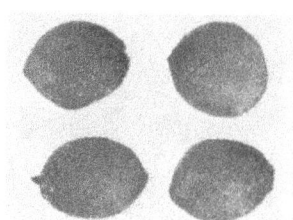

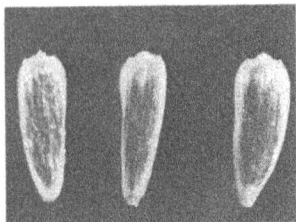

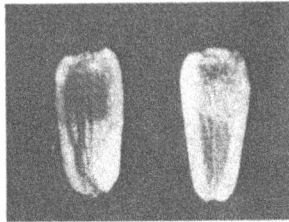

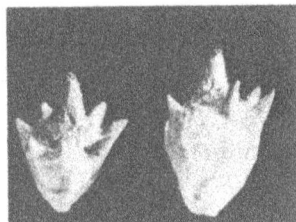

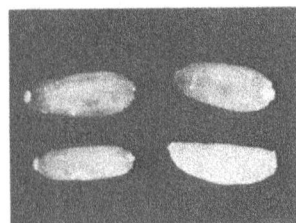

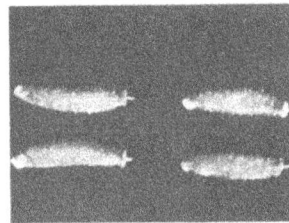

The achenes of most Compositae have a remnant style base and a pappus, or if the pappus is absent, there is usually a remnant pappus rim. Because of their compact arrangement in the head, the majority of the achenes are much longer than they are wide or thick. The ovules are anatropous.

1 **Achillea millefolium** L. var. **lanulosa** (Nutt.) Piper (x12) Yarrow. Achenes 1.6 x 0.7 x 0.3 mm, obovate 46–47 in l.s., elliptic 9–10 in c.s., margin with a narrow gray wing; surface finely longitudinally striate, slightly glossy and light brown; pappus none.

2 **Achillea ptarmica** L. (x12) Achenes 1.7 x 0.9 x 0.3 mm, obovate 47–48 in l.s., elliptic 10 in c.s., margin with a broad grayish wing; body brown and striate; pappus none.

3 **Ambrosia artemisiifolia** L. (x6) Common Ragweed. Achenes 3.0 x 2.0 x 2.0 mm, irregular or obovate 48 in l.s., crowned with 4–7, sharp, conical spines surrounding a central subulate beak, elliptic 6 in c.s.; surface gray to brownish or purplish and pubescent with inconspicuous rather stiff hairs.

4 **Ambrosia trifida** L. (x3) Giant Ragweed. Achenes 8.0 x 8.0 x 5.0 mm, irregular or obovate 50 in l.s., irregularly 5–7 angled in c.s.; apex crowned with 5–7 conical spines surrounding a conical, subulate beak; surface brown, somewhat wrinkled and pubescent.

5 **Anaphalis margaritacea** (L.) Benth. & Hook. (x20) Pearly Everlasting. Achenes 0.7 x 0.3 x 0.2 mm, elliptic 2–3 in l.s., elliptic 8 in c.s., the margins slightly ribbed; surface indistinctly striate grayish or brown; pappus of capillary bristles.

6 **Antennaria neglecta** Greene (x10) Pussy's-toes. Achenes 1.5 x 0.4 x 0.25 mm, elliptic 1–2 in l.s., elliptic 8–9 in c.s.; surface olivaceous and with white, resinous dots and striate; pappus of capillary bristles slightly united at the base.

1 **Antennaria plantaginifolia** (L.) Richards (x10) Achenes 1.3 x 0.4 x 0.25 mm, similar in form to the previous species; surface faintly mottled, brown with white, resinous dots; pappus of capillary bristles.

2 **Anthemis arvensis** L. (x8.5) Corn Chamomile. Disk achenes 1.8 x 0.8 x 0.8 mm, obtriangular 86–87 to oblong 14–15 in l.s., oblong 18 to elliptic 6 in c.s., truncate at the apex; surface longitudinally 10-ribbed, white to gray; pappus not conspicuous. Ray achenes larger, 2.2 x 1.1 x 1.1 mm and truncate at both the apex and the base; they may be flared at the apex and slightly curved.

3 **Anthemis cotula** L. (x14) Dog Fennel. Achenes 1.5 x 0.8 x 0.8 mm, obovate 47–48 in l.s., elliptic 6 in c.s.; surface with 10 longitudinal, verrucose ridges, gray, brown, or black; pappus not evident.

4 **Anthemis tinctoria** L. (x9) Yellow Chamomile. Achenes 2.5 x 0.9 x 0.6 mm, obtriangular 86–87 to oblong 14–15 in l.s., more or less rhombic 32 in c.s., with the margin ridged and a median ridge on each surface; surfaces obscurely longitudinally nerved and faintly rugulose; pappus a mere ridge.

5 **Arctium lappa** L. (x4.5) Great Burdock. Achenes 5.6 x 2.2 x 1.3 mm, obliquely obovate 46–47 in l.s., elliptic 8–9 in c.s.; each face prominently ribbed and faintly nerved, rugulose, mottled gray and black; pappus of deciduous membranous scales.

6 **Arctium minus** Schk. (x4.5) Common Burdock. Achenes 5.3 x 2.4 x 1.5 mm, obovate 46–47 in l.s., elliptic 8–9 in c.s.; surface with a prominent rib on each face and also less prominent ribs or nerves, rugulose, mottled gray and black; pappus of deciduous scales.

7 **Artemisia absinthium** L. (x15) Absinthium. Achenes 1.2 x 0.6 x 0.3 mm, obovate 47 in l.s., elliptic 9 in c.s.; surface faintly longitudinally striate, grayish or brown; pappus none.

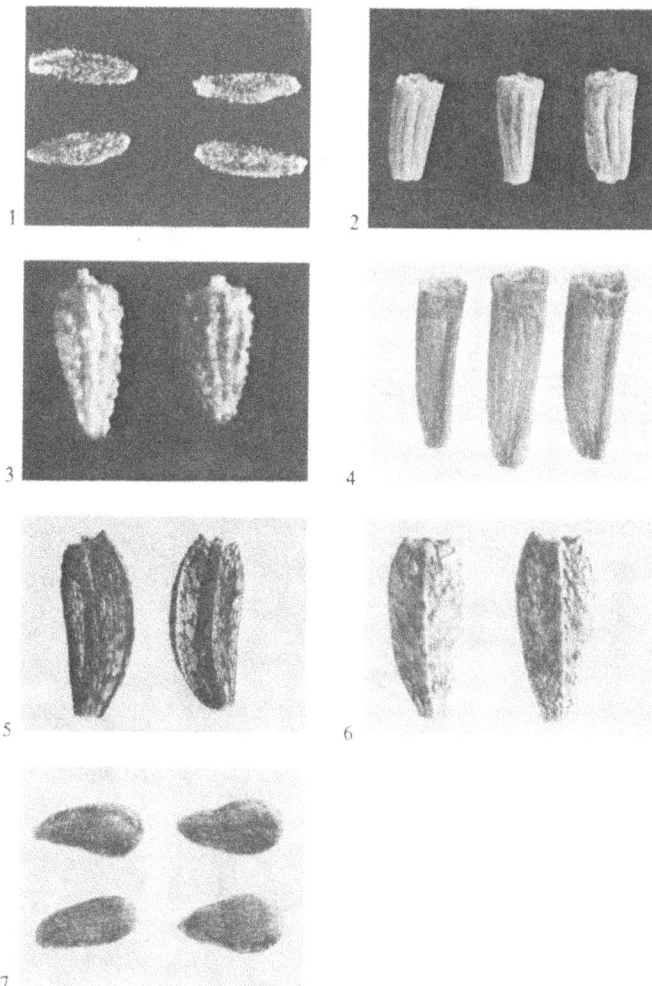

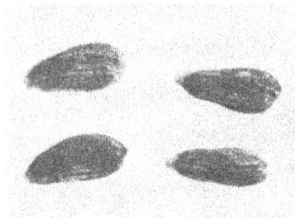

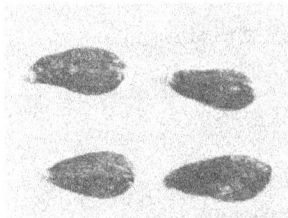

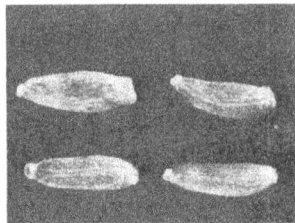

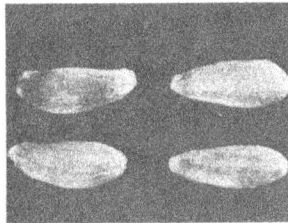

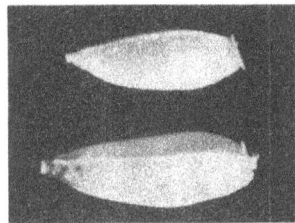

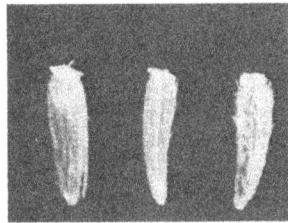

1 **Artemisia biennis** Willd. (x15) Biennial Wormwood. Achenes 1.0 x 0.4 x 0.25 mm, obovate 46–47 in l.s., elliptic 8–9 in c.s.; surface longitudinally striate, gray, and slightly shiny; pappus none.

2 **Artemisia campestris** L. (x15) Achenes 1.0 x 0.5 x 0.3 mm, obovate 47 in l.s., elliptic 8–9 in c.s.; surface faintly rugulose; pappus none.

3 **Artemisia ludoviciana** Nutt. (x11) White Sage. Achenes 1.3 x 0.5 x 0.3 mm, obovate 46–47 in l.s., elliptic 8–9 in c.s.; surface faintly longitudinally striate, brown; pappus none.

4 **Artemisia vulgaris** L. (x11) Common Mugwort. Achenes 1.4 x 0.5 x 0.4 mm, obovate 46–47 in l.s., elliptic 7–8 in c.s.; surface slightly ribbed and striate, dark gray to brown, and slightly glossy; pappus none.

5 **Aster cordifolius** L. (x13) Heart-leaved Aster. Achenes 1.8 x 0.5 x 0.3 mm, obovate 45–46 to oblong 13–14 in l.s., tapering at the base, elliptic 8–9 in c.s.; margins ridged and each face with one rib; pappus of capillary bristles.

6 **Aster ericoides** L. (x12) Heath Aster. Achenes 1.0 x 0.4 x 0.2 mm, obovate 46–47 in l.s., elliptic 9 in c.s.; margins ribbed and each surface with one rib and also nerved, slightly pubescent; pappus of capillary bristles.

7 **Aster macrophyllus** L. (x7) Achenes 3.2 x 0.8 x 0.6 mm, obovate 45–46 in l.s., elliptic 7–8 in c.s.; surface prominently ribbed and with several nerves, glabrous; pappus of capillary bristles.

8 **Aster novae-angliae** L. (x5) New England Aster. Achenes 2.2 x 0.7 x 0.3 mm, more or less oblong 13–14 in l.s., elliptic 9–10 in c.s.; longitudinally ribbed and densely pubescent; pappus of capillary bristles.

1 **Aster ptarmicoides** (Nees.) T.&G. (x10) Achenes 2.0 x 0.8 x 0.6 mm, obovate 46–47 in l.s., elliptic 7–8 in c.s.; surface ribbed and glabrous; pappus of capillary bristles.

2 **Aster puniceus** L. (x5.5) Achenes 3.5 x 0.8 x 0.6 mm, obovate 45–46 in l.s., elliptic 7–8 in c.s.; margin ribbed, surface longitudinally ribbed and faintly nerved, glabrous or slightly pubescent; pappus of capillary bristles.

3 **Aster umbellatus** Mill. (x10) Achenes 2.5 x 0.9 x 0.4 mm, obovate 46–47 in l.s., elliptic 9–10 in c.s.; surface ribbed and nerved, glabrous or slightly appressed pubescent; pappus of capillary bristles.

4 **Bellis perennis** L. (x10) English Daisy. Achenes 1.5 x 0.6 x 0.2 mm, obovate 46–47 in l.s., elliptic 10 in c.s., margin narrowly winged; surface striate, slightly pubescent or glabrous, gray to brown; pappus none.

5 **Bidens cernua** L. (x3.5) Bur Marigold. Outer and inner series of achenes are present, they average 5.5 x 2.4 x 1.0 mm, obtriangular 86–87 in l.s., elliptic 9–10 in c.s.; each surface with a prominent, retrorsely barbed rib, surfaces striate, faintly areolate, gray, brown, or purplish; pappus of 2–4 retrorsely barbed awns 2–3 mm long.

6 **Bidens coronata** (L.) Britt. (x4) Outer and inner series of achenes are present; they average 4.9 x 2.0 x 0.8 mm, obovate 46–47 in l.s., rhombic 33–34 in c.s.; surface nodulose and more or less strigose, margin with upwardly pointed cilia, brown to black; pappus of 2 short setose awns 1–2 mm long.

7 **Bidens frondosa** L. (x3) Achenes 9.8 x 4.0 x 0.5 mm, obovate 46–47 in l.s., oblong 23–24 in c.s.; each face 1-nerved, strigose and minutely scalariform, margin ciliate and slightly toothed; pappus of 2 retrorsely barbed awns 3–4 mm, long.

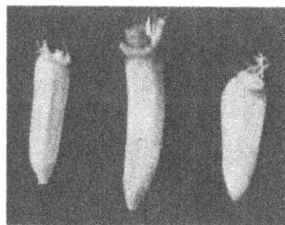

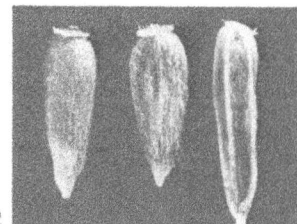

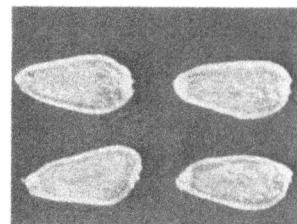

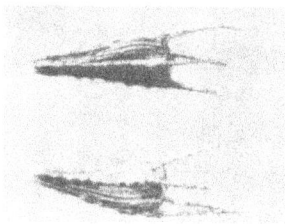

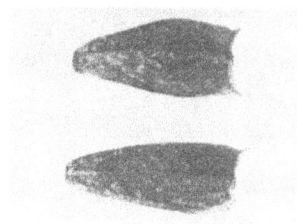

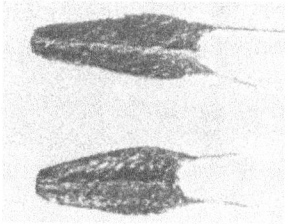

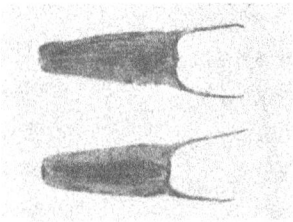

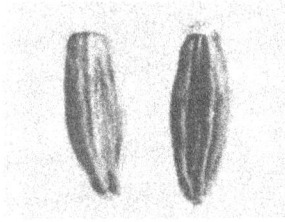

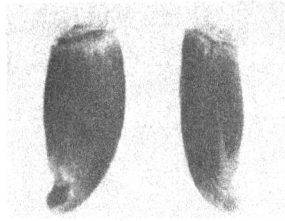

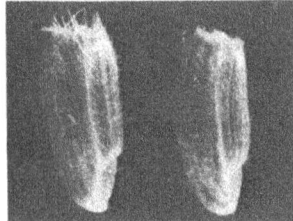

1 **Bidens vulgata** Greene (x2) Beggar's-ticks. Achenes 8.7 x 4.3 x 0.5 mm, obovate 46–47 in l.s., oblong 23–24 in c.s., tending to be concave-convex or plano-convex; midrib on each face prominent and the interveins obscure; margins upwardly ciliate, surface with scattered nodules, brownish or olivaceous; pappus of 2 awns 3–4 mm long and retrorsely barbed.

2 **Cacalia tuberosa** Nutt. (x4.5) White Snakeroot. Achenes 4.8 x 1.9 x 1.1 mm, obliquely elliptic 2–3 to oblong 14–15 in l.s., elliptic 8–9 in c.s.; surface strongly longitudinally ribbed, areolate and dark brown; pappus of capillary barbellate bristles.

3 **Carduus acanthoides** L. (x9) Welted Thistle. Achenes 2.8 x 1.2 x 0.6 mm, obliquely elliptic 2–3 to obovate 46–47 in l.s. with the axis slightly curved, obliquely elliptic 9 in c.s.; surface longitudinally nerved and transversely colliculose, gray to brown; pappus of finely barbed, capillary bristles.

4 **Carduus nutans** L. (x6.5) Nodding Thistle. Achenes 3.7 x 1.6 x 1.0 mm, obliquely obovate 46–47 in l.s., with a slightly curved axis, elliptic 8–9 in c.s.; surface longitudinally striate or nerved and with obscure transverse wavy cross-markings; pappus of finely barbed, capillary bristles.

5 **Centaurea jacea** L. (x8.5) Knapweed. Achenes 2.9 x 1.4 x 0.9 mm, obovate 46–47 to elliptic 2–3 in l.s., the hilum lateral in a crescent-shaped indentation at the base, elliptic 8–9 in c.s; surface faintly longitudinally ribbed and nerved, finely pubescent; pappus none.

6 **Centaurea maculosa** Lam. (x10) Spotted Knapweed. Achenes 2.3 x 1.2 x 0.6 mm, obovate 47–48 in l.s., hilum in a lateral, basal indentation, elliptic 9 in c.s.; surface dark brown with whitish ribs and nerves; pappus a series of unequal scales.

7 **Centaurea nigra** L. (x9) Black Knapweed. Achenes 2.8 x 1.3 x 0.9 mm, obovate 46–47 in l.s., hilum in a lateral, basal indentation, elliptic 7–8 in c.s.; surface brown to purplish with whitish ribs and nerves, slightly hirsute; pappus of short fine unequal scales.

1 **Centaurea repens** L. (x8.5) Russian Knapweed. Achenes 2.7 x 1.9 x 1.1 mm, elliptic 4–5 in l.s., with the hilum sub-basal, elliptic 8–9 in c.s.; surface white, inconspicuously ribbed and nerved, glabrous; pappus of short bristles or none.

2 **Chrysanthemum balsamita** L. (x12) Costmary. Achenes 1.3 x 0.4 x 0.2 mm, obtriangular 85–86 to obovate 45–46 in l.s., oblong 21 in c.s.; surface longitudinally ribbed, slightly pubescent, and glandular dotted; pappus a short membranous crown.

3 **Chrysanthemum leucanthemum** L. (x11) Ox-eye Daisy. Achenes 1.8 x 0.7 x 0.5 mm, obovate 46–47 in l.s., elliptic 7–8 in c.s.; surface with 10 longitudinal white ribs alternating with brown or black sulci, white-papillose; pappus none.

4 **Chrysanthemum parthenium** (L.) Bernh. (x13) Feverfew. Achenes 1.4 x 0.4 x 0.4 mm, obtriangular 85–86 in l.s., oblong 18 in c.s.; surface longitudinally 8–10-ribbed and glandular dotted or glabrous; pappus merely a short toothed crown or absent.

5 **Cichorium intybus** L. (x8) Chicory. Achenes 2.6 x 1.2 x 0.8 mm, obliquely obovate 46–47 in l.s., apex truncate, elliptic 8 in c.s., the strong ribs giving a somewhat angular appearance; surface longitudinally striate or finely ribbed and transversely rugulose; pappus a ring of white chaffy scales.

6 **Cirsium arvense** (L.) Scop. (x7) Canada Thistle. Achenes 3.2 x 1.1 x 0.8 mm, obliquely oblong 14–15 to elliptic 2–3 in l.s., tapering slightly at the base, axis slightly curved, elliptic 7–8 in c.s.; surface finely striate, apex with a cartilaginous rim, gray to brown; pappus of plumose bristles.

7 **Cirsium muticum** Michx. (x4) Swamp Thistle. Achenes 4.5 x 1.7 x 1.0 mm, obliquely obovate 46–47 in l.s., elliptic 8–9 in c.s.; surface with rather prominent ribs tending to give a 4-angled or rhombic appearance, surface black or brown with purplish longitudinal lines; pappus of plumose bristles.

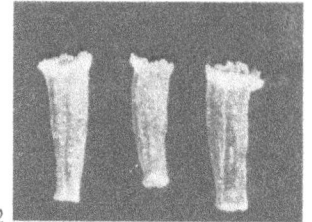

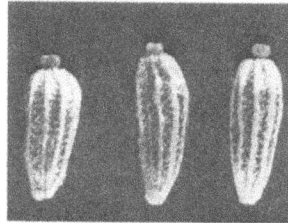

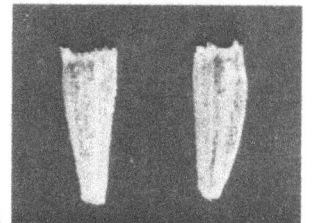

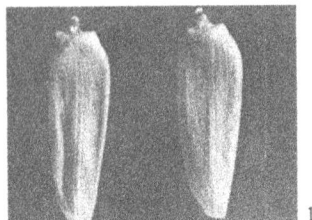

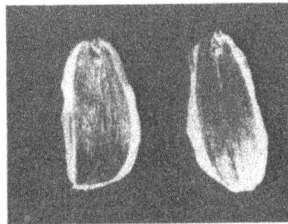

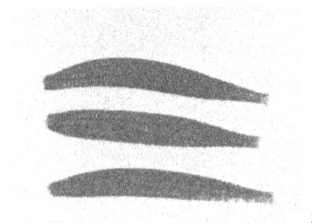

1 **Cirsium pitcheri** (Torr.) T.&G. (x3.6) Achenes 6.7 x 2.3 x 1.2 mm, obovate 46–47 in l.s., elliptic 8–9 in c.s.; surface faintly nerved, apex with a cartilaginous rim, gray to brown; pappus of plumose bristles.

2 **Cirsium vulgare** (Savi) Tenore (x6.5) Bull Thistle. Achenes 3.5 x 1.5 x 1.0 mm, obliquely obovate 46–47 in l.s. with a slightly curved axis, elliptic 8 in c.s., faces rather prominently ribbed tending to give a 4-angled appearance; surface with longitudinal dark lines on a grayish background, apex with a narrow, cartilaginous rim; pappus of plumose bristles.

3 **Conyza canadensis** (L.) Cronq. (x17) Canada Fleabane. Achenes 1.3 x 0.4 x 0.2 mm, oblong 13–14 in l.s., elliptic 9 in c.s., margin with a prominent rib; surface pubescent, gray to straw-coloured; pappus of capillary bristles.

4 **Coreopsis tripteris** L. (x3.5) Tall Coreopsis. Achenes 5.8 x 2.3 x 0.5 mm, oblong 14–15 in l.s., with a rounded apex and base, elliptic 10–11 or concave-convex in c.s.; both surfaces with a median rib and numerous, obscure striations, faintly areolate and papillate, margins with a light-coloured, coriaceous wing, central portion dark brown to black, apex may be minutely setose; pappus a few short bristles and occasionally barbed awns.

5 **Crepis tectorum** L. (x8) Hawk's-beard. Achenes 3.5 x 0.5 x 0.3 mm, elliptic 1 in l.s., elliptic 8–9 in c.s.; surface longitudinally 10-ribbed and transversely rugulose or spiculose, brown or purplish in colour; pappus of capillary bristles.

6 **Erechtites hieracifolia** (L.) Raf. (x7.5) Fireweed. Achenes 3.3 x 0.6 x 0.5 mm, elliptic 1–2 in l.s., elliptic 7 in c.s. or slightly angled; surface with 10–13 pale longitudinal ribs, strigose between the ribs; pappus of capillary bristles.

1 **Erigeron annuus** (L.) Pers. (x17) Annual Daisy Fleabane. Achenes 0.8 x 0.3 x 0.2 mm, oblong 14–15 in l.s., elliptic 8 to slightly angular in c.s.; surface 2-nerved and slightly pubescent; pappus of the disk florets of bristles and delicate scales, that of the ray florets consisting of scales only.

2 **Erigeron philadelphicus** L. (x17) Philadelphia Daisy Fleabane. Achenes 0.9 x 0.4 x 0.2 mm, oblong 14–15 in l.s., elliptic 9 to slightly angular in c.s.; surface 2-nerved and slightly pubescent; pappus of simple bristles.

3 **Erigeron pulchellus** Michx. (x14) Robin's Plantain. Achenes 1.4 x 0.5 x 0.15 mm, oblong 14–15 in l.s., elliptic 10–11 to slightly angular in c.s.; surface 2-nerved, glabrous or almost so; pappus of simple bristles.

4 **Erigeron strigosus** Muhl. (x14) Rough Daisy Fleabane. Achenes 1.0 x 0.4 x 0.2 mm, oblong 14–15 in l.s., elliptic 9 or slightly angular in c.s.; surface 2-nerved, pubescent; pappus double, the outer series of delicate scales and the inner of fragile bristles, bristles lacking on the pistillate florets.

5 **Eupatorium maculatum** L. (x5.5) Joe-pye-weed. Achenes 3.5 x 0.5 x 0.5 mm, oblong 12–13 in l.s., tapering to an acute base, oblong 18 in c.s. or 5-angled; surface with intermediate ribs between the angles, colour black with minute, light-coloured, glandular dots and obscurely transversely rugulose; pappus of slightly barbed capillary bristles.

6 **Eupatorium perfoliatum** L. (x13) Boneset. Achenes 1.6 x 0.4 x 0.4 mm, oblong 13–14 in l.s., tapering at the base, oblong 18 or 5-angled in c.s.; surface with minute white glandular dots, very finely striate and transversely rugulose; pappus of slightly barbed bristles.

7 **Eupatorium purpureum** L. (x6) Green-stemmed Joe-pye-weed. Achenes 4.1 x 0.5 x 0.5 mm, oblong 12–13 in l.s., pointed at the base, oblong 18 or 5-angled in c.s.; surface black with white or brownish glandular dots, very finely longitudinally striate, and transversely rugulose; pappus of slightly barbed bristles.

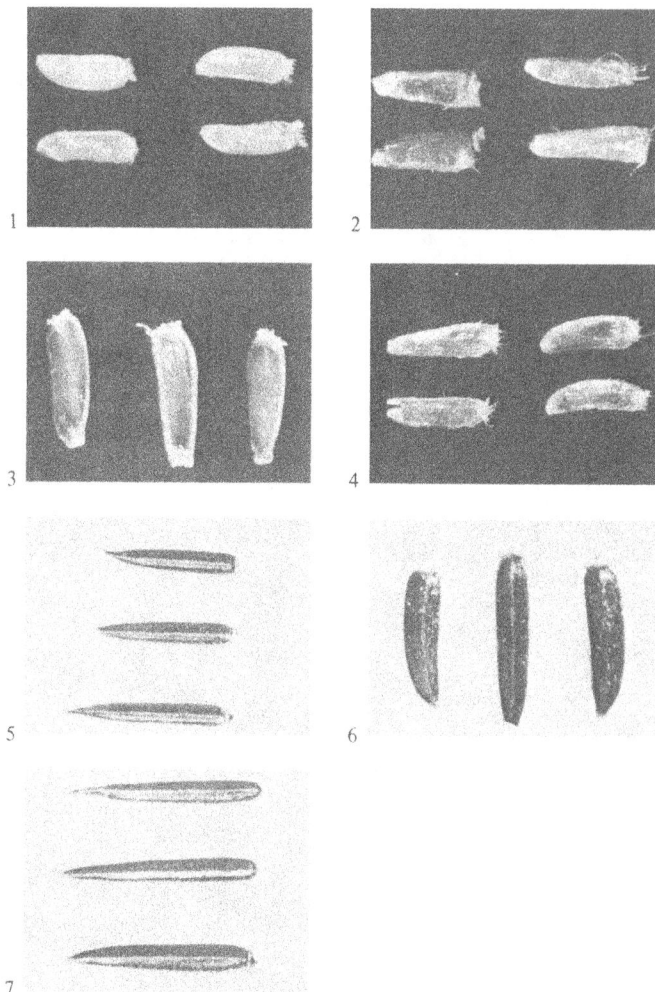

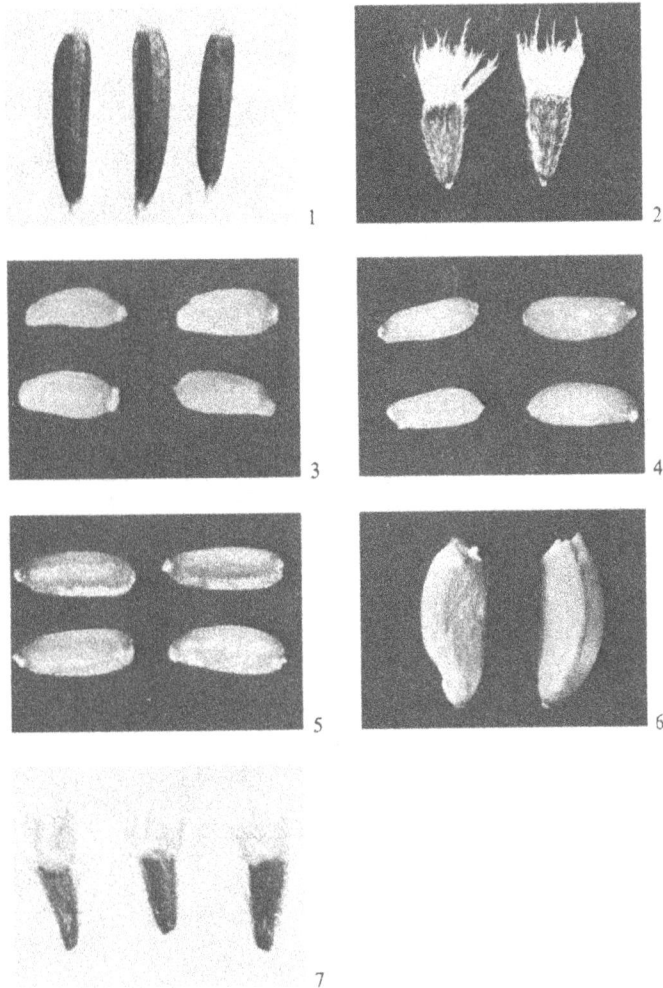

1 **Eupatorium rugosum** Houtt. (x9) White Snakeroot. Achenes 2.6 x 0.4 x 0.4 mm, oblong 12–13 in l.s., pointed at the base, oblong 18 or 5-angled in c.s.; surface black, not obviously glandular-dotted, faintly areolate; pappus of slightly barbed bristles.

2 **Galinsoga ciliata** (Raf.) Blake (x10) Achenes 1.4 x 0.6 x 0.5 mm, obtriangular 86–87 in l.s., elliptic 7 or obtriangular 91 in c.s.; surface black and strigose; pappus of fimbriate or awn-tipped scales or both.

3 **Gnaphalium obtusifolium** L. (x17) Cat's-foot. Achenes 0.7 x 0.3 x 0.2 mm, oblong 14–15 to elliptic 2–3 in l.s., elliptic 8 in c.s., slightly angular; surface smooth and grayish; pappus of capillary bristles.

4 **Gnaphalium uliginosum** L. (x25) Low Cudweed. Achenes 0.6 x 0.2 x 0.1 mm, oblong 14 to elliptic 2 in l.s., elliptic 9 in c.s.; surface smooth or faintly longitudinally ribbed, light brown; pappus of capillary bristles.

5 **Gnaphalium viscosum** H.B.K. (x20) Achenes 0.7 x 0.3 x 0.1 mm, oblong 14–15 to elliptic 2–3 in l.s., elliptic 10 in c.s.; margin ridged and each face with a low rib; surface gray to brownish, slightly roughened; pappus of capillary bristles.

6 **Grindelia squarrosa** (Pursh) Dunal (x7) Gum Plant. Achenes 2.9 x 1.1 x 0.5 mm, obliquely obovate 46–47 in l.s., elliptic 9–10 to obtriangular 86–87 in c.s.; surface glabrous, longitudinally ridged and nerved, white to brown; pappus of 2–6 deciduous awns. The axes of the ray achenes are more curved than those of the disk achenes.

7 **Helenium autumnale** L. (x8) Sneezeweed. Achenes 1.5 x 0.6 x 0.4 mm, obovate 46–47 in l.s., elliptic 8 in c.s. with 5, well-developed ribs and 5 delicate intermediate nerves; main ribs strigose, nerves glabrous, surface gray with brownish, resinous dots; pappus of papery, awn-tipped scales.

1 **Helenium flexuosum** Raf. (x11) Achenes 1.0 x 0.5 x 0.4 mm, obtriangular 87 in l.s., 5-angled in c.s.; angles white-strigose, surface brownish with brownish resinous dots; pappus of scarious, awn-tipped scales.

2 **Helianthus divaricatus** L. (x5) Achenes 4.2 x 2.3 x 1.6 mm, obovate 47–48 in l.s., elliptic 7–8 in c.s.; surface longitudinally striate, glabrous, black or slightly mottled; pappus of 2 thin deciduous, and chaffy awns.

3 **Helianthus tuberosus** L. (x5.5) Jerusalem Artichoke. Achenes 4.5 x 1.2 x 0.8 mm, oblong 13–14 in l.s., rhombic 32 in c.s.; lateral angles prominent and slightly winged; surface grayish and slightly pubescent, particularly at the apex; pappus of 2 deciduous awns. All achenes collected appeared to be sterile.

4 **Hieracium aurantiacum** L. (x11) Devil's Paintbrush. Achenes 1.9 x 0.4 x 0.4 mm, oblong 13–14 in l.s., abruptly tapering at the base, oblong 18 in c.s.; surface with 10 longitudinal ribs, scalariform, dull, and black; pappus of capillary bristles.

5 **Hieracium canadense** Michx. (x9) Achenes 3.0 x 0.6 x 0.6 mm, oblong 13–14 in l.s., tapering slightly at the base, oblong 18 in c.s.; surface longitudinally ribbed and scalariform, brown; pappus of capillary bristles.

6 **Hieracium pilosella** L. (x10) Mouse-ear Hawkweed. Achenes 2.0 x 0.4 x 0.4 mm, oblong 13–14 in l.s., tapering abruptly at the base, oblong 18 in c.s.; surface with 10 longitudinal ribs, scalariform, black; pappus of capillary bristles.

7 **Hieracium scabrum** Michx. (x12) Achenes 2.5 x 0.6 x 0.6 mm, oblong 13–14 in l.s., tapering at the base, oblong 18 in c.s.; faces longitudinally ribbed and scalariform, black; pappus of capillary bristles.

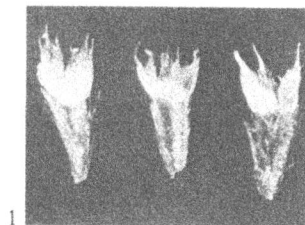

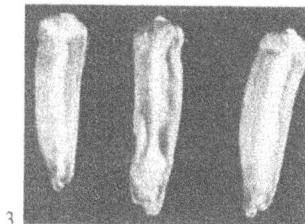

COMPOSITAE

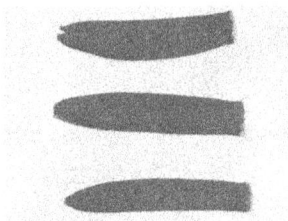

1

2

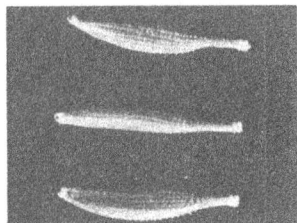
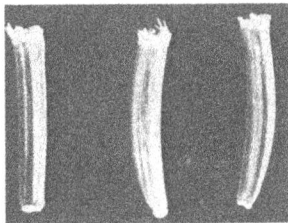

3

4

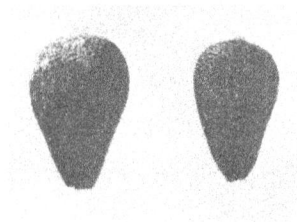

5

6

7

1 **Hieracium umbellatum** L. (x10) Achenes 2.8 x 0.7 x 0.7 mm, oblong 13–14 in l.s., tapering at the base, oblong 18 in c.s.; surface longitudinally ribbed and scalariform, brown to black; pappus of capillary bristles.

2 **Hieracium vulgatum** Fries (x8) Achenes 3.0 x 0.5 x 0.5 mm, oblong 13 in l.s., pointed at the base, oblong 18 in c.s.; surface longitudinally ribbed, scalariform or rugulose, brown to black; pappus of capillary bristles.

3 **Hypochaeris radicata** L. (x5.5) Cat's-ear. Body of the achene 5.0 x 0.6 x 0.4 mm, elliptic 1 in l.s. with a beak about 6.0 mm long, elliptic 8 in c.s.; surface longitudinally 15-ribbed and transversely rugulose or spiculose, rather muricate at the apex; pappus of barbellate bristles, often of two different lengths.

4 **Inula helenium** L. (x6) Elecampane. Achenes 4.5 x 0.7 x 0.7 mm, oblong 12 in l.s., oblong 18 in c.s.; faces ribbed, faintly areolate and brown; pappus of capillary bristles.

5 **Iva xanthifolia** Nutt. (x12) Marsh Elder. Achenes 2.2 x 1.5 x 1.0 mm, obovate 48–49 in l.s., elliptic 8 in c.s.; surface striate, finely papillose, brown to black; pappus none.

6 **Krigia biflora** (Walt.) Blake (x10) Dwarf Dandelion. Achenes 1.8 x 0.7 x 0.7 mm, oblong 14–15 in l.s., elliptic 6 in c.s.; surface longitudinally ribbed and transversely rugulose, brown to black; pappus of scales and capillary bristles.

7 **Lactuca biennis** (Moench.) Fern. (x5) Achenes 4.7 x 1.3 x 0.5 mm, obliquely elliptic 1–2 in l.s. terminating in a short beak, elliptic 9–10 in c.s.; surface longitudinally ribbed and transversely rugulose, mottled gray and brown; pappus of brownish capillary bristles.

1 **Lactuca canadensis** L. (x6) Achenes 5.6 x 1.5 x 0.2 mm, elliptic 1–2 in l.s. with a slender beak about 2.0 mm long, oblong 23–24 in c.s. and slightly winged; each surface with one prominent rib and two less conspicuous ones, finely transversely rugulose, black, or mottled brown and black; pappus of capillary bristles.

2 **Lactuca pulchella** (Pursh) DC. (x5) Blue Lettuce. Achenes 5.0 x 1.2 x 0.5 mm, elliptic 1–2 in l.s. with a short beak, elliptic 9–10 in c.s.; surface longitudinally ribbed and faintly transversely rugulose, dull, black with the tip of the beak whitish; pappus of capillary bristles.

3 **Lactuca serriola** L. (x6) Prickly Lettuce. Body of the achenes 3.5 x 1.0 x 0.5 mm, obovate 45 in l.s. with a long, fragile, and filiform beak about 3.5 mm long, elliptic 9 in c.s.; faces with 5–7 ribs and faintly transversely rugulose, ribs slightly strigose at the apex, mottled gray to brown; pappus of white capillary bristles.

4 **Lapsana communis** L. (x6) Nipplewort. Achenes 4.0 x 1.0 x 0.6 mm, obliquely obovate 45–46 in l.s., elliptic 8–9 in c.s. with prominent marginal ribs; surface finely ribbed with the central rib being the most prominent, light brown; pappus none.

5 **Leontodon autumnalis** L. (x5) Fall Dandelion. Achenes 4.7 x 0.6 x 0.3 mm, oblong 12–13 in l.s., oblong 21 in c.s.; surface longitudinally 5-ribbed and transversely rugulose or spiculose, straw-coloured with the spicules a darker brown; pappus of plumose bristles.

6 **Liatris asper** Michx. (x5) Achenes 4.5 x 1.2 x 0.8 mm, obtriangular 85–86 in l.s., rhombic 32 in c.s.; faces longitudinally nerved and densely pubescent with white hairs, body minutely transversely rugulose and black; pappus of barbellate bristles.

7 **Liatris cylindracea** Michx. (x4) Achenes 5.5 x 1.0 x 0.7 mm, obtriangular 85–86 in l.s., rhombic 31–32 in c.s.; surface longitudinally ribbed and finely transversely rugulose, densely pubescent, black; pappus of plumose bristles.

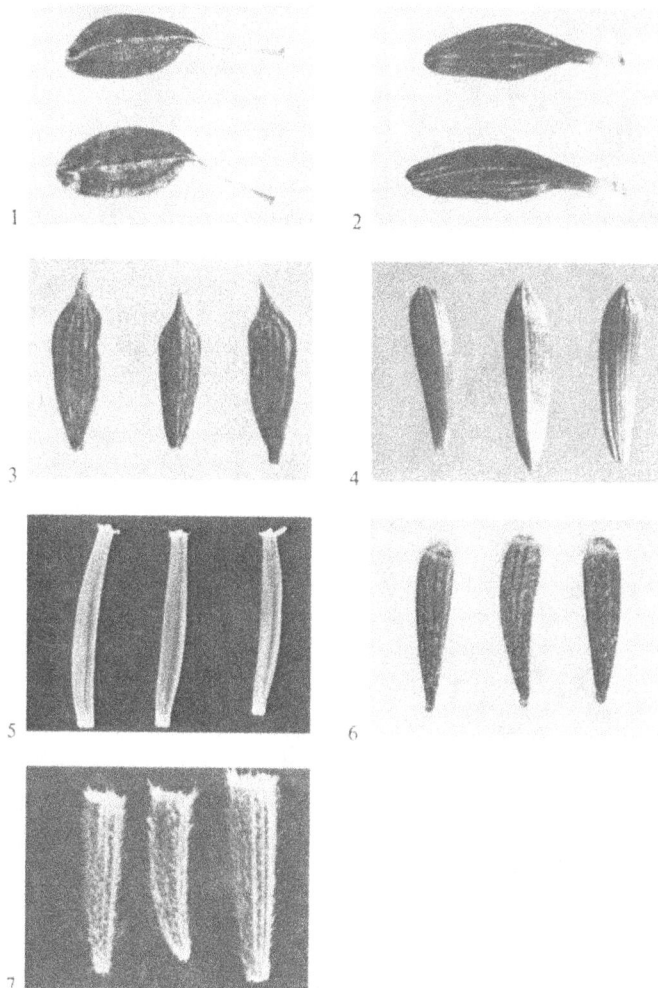

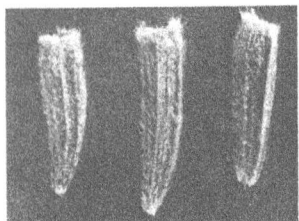

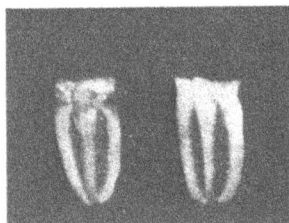

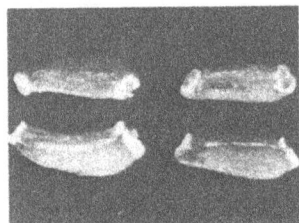

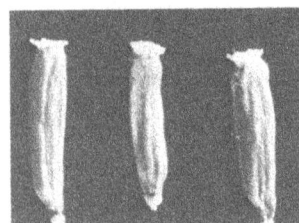

1 **Liatris spicata** (L.) Willd. (x5.5) Achenes 4.5 x 1.3 x 0.9 mm, obtriangular 85–86 in l.s., rhombic 31–32 in c.s.; faces strongly nerved with striations between the nerves, finely transversely rugulose, white pubescent, body black; pappus of barbellate bristles.

2 **Matricaria maritima** L. (x10) Achenes 1.6 x 0.8 x 0.5 mm, obovate 47 in l.s., obliquely triangular 92–93 in c.s.; surface with 3 broad whitish ribs, body black and transversely rugulose; pappus a short, toothed crown.

3 **Matricaria matricarioides** (Less.) Porter (x15) Pineapple-weed. Achenes 1.0 x 0.3 x 0.2 mm, obliquely obovate 45–46 in l.s., elliptic 8 or slightly 4-angled in c.s.; surface with 2–4 longitudinal nerves, smooth, olive or brown in colour; pappus a short crown or none.

4 **Onopordum acanthium** L. (x5) Scotch Thistle. Achenes 4.5 x 2.4 x 1.6 mm, obovate 47–48 in l.s., elliptic 8 in c.s. with a strong rib on the surfaces tending to give a slightly angular appearance; surface faintly longitudinally nerved and transversely rugose; gray to brown with irregular black dotting; pappus of barbellate bristles.

5 **Petasites frigidus** (L.) Fries (x12) Sweet Colt's-foot. Achenes 2.2 x 0.5 x 0.5 mm, oblong 13–14 in l.s., elliptic 6 in c.s.; surface longitudinally ribbed, light brown; pappus of capillary bristles up to 15 mm long.

6 **Petasites sagittatus** (Pursh) Gray (x12) Achenes 2.8 x 0.5 x 0.5 mm, oblong 13–14 in l.s., elliptic 6 in c.s.; surface longitudinally ribbed, brown; pappus of capillary bristles up to 18 mm long.

7 **Picris hieracioides** L. (x5.5) Ox-tongue. Achenes 3.9 x 0.8 x 0.8 mm, elliptic 1–2 in l.s., elliptic 6 in c.s.; surface longitudinally ribbed and transversely rugulose, black or dark brown; pappus of plumose bristles.

1 **Polymnia canadensis** L. (x6) Leafcup. Achenes 3.9 x 2.5 x 1.9 mm, obovate 47–48 in l.s., elliptic 7–8 in c.s. with the margins strongly ribbed and a prominent rib on one face; surface finely striate, areolate, pubescent, and black; pappus none.

2 **Prenanthes alba** L. (x6) White Lettuce. Achenes 4.9 x 1.0 x 0.7 mm, oblong 13–14 in l.s., elliptic 7–8 in c.s. with prominent ribs giving a rather angular appearance; surface longitudinally nerved, obscurely spiculose and brownish; pappus of cinnamon-coloured barbellate bristles.

3 **Prenanthes altissima** L. (x5) Tall White Lettuce. Achenes 4.7 x 0.9 x 0.6 mm, oblong 13–14 in l.s., elliptic 8 in c.s. with prominent ribs giving an angular appearance; surface strongly nerved and brown; pappus of whitish barbellate bristles.

4 **Prenanthes racemosa** Michx. (x5.5) Rattlesnake Root. Achenes 4.2 x 0.9 x 0.6 mm, oblong 13–14 in l.s., elliptic 8 in c.s.; surface longitudinally ribbed and nerved giving a somewhat angular appearance; pappus of light brown barbellate bristles.

5 **Ratibida pinnata** (Vent.) Barnh. (x9) Achenes 2.2 x 1.1 x 0.7 mm, obovate 47 in l.s., elliptic 8–9 in c.s., margins ribbed; surface longitudinally striate and minutely spiculose, brown or black; pappus none.

6 **Rudbeckia hirta** L. var. **pulcherrima** Farw. (x15) Black-eyed Susan. Achenes 1.7 x 0.4 x 0.4 mm, oblong 13–14 in l.s., tapering at the base, oblong 18 in c.s.; surface finely nerved and transversely rugulose, black; pappus none. Only the disk florets are fertile.

7 **Rudbeckia laciniata** L. (x5) Cut-leaved Coneflower. Achenes 4.9 x 1.5 x 1.0 mm, obliquely oblong 13–14 in l.s., elliptic 8 to rhombic 32 in c.s.; surfaces longitudinally ribbed and finely nerved, obscurely transversely rugulose; pappus a short, toothed crown. Only the disk florets are fertile.

8 **Senecio aureus** L. (x10) Golden Ragwort. Achenes 2.4 x 0.5 x 0.5 mm, oblong 13–14 in l.s., elliptic 6 in c.s.; surface longitudinally ribbed and nerved, glabrous, and brown; pappus of white bristles.

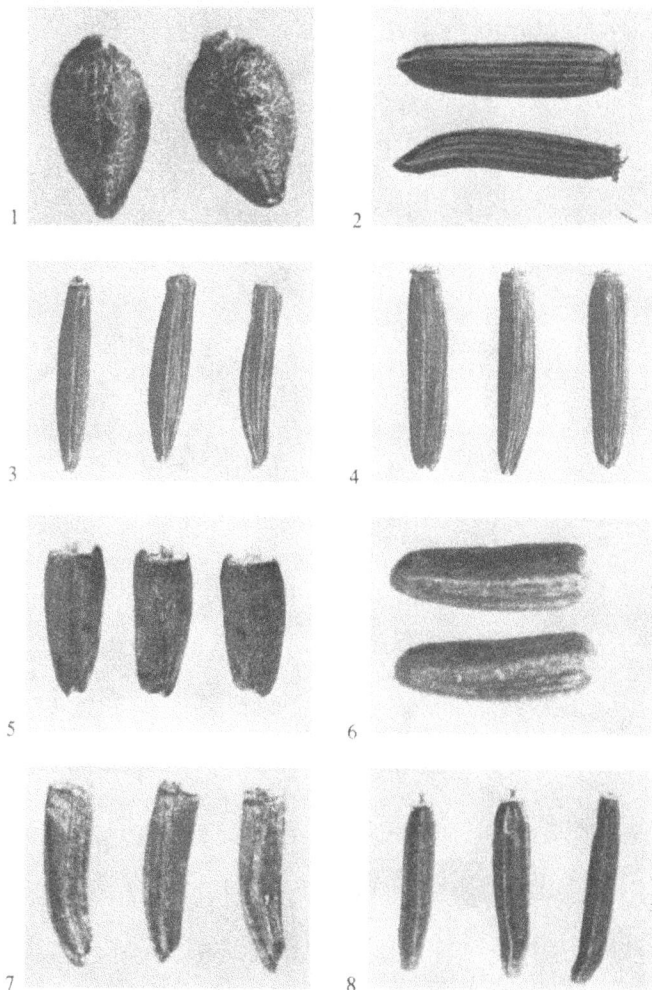

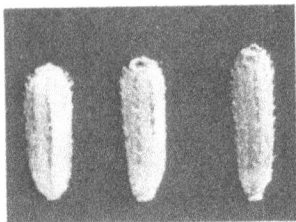

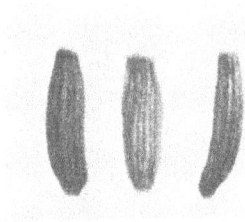

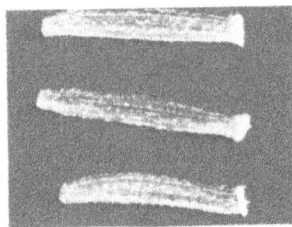

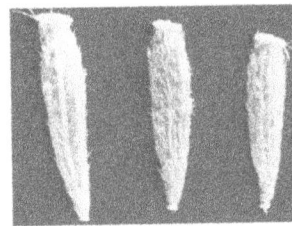

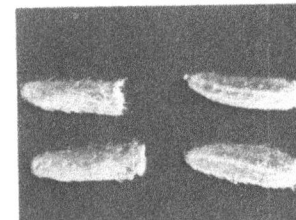

1 **Senecio jacobaea** L. (x10) Ragwort. Achenes 1.7 x 0.6 x 0.5 mm, oblong 14–15 in l.s., elliptic 7 in c.s.; surface longitudinally ribbed, pubescent and grayish; pappus of white bristles.

2 **Senecio pauperculus** Michx. (x10) Achenes 2.0 x 0.4 x 0.4 mm, oblong 13–14 in l.s., elliptic 6 in c.s.; surface strongly longitudinally ribbed and nerved, glabrous or sometimes hispidulous on the ridges; pappus of white bristles.

3 **Senecio plattensis** Nutt. (x10) Achenes 1.8 x 0.4 x 0.4 mm, oblong 13–14 in l.s., elliptic 6 to oblong 18 in c.s.; surface longitudinally ribbed and glabrous or hispidulous; pappus of white bristles.

4 **Senecio vulgaris** L. (x12) Common Groundsel. Achenes 2.3 x 0.4 x 0.4 mm, oblong 13–14 in l.s., elliptic 6 in c.s.; surface longitudinally ribbed with the ridges hirtellous or puberulent, intermediate sulci white-strigose; pappus of white bristles.

5 **Silphium perfoliatum** L. (x2) Prairie Dock. Achenes 10.0 x 5.9 x 1.0 mm, obovate 47–48 in l.s., very thin and linear in c.s., margin winged; surface with a prominent central rib, striate and finely areolate; pappus none. Only the ray florets produce achenes.

6 **Solidago caesia** L. (x13) Blue-stem Goldenrod. Achenes 2.3 x 0.5 x 0.3 mm, oblong 13–14 in l.s., tapering below the middle, elliptic 8–9 in c.s.; surface longitudinally ribbed and pubescent; pappus of capillary bristles.

7 **Solidago canadensis** L. (x11) Canada Goldenrod. Achenes 1.3 x 0.5 x 0.5 mm, oblong 14–15 in l.s., elliptic 6 in c.s.; surface longitudinally ribbed and slightly pubescent; pappus of capillary barbellate bristles.

8 **Solidago flexicaulis** L. (x7) Zig-zag Goldenrod. Achenes 2.5 x 0.5 x 0.5 mm, oblong 13–14 in l.s., tapering below the middle, elliptic 6 in c.s.; longitudinally ribbed and pubescent; pappus of capillary bristles.

1 **Solidago graminifolia** (L.) Salisb. (x20) Grass-leaved Goldenrod. Achenes 0.8 x 0.3 x 0.3 mm, obovate 46–47 in l.s., elliptic 6 in c.s.; faintly ribbed and pubescent; pappus of capillary bristles.

2 **Solidago nemoralis** Ait. (x10) Field Goldenrod. Achenes 1.5 x 0.5 x 0.3 mm, oblong 14 in l.s., elliptic 8–9 in c.s.; surface ribbed and obscurely pubescent; pappus of capillary bristles.

3 **Solidago ohioensis** Riddell (x16) Achenes 1.2 x 0.5 x 0.4 mm, oblong 14–15 in l.s., elliptic 7–8 in c.s.; surface glabrous, grayish, or light brown; pappus of capillary bristles.

4 **Sonchus arvensis** L. (x9) Perennial Sow Thistle. Achenes 2.6 x 0.9 x 0.6 mm, elliptic 2–3 to oblong 14–15 in l.s., elliptic 8 in c.s.; surface longitudinally ribbed and transversely rugulose, brown to black; pappus of capillary bristles.

5 **Sonchus asper** (L.) Mill. (x8) Spiny Annual Sow Thistle. Achenes 2.6 x 1.0 x 0.3 mm, elliptic 2–3 in l.s., elliptic 10–11 in c.s., each face with 5–7 ribs and striate, not transversely rugulose; pappus of capillary bristles.

6 **Sonchus oleraceus** L. (x7.5) Common Annual Sow Thistle. Achenes 3.0 x 0.8 x 0.4 mm, obovate 45–46 in l.s., elliptic 9 in c.s.; each face with 3–5 ribs and finely transversely rugulose; pappus of capillary bristles.

7 **Sonchus uliginosus** Bieb. (x12) Achenes 3.4 x 1.1 x 0.6 mm, obliquely elliptic 2–3 to oblong 14–15 in l.s., elliptic 8–9 in c.s.; surface longitudinally ribbed and transversely rugulose, dark brown to black; pappus of capillary bristles.

8 **Tanacetum vulgare** L. (x12) Tansy. Achenes 1.6 x 0.5 x 0.5 mm, obtriangular 85–86 in l.s., rhombic 30 in c.s., sometimes 5-angled; surface slightly glandular and gray; pappus none or a mere 3- to 5-toothed crowning ridge.

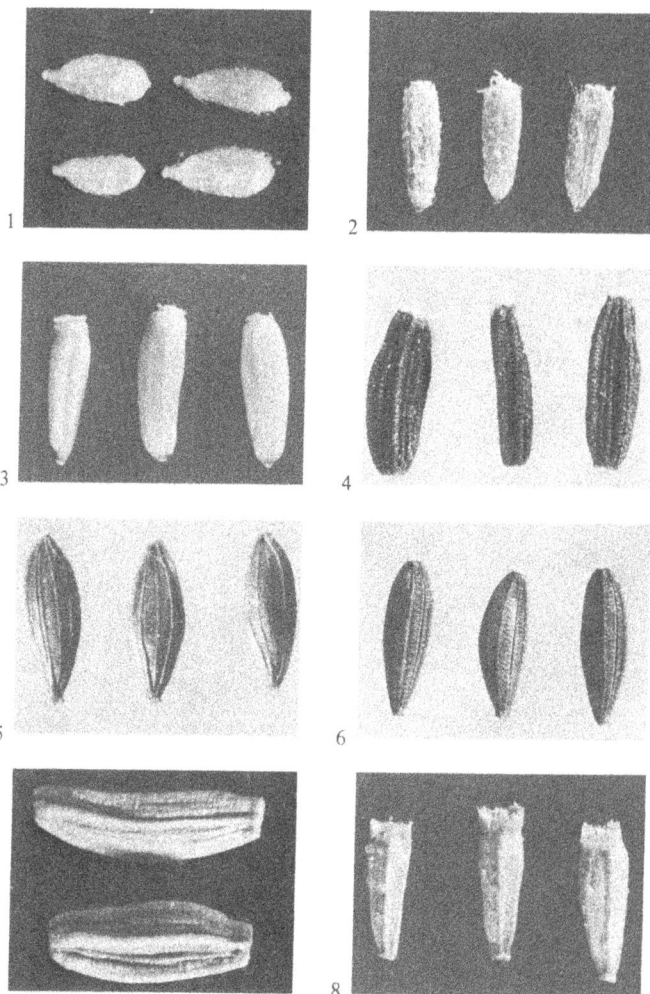

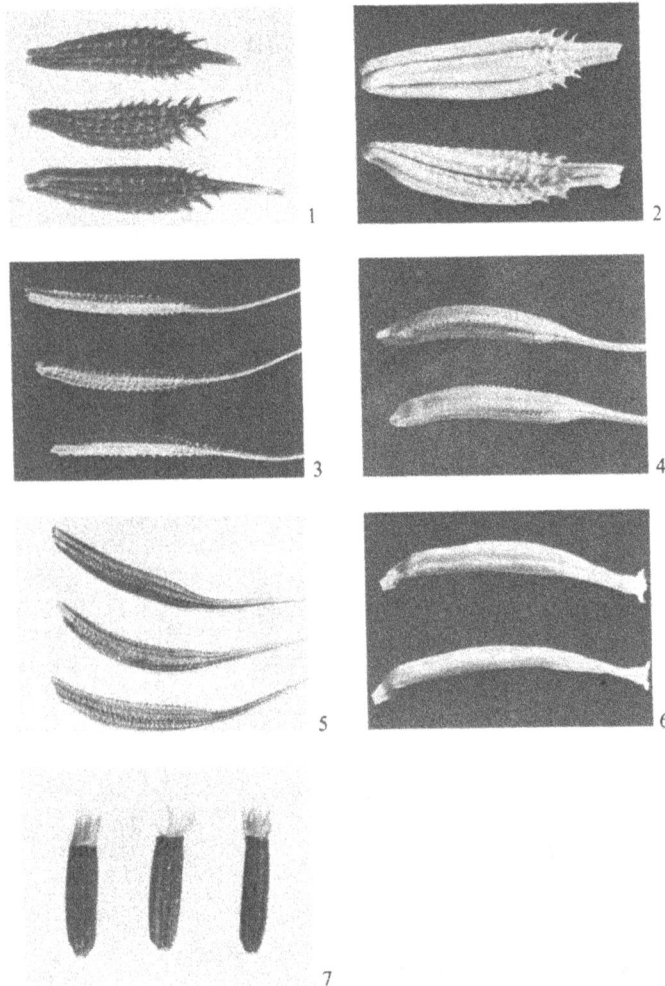

1 **Taraxacum laevigatum** (Willd.) DC. (x6) Red-seeded Dandelion. Achenes 3.0 x 0.8 x 0.5 mm, obovate 45–46 in l.s. tapering to a beak up to three times as long as the body, elliptic 8–9 in c.s.; surface longitudinally 12-ribbed with the ribs rugulose below the middle and spiny or muricate above, reddish-brown in colour; pappus of capillary bristles.

2 **Taraxacum officinale** Weber (x8.5) Common Dandelion. Achenes 3.5 x 0.8 x 0.6 mm, obovate 45–46 in l.s. tapering to a beak 2.5–4 times as long as the body, elliptic 7–8 in c.s.; surface 10–15 ribbed and smooth below the middle, spiny or muricate above, gray to brown; pappus of capillary bristles.

3 **Tragopogon dubius** Scop. (x1.4) Body of the achene 12.0 x 1.3 x 1.3 mm, obliquely elliptic 1 in l.s. and terminating in a beak about as long as the body, oblong 18 or 5-angled in c.s.; surface longitudinally 10-ribbed and transversely rugulose or muricate; pappus of white plumose bristles.

4 **Tragopogon porrifolius** L. (x2.3) Salsify. Achene body 12.0 x 1.3 x 1.3 mm, obliquely elliptic 1 in l.s., tapering to a beak about twice as long as the body, elliptic 6 in c.s. or slightly angled; surface longitudinally 10-ribbed and rugulose or muricate; pappus of brownish plumose bristles.

5 **Tragopogon pratensis** L. (x2) Goat's-beard. Body of the achene 13.0 x 1.8 x 1.8 mm, obliquely elliptic 1 in l.s. and tapering to a beak about 10 mm long, elliptic 6 in c.s.; surface with 10 prominent ribs and transversely rugulose; pappus of whitish plumose bristles.

6 **Tussilago farfara** L. (x10) Colt's-foot. Achenes 3.0 x 0.5 x 0.3 mm, oblong 13 in l.s., elliptic 8–9 in c.s.; surface faintly ribbed; pappus of white capillary bristles about 10 mm long.

7 **Vernonia missurica** Raf. (x5) Ironweed. Achenes 3.1 x 0.8 x 0.6 mm, oblong 13–14 in l.s., oblong 19–20 in c.s.; surface strongly ribbed, finely pubescent and slightly glandular, particularly in the sulci; pappus of purplish or tawny bristles.

1 **Xanthium strumarium** L. (x1) Cocklebur. Fruit an elliptic bur formed from the involucre of the pistillate flowers, achenes enclosed by the bur; bur 30 x 20 x 20 mm, ovate 39 in l.s., elliptic 6 in c.s.; body covered with long, hooked prickles, apex with two stout incurved beaks, surface and bases of the prickles pubescent and with stalked glands; achenes elongate, longitudinally ribbed, rugulose, and black; pappus none.

CONVOLVULACEAE

2 **Convolvulus arvensis** L. (x5.5) Field Bindweed. Seeds 3.8 x 2.6 x 2.4 mm, obovate 48–49 in l.s., obtriangular 90–91 to elliptic 6–7 in c.s.; inner surface slightly angular, outer surface rounded, rugulose and papillose, dark brown.

3 **Convolvulus sepium** L. (x4) Hedge Bindweed. Seeds 4.8 x 4.3 x 3.5 mm, obliquely elliptic 5–6 in l.s., obtriangular 91–92 in c.s.; inner surface slightly angular and faces concave, outer surface rounded, slightly rugulose, papillose, and black.

4 **Convolvulus spithamaeus** L. (x5) Low Bindweed. Seeds 3.8 x 3.2 x 2.7 mm, obovate 49–50 in l.s., elliptic 6–7 in c.s.; slightly angled on the inner surface, smooth, dull, and black.

5 **Cuscuta gronovii** Willd. (x10) Dodder. Seeds 1.5 x 1.4 x 1.2 mm, obovate 49–50 in l.s., with the hilum at the slightly oblique base, elliptic 6–7 in c.s.; surface rather granular, rugulose, and brownish.

CORNACEAE

6 **Cornus alternifolia** L.f. (x3.2) Alternate-leaved Dogwood. Stone 5.0 x 5.5 x 5.5 mm, elliptic 6–7 in l.s., elliptic 6 in c.s.; surface veined and sulcate, looking somewhat like a peeled orange.

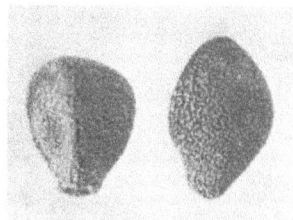
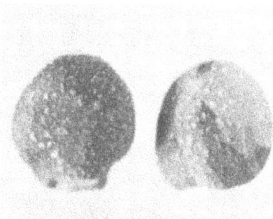

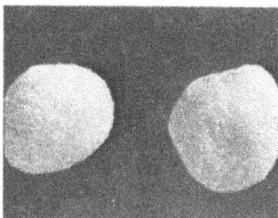

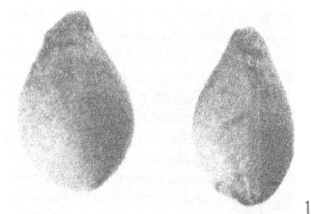
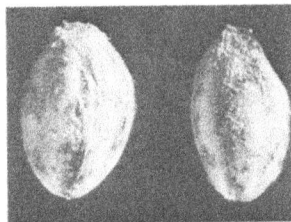
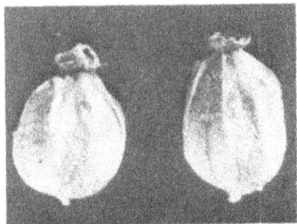

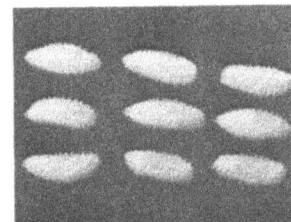

CORNACEAE

1 **Cornus canadensis** L. (x8.5) Bunchberry. Stone 3.0 x 2.0 x 2.0 mm, elliptic 4 or ovate 39 in l.s., elliptic 6 in c.s.; surface smooth and light brown.

2 **Cornus florida** L. (x3.5) Flowering Dogwood. Stone 7.0 x 4.8 x 4.8 mm, elliptic 4–5 in l.s., elliptic 6 in c.s.; surface with 4 or 5 prominent longitudinal sulci and nerves, slightly rugulose, gray to brown.

3 **Cornus purpusi** Koehne (x4) Silky Dogwood. Stone 4.5 x 3.9 x 3.9 mm, elliptic 5–6 in l.s., elliptic 6 in c.s.; surface strongly ribbed, green, and with a prominent terminal projection.

4 **Cornus racemosa** Lam. (x3.5) Gray Dogwood. Stone 4.1 x 4.0 x 4.0 mm, elliptic 6 in l.s. and in c.s.; surface green with white veins or ribs and prominent apex.

5 **Cornus rugosa** Lam. (x5) Round-leaved Dogwood. Stone 3.0 x 4.3 x 4.3 mm, elliptic 7–8 in l.s., elliptic 6 in c.s.; surface green with white veins or ribs, apex not prominent.

6 **Cornus stolonifera** Michx. (x5) Red-osier Dogwood. Stone 4.0 x 4.0 x 3.3 mm, elliptic 6 in l.s., elliptic 7–8 in c.s.; surface smooth, dull, gray, and longitudinally white-nerved.

7 **Nyssa sylvatica** Marsh (x2.5) Sour Gum. Stone 7.5 x 5.8 x 4.1 mm, elliptic 4–5 in l.s., elliptic 7–8 in c.s.; surface with 10–12 prominent, rounded, pale ridges with intervening nerves, gray in colour.

CRASSULACEAE

8 **Penthorum sedoides** L. (x20) Ditch Stonecrop. Seeds 0.5 x 0.2 x 0.2 mm, elliptic 2–3 in l.s., elliptic 6 in c.s.; surface with longitudinal rows of spicules or spines, white or brown.

1 **Sedum acre** L. (x20) Mossy Stonecrop. Seeds 0.6 x 0.3 x 0.3 mm, obovate 47 in l.s., elliptic 6 in c.s.; surface longitudinally ridged and obscurely areolate, slightly glossy and brown.

2 **Sedum ternatum** Michx. (x25) Seeds 0.8 x 0.4 x 0.4 mm, obovate 47 in l.s., elliptic 6 in c.s.; surface with about 16 longitudinal ribs, finely areolate between the ribs, brown to black.

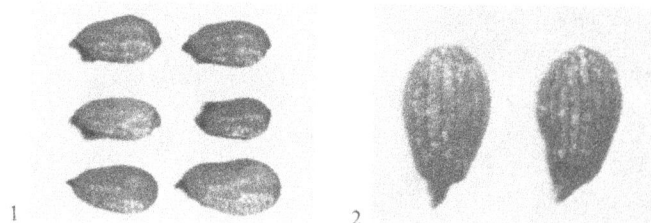

CRUCIFERAE

The embryo of the seeds of this family may be folded in one of three ways, (a) the cotyledons accumbent, (b) the cotyledons incumbent, or (c) the cotyledons conduplicate. Very frequently the folding is externally indicated by a ridge or sulcus and the position of the cotyledons and radicle evident.

3 **Alliaria officinalis** Andrz. (x5.5) Garlic Mustard. Seeds 3.0 x 1.0 x 1.0 mm, irregular in form or oblong 14 in l.s., oblique at both ends, elliptic 6 in c.s.; surface longitudinally ridged with the ridges occasionally branching, dark brown to black; cotyledons incumbent.

4 **Alyssum alyssoides** L. (x13) Alyssum. Seeds 1.4 x 1.2 x 0.5 mm, obovate 49–50 to elliptic 5–6 in l.s., elliptic 9–10 in c.s., with a conspicuous marginal wing; surface slightly roughened, wing whitish and areolate, body light brown; cotyledons accumbent, folding apparent by the small lateral sulcus.

5 **Arabis canadensis** L. (x11) Sicklepod. Seeds 2.0 x 1.0 x 0.3 mm, elliptic 4–5 in l.s., elliptic 11 in c.s. with a membranous wing; surface brown, slightly roughened or rugulose and the wing areolate and translucent; cotyledons accumbent and folding apparent by the sulcus.

6 **Arabis divaricarpa** A. Nels. (x8.5) Seeds 1.3 x 1.0 x 0.2 mm, elliptic 4–5 in l.s., elliptic 10–11 in c.s. and with a narrow wing; surface areolate or reticulate and brown; cotyledons accumbent and folding apparent.

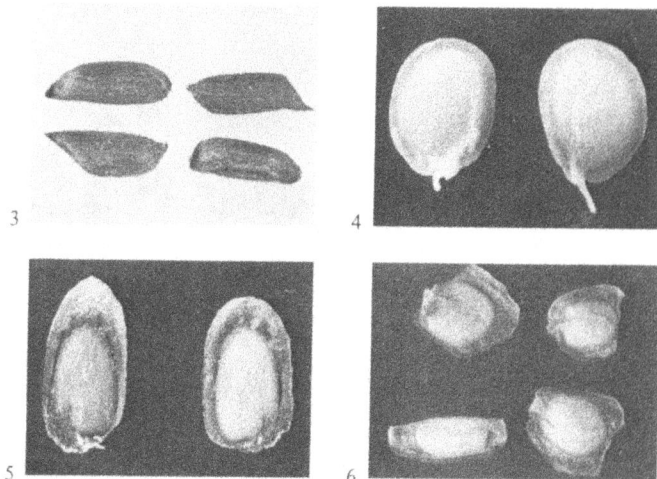

CRUCIFERAE

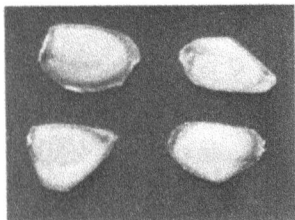
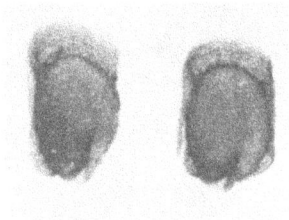
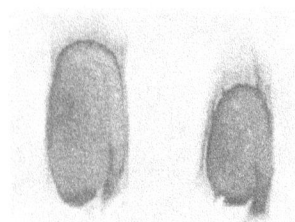
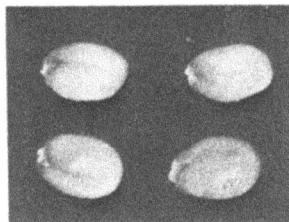
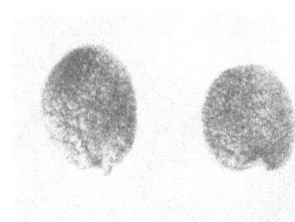
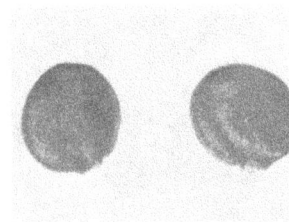
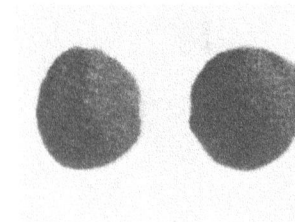

1 **Arabis glabra** (L.) Bernh. (x14) Smooth Arabis. Seeds 0.9 x 0.5 x 0.2 mm, elliptic 3–4 in l.s., elliptic 9–10 in c.s., margin narrowly winged; surface brown and areolate; cotyledons accumbent and folding apparent by the sulcus.

2 **Arabis hirsuta** (L.) Scop. (x20) Seeds 0.8 x 0.6 x 0.2 mm, elliptic 4–5 in l.s., elliptic 10 in c.s., broadly winged; surface areolate or reticulate; cotyledons accumbent and lateral sulcus present.

3 **Arabis laevigata** (Muhl.) Poir. (x15) Seeds 1.5 x 0.7 x 0.3 mm, oblong 14–15 in l.s., elliptic 9–10 in c.s., margin winged, the wing larger at the apex; surface reticulate or areolate; lateral sulcus present and cotyledons accumbent.

4 **Arabis lyrata** L. (x12) Seeds 1.0 x 0.6 x 0.3 mm, elliptic 3–4 in l.s., elliptic 9 in c.s., margin with a narrow wing or wingless; surface areolate and light brown; cotyledons accumbent.

5 **Barbarea vulgaris** R. Br. (x11) Yellow Rocket. Seeds 1.5 x 1.0 x 0.6 mm, obliquely elliptic 4–5 in l.s., obliquely elliptic 8–9 in c.s.; surface alveolate or rugulose, light brown to black, and glossy; cotyledons accumbent.

6 **Berteroa incana** (L.) DC. (x10) Hoary Alyssum. Seeds 1.4 x 1.4 x 0.6 mm, elliptic 6 in l.s., elliptic 9–10 in c.s. with a narrow marginal wing or ridge; surface areolate, dull to glossy, and brown; cotyledons accumbent and folding evident.

7 **Brassica juncea** (L.) Cosson (x10) Indian Mustard. Seeds 1.6 x 1.3 x 1.0 mm, elliptic 4–5 in l.s., elliptic 7–8 in c.s.; surface reticulate and dark brown; cotyledons conduplicate.

8 **Brassica kaber** (DC.) L.C. Wheeler (x15) Common Mustard. Seeds 1.2 x 1.2 x 1.2 mm, elliptic 6 in l.s. and in c.s.; surface areolate, base with striations radiating from the hilum, brown to black; cotyledons conduplicate.

1 **Brassica nigra** (L.) Koch (x9) Black Mustard. Seeds 1.3 x 1.2 x 1.2 mm, obovate 49–50 to elliptic 6 in l.s., elliptic 6 in c.s.; surface areolate, brown to black; cotyledons conduplicate.

2 **Cakile edentula** (Bigel.) Hook. (x4.6) Sea Rocket. Seeds enclosed in an indehiscent, cylindrical, ridged pod with a transverse constriction or joint forming two internal cavities (torose), the upper one being the larger; seeds 5.0 x 2.3 x 1.3 mm, irregular in shape or elliptic 2–3 in l.s., elliptic 8–9 in c.s.; surface brown and smooth or almost so; lateral sulcus showing the accumbent cotyledons.

3 **Camelina microcarpa** Andrz. (x12) Small-seeded False Flax. Seeds 1.1 x 0.8 x 0.5 mm, obliquely elliptic 4–5 in l.s., obliquely elliptic 8–9 in c.s.; surface obscurely roughened and reticulate, brown; cotyledons incumbent and indicated by a sulcus.

4 **Capsella bursa-pastoris** (L.) Medic. (x20) Shepherd's Purse. Seeds 1.0 x 0.6 x 0.4 mm, elliptic 3–4 in l.s., elliptic 8 in c.s.; surface slightly roughened, scalariform; cotyledons incumbent.

5 **Cardamine bulbosa** (Schreb.) BSP. (x7) Spring Cress. Seeds 1.6 x 1.2 x 0.6 mm, elliptic 4–5 in l.s., elliptic 9 in c.s.; surface slightly rugulose and greenish; cotyledons accumbent, radicle prolonged under the cotyledons and below the hilum.

6 **Cardamine parviflora** L. (x21) Seeds 0.9 x 0.5 x 0.3 mm, oblong 15–16 in l.s., elliptic 8–9 in c.s.; surface faintly striate from rows of angular cells; cotyledons accumbent, position of the radicle evident.

7 **Cardamine pensylvanica** Muhl. (x12) Seeds 1.0 x 0.7 x 0.3 mm, elliptic 4–5 in l.s., elliptic 9–10 in c.s. with a narrow marginal ridge or wing; surface alveolate, light brown; cotyledons accumbent.

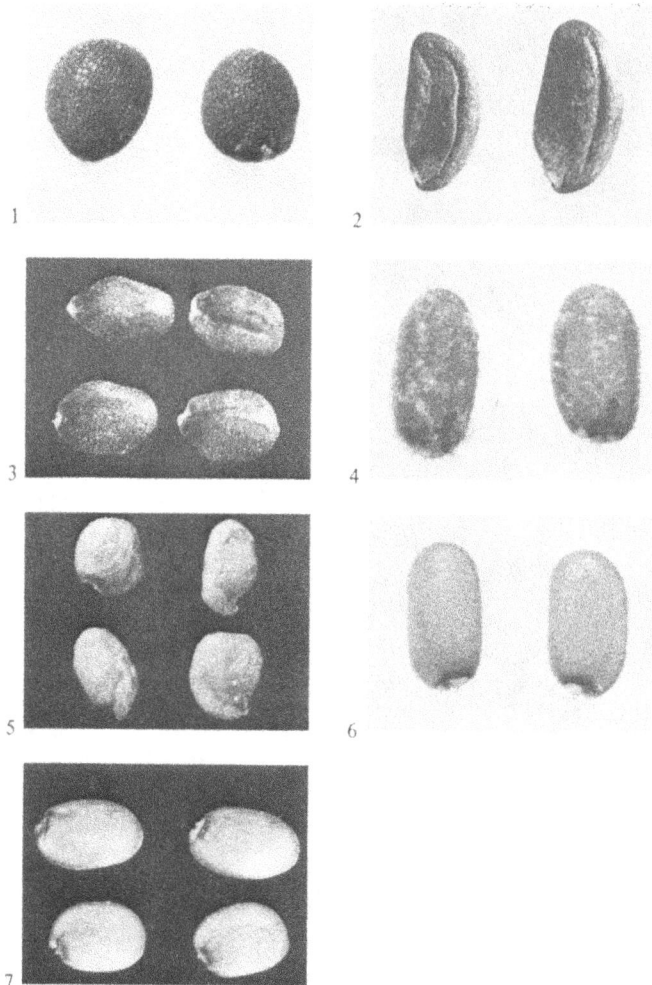

94

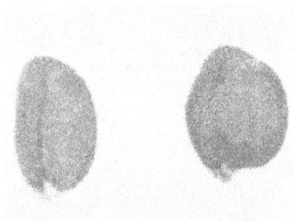

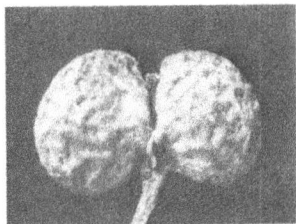
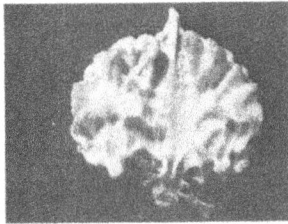

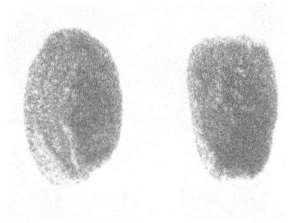

1 **Cardaria draba** (L.) Desv. (x8) Seeds 2.2 x 1.5 x 1.0 mm, obovate 48–49 in l.s., elliptic 8 in c.s.; surface areolate and scalariform; cotyledons incumbent, radicle prolonged at the base.

2 **Conringia orientalis** (L.) Andrz. (x8) Hare's-ear Mustard. Seeds 2.6 x 1.4 x 1.4 mm, elliptic 3–4 in l.s., elliptic 6 in c.s.; remains of the funiculus rather prominent, surface with minute, rather eye-like pittings (ocellate); cotyledons incumbent.

3 **Coronopus didymus** (L.) Smith (x12) Swine Cress. Fruit an indehiscent silicle, bilocular, indented at the apex and the base, 1.5 x 2.5 x 0.9 mm, surface transversely rugose; seeds obovate in l.s. with the axis slightly curved, elliptic in c.s.; surface areolate; cotyledons incumbent.

4 **Coronopus procumbens** Gilib. (x12) Fruit similar to the previous species; silicle 2.7 x 3.5 x 1.0 mm, not indented at the apex and base, style prominent and persistent, surface roughly transversely ridged and ridges acute, often appearing reticulate; cotyledons incumbent.

5 **Dentaria diphylla** Michx. (x7.5) Two-leaved Pepperroot. Seeds 2.8 x 1.8 x 1.0 mm, elliptic 3–4 in l.s., elliptic 8–9 in c.s.; surface slightly wrinkled or rugulose, green to brown.

6 **Dentaria laciniata** Muhl. (x9) Cut-leaved Pepperroot. Seeds 2.2 x 1.5 x 1.0 mm, elliptic 4–5 in l.s., elliptic 8 in c.s.; surface rugulose, green to brown.

7 **Descurania sophia** (L.) Webb. (x12) Flixweed. Seeds 1.2 x 0.6 x 0.5 mm, irregular in l.s. and in c.s. or obliquely ovate 38 in l.s., elliptic 7 in c.s.; surface slightly rugulose; cotyledons incumbent.

8 **Diplotaxis muralis** (L.) DC. (x15) Seeds 1.2 x 0.8 x 0.7 mm, elliptic 4 in l.s., elliptic 6–7 in c.s.; surface light brown, obscurely areolate or puncticulate; cotyledons conduplicate.

1 **Diplotaxis tenuifolia** (L.) DC. (x12) Seeds 1.1 x 0.8 x 0.6 mm, elliptic 4–5 in l.s., elliptic 7–8 in c.s.; surface light brown, obscurely pitted or puncticulate; cotyledons conduplicate.

2 **Draba arabisans** Michx. (x16) Seeds 1.1 x 0.7 x 0.3 mm, elliptic 3–4 in l.s., elliptic 9–10 in c.s., margin slightly ridged; surface puncticulate and dark brown; cotyledons accumbent and folding evident by the lateral sulcus.

3 **Erucastrum gallicum** (Willd.) O.E. Schulz (x12) Dog Mustard. Seeds 1.2 x 0.8 x 0.6 mm, elliptic 4 in l.s., elliptic 7–8 in c.s.; surface finely alveolate and areolate; cotyledons conduplicate.

4 **Erysimum cheiranthoides** L. (x13) Wormseed Mustard. Seeds 1.2 x 0.6 x 0.5 mm, irregular or obliquely ovate 38 in l.s., obliquely elliptic 7 in c.s.; surface inconspicuously areolate and light brown; cotyledons incumbent.

5 **Erysimum hieracifolium** L. (x13) Seeds 1.3 x 0.6 x 0.4 mm, irregular or obliquely ovate 37–38 in l.s., apex slightly winged, obliquely elliptic 8 in c.s.; surface areolate and light brown; cotyledons incumbent and folding evident.

6 **Hesperis matronalis** L. (x7) Dame's-violet. Seeds 1.9 x 1.0 x 1.0 mm, oblong 15–16 to elliptic 3–4 in l.s., elliptic 6 in c.s.; surface ocellate and brown; cotyledons incumbent.

7 **Lepidium campestre** (L.) R. Br. (x7.5) Field Peppergrass. Seeds 2.5 x 1.5 x 1.5 mm, obovate 47–48 in l.s., elliptic 6 in c.s.; remnant of the funiculus prominent; surface slightly roughened or papillose and ocellate, light to dark brown; cotyledons incumbent and sulcus present.

8 **Lepidium densiflorum** Schrader (x11) Poor Man's Pepper. Seeds 1.3 x 0.9 x 0.4 mm, obovate 48–49 in l.s., elliptic 9–10 in c.s. with a narrow marginal wing; surface obscurely areolate and slightly rugulose, light brown; cotyledons incumbent or accumbent.

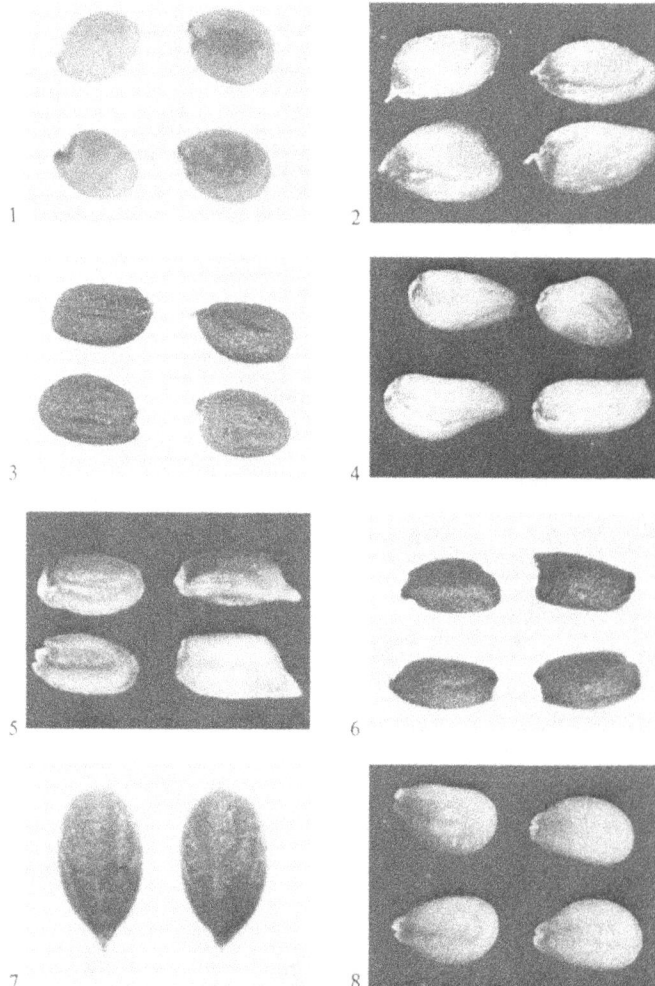

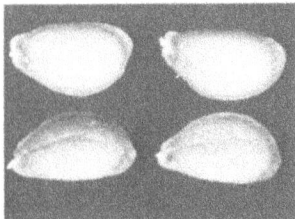

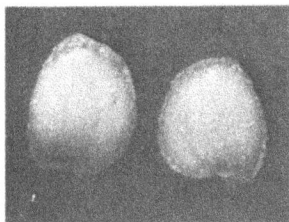

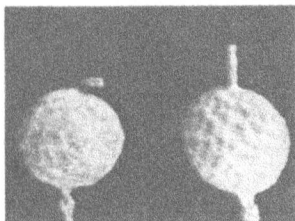

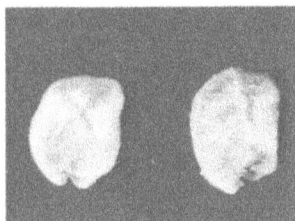

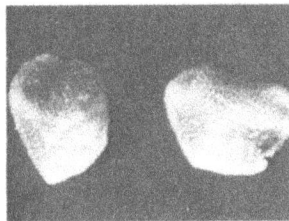

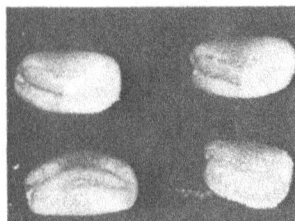

1 **Lepidium virginicum** L. (x8.5) Seeds 1.9 x 1.1 x 0.5 mm, obovate 47–48 in l.s., elliptic 9–10 in c.s. with a narrow marginal wing; surface areolate, light brown; cotyledons incumbent or accumbent, lateral sulcus shows the folding.

2 **Nasturtium officinale** R. Br. (x19) Water Cress. Seeds 0.9 x 0.8 x 0.3 mm, irregular in shape or elliptic 5–6 in l.s., elliptic 9–10 in c.s., margin slightly winged by an enlargement of the marginal cells; surface reticulate, slightly glossy, and dark brown; cotyledons accumbent.

3 **Neslia paniculata** (L.) Desv. (x7) Ball Mustard. Fruits indehiscent, 2.2 x 2.5 x 1.7 mm, elliptic 6–7 in l.s., elliptic 7–8 in c.s.; surface alveolate, style persistent; cotyledons incumbent.

4 **Raphanus raphanistrum** L. (x7.5) Wild Radish. Fruit an indehiscent silique, cylindrical, strongly beaked, longitudinally ridged and transversely constricted into 2 or more locules (torose); seeds 2.4 x 1.9 x 1.4 mm, irregular in form or elliptic 4–5 in l.s., elliptic 7–8 in c.s.; surface alveolate, dark brown to black; cotyledons conduplicate.

5 **Rorippa islandica** (Oeder) Borbas (x20) Seeds very irregular in shape, 0.8 x 0.6 x 0.4 mm, surface reticulate, slightly glossy, and gray to brown; cotyledons accumbent.

6 **Rorippa sylvestris** (L.) Besser (x30) Creeping Yellow Cress. Seeds 0.5 x 0.4 x 0.3 mm, irregular in shape, surface finely reticulate, slightly glossy, and brown; cotyledons accumbent.

7 **Sisymbrium altissimum** L. (x13) Tumble Mustard. Seeds 1.1 x 0.6 x 0.6 mm, irregular in form or obliquely oblong 15–16 in l.s., obliquely oblong 18 in c.s.; surface inconspicuously roughened or granular and striate; cotyledons incumbent, folding evident.

1 **Sisymbrium loeselii** L. (x15) Seeds 0.8 x 0.5 x 0.4 mm, irregular or obliquely elliptic 3–4 in l.s., obliquely elliptic 7–8 in c.s.; surface obscurely roughened and faintly areolate; cotyledons incumbent.

2 **Sisymbrium officinale** (L.) Scop. (x11) Hedge Mustard. Seeds 1.3 x 0.7 x 0.5 mm, irregular in form or obliquely oblong 15–16 to ovate 38–39 in l.s., elliptic 7–8 in c.s. or somewhat angular; surface slightly striate and faintly areolate; cotyledons incumbent.

3 **Thlaspi arvense** L. (x11) Penny Cress. Seeds 1.8 x 1.2 x 0.6 mm, obovate 48 to elliptic 4 in l.s., oblong 21 in c.s.; surface with distinct concentric ridges or loops like a finger print, inconspicuously pitted or transversely rugulose, black, and slightly glossy; cotyledons accumbent.

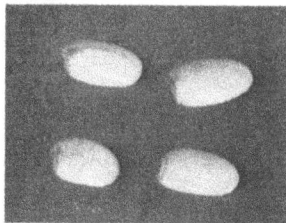

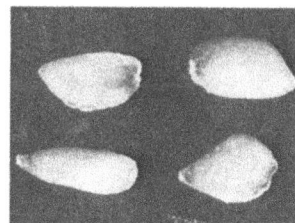

CUCURBITACEAE

4 **Echinocystis lobata** (Michx.) T.&G. (x2) Wild Balsam Apple. Seeds 17.0 x 8.0 x 3.4 mm, elliptic 2–3 in l.s., elliptic 9–10 in c.s. with the margins ridged; surface slightly roughened or granular and black.

5 **Sicyos angulatus** L. (x3) Bur Cucumber. Seeds 8.6 x 6.0 x 3.0 mm, obovate 48–49 in l.s., elliptic 9 in c.s.; base of the seed with a swelling on each side of the hilum; surface smooth, dull, and gray to brown.

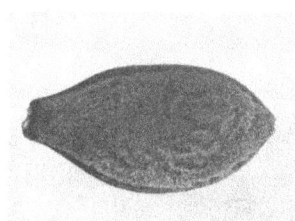

CUPRESSACEAE

6 **Juniperus communis** L. (x5.5) Common Juniper. Seeds 4.6 x 3.0 x 2.5 mm, ovate 38–39 in l.s., elliptic 7 in c.s.; surface with a lateral, rounded scar at the base, rugose and with large resin glands, rather pitted at the base.

7 **Juniperus horizontalis** Moench. (x5.5) Creeping Juniper. Seeds 3.6 x 2.4 x 2.0 mm, ovate 39 in l.s., elliptic 7 in c.s.; surface with a rounded, whitish, pitted scar at the base, rugose and with large resin glands.

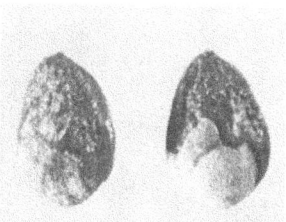

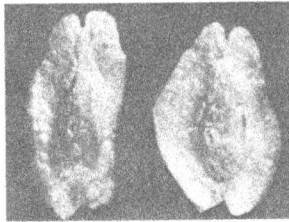

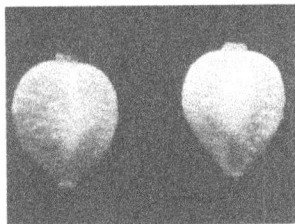

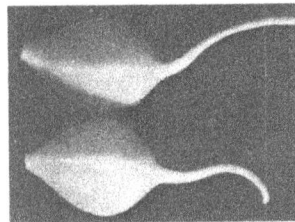

CUPRESSACEAE

1 **Juniperus virginiana** L. (x5.5) Red Cedar. Seeds 3.4 x 2.7 x 2.2 mm, ovate 39–40 in l.s., elliptic 7–8 in c.s.; surface with a whitish pitted scar at the base, rugulose and with large resin glands.

2 **Thuja occidentalis** L. (x6) White Cedar. Seeds 5.0 x 3.0 x 0.9 mm, elliptic 3–4 to ovate 38–39 in l.s., elliptic 10–11 in c.s.; margin with a membranous wing notched at the apex and the base, slightly inflated and rugulose, surface with resin glands.

CYPERACEAE

The achenes in this family are either lenticular or trigonous in c.s. The base of the style, or in some species a large portion of the style, is persistent.

In the genus *Carex* the achenes are enclosed in a saclike structure called the perigynium. The style remnant may be very small or long and contorted.

Eleocharis species are characterized by the comparatively large style base or tubercle, the form of which may be important in identification. The achenes are subtended by barbed bristles.

The achenes of *Scirpus* are subtended by up to 6 bristles which are shorter or longer than the achenes. Some other genera have no subtending structures.

3 **Bulbostylis capillaris** (L.) Clarke (x18) Achenes 0.8 x 0.6 x 0.6 mm, obovate 48–49 in l.s., triangular 78 in c.s. with the faces slightly convex; angles prominently ridged, faces transversely rugose, tubercle minute, white or gray.

4 **Carex aurea** Nutt. (x11) Achenes 1.5 x 1.4 x 0.6 mm, ovate 40–41 in l.s., elliptic 9–10 in c.s.; surface areolate, dark brown to black.

5 **Carex comosa** Boot. (x11) Achenes 1.5 x 1.0 x 1.0 mm, elliptic 4 in l.s. with a persistent, contorted style, triangular 78 in c.s.; surface obscurely areolate, light brown.

1 **Carex lupulina** Muhl. (x4.5) Achenes 3.7 x 2.3 x 2.2 mm, elliptic 3–4 in l.s., tapering to a rather broad, stipitate base, triangular 78–79 in c.s. with the faces slightly concave; surface areolate, style long and contorted or spiralling.

2 **Carex pseudo-cyperus** L. (x10) Achenes 1.7 x 1.5 x 1.5 mm, obovate 49–50 in l.s. with a persistent, contorted style, triangular 78 in c.s.; surface faintly areolate and brown.

3 **Carex stipata** Muhl. (x11) Achenes 1.3 x 1.2 x 0.6 mm, ovate 40–41 in l.s., elliptic 9 in c.s.; surface irregularly alveolate and finely areolate.

4 **Cladium marsicoides** (Muhl.) Torr. (x6.5) Nut-rush. Achenes 3.0 x 1.7 x 1.7 mm, ovate 38–39 in l.s., truncate at the base and acuminate at the apex, elliptic 6 in c.s.; surface irregularly rugulose or ridged and green to brown, style base small.

5 **Cyperus esculentus** L. (x10) Yellow Nut-grass. Achenes 1.1 x 0.6 x 0.6 mm, elliptic 3–4 to obovate 47–48 in l.s., triangular 78 in c.s.; angles rounded and sides slightly concave, surface reticulate, yellowish or cream colour, style base small.

6 **Cyperus filiculmis** Vahl (x11) Achenes 1.8 x 0.8 x 0.7 mm, elliptic 2–3 in l.s., triangular 78–79 in c.s. with the faces slightly convex; surfaces areolate, gray to brown, style base short.

7 **Cyperus schweinitzii** Torr. (x10) Achenes 2.5 x 1.2 x 1.1 mm, obovate 46–47 in l.s., triangular 78–79 in c.s., angles sharp and faces flat or slightly concave; surface areolate or ocellate, dull, green to brown, style base short.

8 **Dulichium arundinaceum** (L.) Britt. (x10) Three-way Sedge. Achenes 2.6 x 0.8 x 0.4 mm, oblong 13–14 in l.s., acuminate at the apex and base and terminating in a long, slender style, triangular 81 in c.s.; base with 6–9 bristles, retrorsely barbed and longer than the achene; surface areolate and light brown.

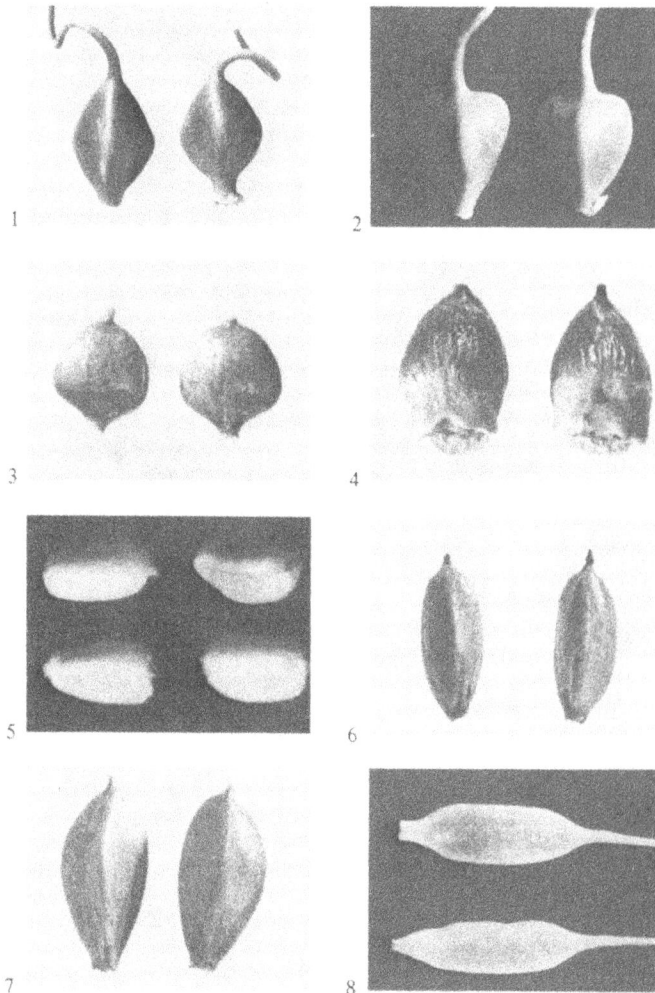

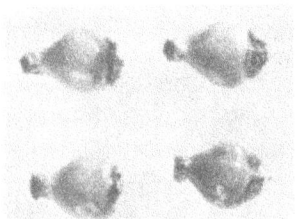

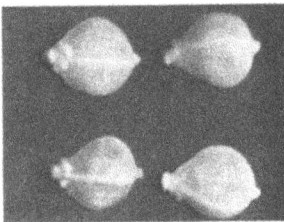

1 **Eleocharis ovata** (Roth) R.&S. (x10) Achenes 1.3 x 0.9 x 0.5 mm, obovate 48–49 in l.s., elliptic 8–9 in c.s.; apex terminating in a broad, triangular tubercle about as wide as the achene; bristles about 6, retrorsely barbed and as long as, or longer than, the achene; surface smooth or very faintly areolate.

2 **Eleocharis palustris** (L.) R.&S. (x10) Achenes 1.7 x 1.3 x 0.9 mm, obovate 48–49 in l.s., elliptic 7–8 in c.s.; apex with a conical tubercle longer than broad; bristles 4–6 of unequal length and retrorsely barbed; surface faintly sulcate, obscurely areolate, and brown.

3 **Eleocharis quadrangulata** (Michx.) R.&S. (x8.5) Achenes 2.2 x 1.5 x 1.0 mm, obovate 48–49 in l.s., obliquely elliptic 8 in c.s., tending to be plano-convex, tubercle triangular, large and flattened dorso-ventrally; surface longitudinally striate and with rows of transverse, linear cells or scalariform, shiny, and brown; bristles retrorsely barbed and as long as, or longer than, the achene.

4 **Eriophorum spissum** Fern. (x10) Achenes 2.4 x 1.4 x 0.9 mm, obovate 47–48 in l.s., triangular 80–81 in c.s., angles acute; surface very finely areolate, white bristles longer than the dark brown achene.

5 **Eriophorum viridi-carinatum** (Engelm.) Fern. (x10) Cottongrass. Achenes 2.8 x 1.3 x 0.7 mm, obovate 46–47 in l.s., triangular 80–81 in c.s., angles acute; surface minutely areolate; bristles long, whitish, achenes brown.

6 **Fimbristylis autumnalis** (L.) R.&S. (x20) Achenes 0.5 x 0.5 x 0.5 mm, obovate 50 in l.s., triangular 78 in c.s., angles prominently ridged, faces convex; surface smooth to very finely areolate, sometimes faintly verrucose at the base or over the whole surface, white to brown.

1 **Rhynchospora alba** (L.) Vahl (x10) Achenes 2.2 x 1.1 x 0.6 mm, obovate 47 in l.s. tapering to a short stipe, obliquely elliptic 8–9 in c.s., more or less plano-convex; terminating in a large, subulate tubercle, margins slightly ridged; surface faintly areolate; bristles 8–14, about as long as the achene plus the tubercle, retrorsely barbed and slightly pubescent at the base.

2 **Rhynchospora capillacea** Torr. (x13) Achenes 1.5 x 0.8 x 0.4 mm, obovate 47–48 in l.s., elliptic 9 in c.s.; base with a prominent gynophore or stalk, tubercle stout and acuminate; surface faintly rugulose and with indistinct reddish lines near the base; bristles 6, retrorsely barbed and longer than the achene plus the tubercle.

3 **Scirpus americanus** Pers. (x7) Achenes 2.9 x 2.1 x 1.2 mm, obovate 48–49 in l.s., elliptic 8–9 in c.s. tending to be plano-convex, apiculate; retrorsely barbed bristles 4 and shorter than, to as long as, the achenes; surface smooth, dull, and brown to black.

4 **Scirpus atrovirens** Willd. (x20) Achenes 1.0 x 0.5 x 0.3 mm, elliptic 3 in l.s., triangular 80–81 in c.s.; retrorsely barbed bristles 4–6 and shorter than or equal to the achene; surface inconspicuously papillose, white, or gray.

5 **Scirpus cyperinus** (L.) Kunth (x17) Achenes 0.7 x 0.4 x 0.2 mm, elliptic 3–4 in l.s., triangular 81 in c.s. with the faces convex; bristles longer than the achenes; surface faintly papillose, white to gray.

6 **Scirpus hudsonianus** (Michx.) Fern. (x15) Achenes 1.2 x 0.5 x 0.4 mm, obovate 46–47 in l.s., triangular 79–80 in c.s., angles prominent; bristles 3–6, white, ligulate, much longer than the achenes; surface smooth or faintly papillose, brown to black.

7 **Scirpus lineatus** Michx. (x20) Achenes 1.0 x 0.5 x 0.3 mm, elliptic 3 in l.s., triangular 80–81 in c.s.; surface faintly areolate, light to dark brown; bristles very twisted and about twice as long as the achene.

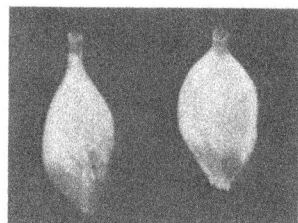

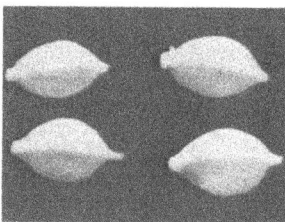

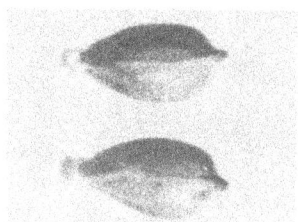

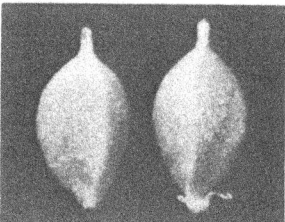

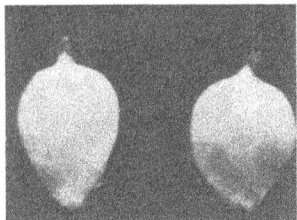
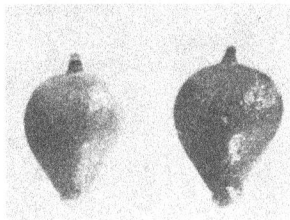
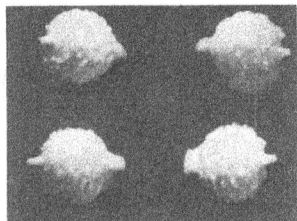
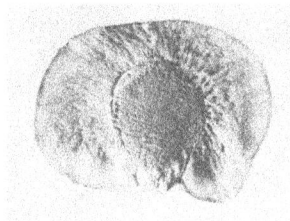
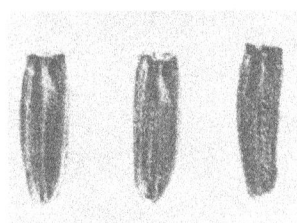
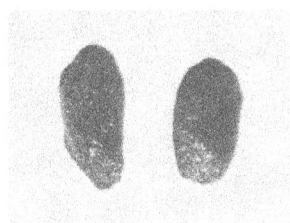

CYPERACEAE

1 **Scirpus rubrotinctus** Fern. (x20) Achenes 0.9 x 0.6 x 0.3 mm, obovate 48 in l.s., elliptic 9 in c.s. or slightly triangular; retrorsely barbed bristles 6 and as long as, or slightly longer than the achene; surface whitish and faintly areolate.

2 **Scirpus validus** Vahl (x7) Great Bulrush. Achenes 1.7 x 1.2 x 0.7 mm, obovate 48–49 in l.s., elliptic 8–9 in c.s., tending to be plano-convex; bristles 4–6, retrorsely barbed and as long as, or longer than the achene; surface brown to black and finely reticulate.

3 **Scleria verticillata** Muhl. (x9) Nut-rush. Achenes 1.2 x 1.2 x 1.2 mm, elliptic 6 in l.s., elliptic 6 or obscurely triangular in c.s.; base large, style base obvious; surface verrucose or coarsely rugose and faintly areolate, glossy and white.

DIOSCORIACEAE

4 **Dioscorea villosa** L. (x3.5) Wild Yam. Fruit a 3-winged capsule enclosing a number of winged seeds; seeds 8.0 x 10.0 x 0.7 mm, elliptic 7–8 in l.s., very thin in c.s. with a spongy wing of varying width; surface areolate and brown.

DIPSACACEAE

5 **Dipsacus sylvestris** Huds. (x5) Teasel. Achenes 3.8 x 1.1 x 1.1 mm, oblong 13–14 in l.s., oblong 18 in c.s., each face with usually one longitudinal vein or rib; surface with a fine, slightly stiff, appressed pubescence.

DROSERACEAE

6 **Drosera intermedia** Hayne (x25) Seeds 0.6 x 0.3 x 0.3 mm, elliptic 3 in l.s., elliptic 6 in c.s.; surface black and covered with papillae terminating with reddish glands.

1 **Drosera linearis** Goldie (x15) Seeds 0.9 x 0.4 x 0.3 mm, elliptic 2–3 in l.s., obliquely elliptic 7–8 in c.s.; surface very finely papillose, the papillae having light-coloured apices, seed body black.

2 **Drosera rotundifolia** L. (x15) Round-leaved Sundew. Seeds 1.3 x 0.3 x 0.2 mm, elliptic 1–2 in l.s., elliptic 8 in c.s., or irregular in l.s. and c.s.; testa forming a network around the embryo and protruding on each side of the seed body as a wing, surface finely striate and reticulate.

ELAEAGNACEAE

3 **Shepherdia argentea** Nutt. (x6) Buffalo Berry. Achenes covered by a pulpy calyx, 3.4 x 2.5 x 1.8 mm, elliptic 4–5 in l.s., oblique at the base, elliptic 7–8 in c.s.; surface with a single sulcus, smooth or slightly rugulose, somewhat glossy and dark brown.

4 **Shepherdia canadensis** (L.) Nutt. (x4.8) Soapberry. Achenes covered by a pulpy calyx, 4.9 x 2.2 x 1.1 mm, elliptic 2–3 in l.s., oblique at the base, elliptic 9 in c.s.; surface with a single median sulcus, somewhat glossy, and brown.

EMPETRACEAE

5 **Empetrum nigrum** L. (x9) Crowberry. Seeds 1.1 x 1.8 x 0.9 mm, obliquely elliptic 3–4 in l.s., obtriangular 88–89 in c.s. with the dorsal surface rounded; surface faintly rugulose and brown.

ERICACEAE

6 **Andromeda glaucophylla** L. (x12) Bog-rosemary. Seeds 1.6 x 0.9 x 0.5 mm, obliquely elliptic 3–4 in l.s., elliptic 8–9 in c.s.; surface smooth and glossy.

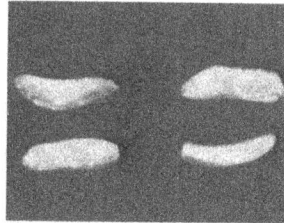
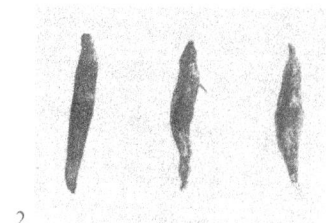
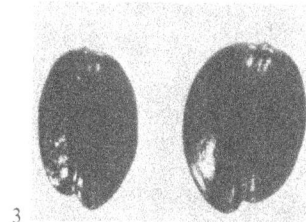
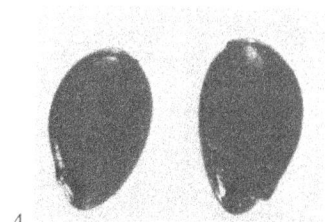
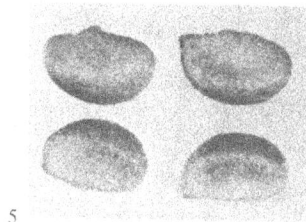
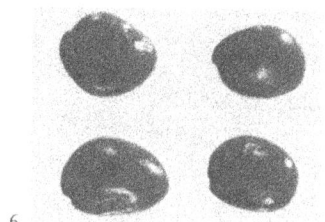

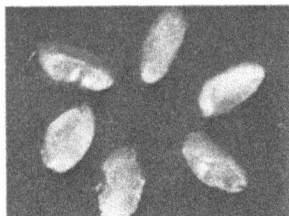

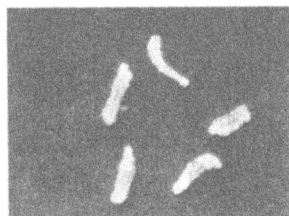
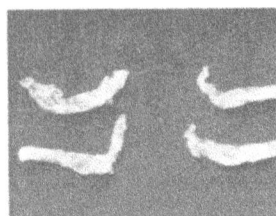
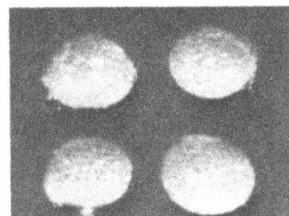
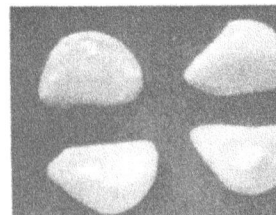

ERICACEAE

1 **Arctostaphylos uva-ursi** (L.) Spreng. (x5) Bearberry. Seeds 4.3 x 2.5 x 2.0 mm, irregular in form or obliquely elliptic 3–4 in l.s., obtriangular 88–89 in c.s. with the dorsal surface rounded, ridged, and rugulose, lateral surfaces smooth.

2 **Cassiope hypnoides** (L.) D. Don (*Harrimanella hypnoides*) (x20) Seeds 0.5 x 0.25 x 0.25 mm, elliptic 3 in l.s., elliptic 6 in c.s.; surface finely rugulose and areolate, brown in colour.

3 **Chamaedaphne calyculata** (L.) Moench. (x12) Leather Leaf. Seeds 1.0 x 0.6 x 0.4 mm, irregular in shape or obliquely elliptic 3–4 in l.s., obtriangular 88 in c.s.; surface rugulose, light brown, and glossy.

4 **Chimaphila maculata** (L.) Pursh. Spotted Pipsissewa. Seeds elongate with the seed coating forming a network around a minute embryo. Very similar in appearance to the seeds of the Orchid Family.

5 **Chimaphila umbellata** (L.) Bart. var. **cisatlantica** Blake Pipsissewa. Seeds similar to the previous species.

6 **Epigaea repens** L. (x25) Trailing Arbutus. Seeds 0.5 x 0.4 x 0.35 mm, elliptic 4–5 in l.s., elliptic 6–7 in c.s.; surface reticulate, brown, and slightly glossy.

7 **Gaultheria hispidula** (L.) Muhl. (x16) Creeping Snowberry. Seeds 0.9 x 0.7 x 0.4 mm, obliquely elliptic 4–5 to obovate 48–49 in l.s., obtriangular 87–88 in c.s. with the outer surface rounded, cream-coloured, and glossy.

8 **Gaultheria procumbens** L. (x11) Wintergreen. Seeds 1.0 x 0.8 x 0.5 mm, irregularly angular to circular in outline, compressed and irregularly angular in c.s.; surface slightly rugulose, shiny, and brown.

1 **Gaylussacia baccata** (Wang.) K. Koch (x11) Black Huckleberry. Seeds 1.9 x 1.5 x 0.8 mm, obliquely elliptic 4–5 in l.s., obtriangular 87–88 in c.s.; surface slightly roughened and light brown.

2 **Kalmia angustifolia** L. Sheep Laurel. Seeds elongated, seed coatings forming a whitish network around the embryo.

3 **Kalmia polifolia** Wang. Laurel. Seeds similar in structure to the previous species.

4 **Ledum groenlandicum** Oeder. Greenland Tea. Seeds irregular in l.s. and in c.s. with the seed coating forming a white network around the embryo.

5 **Moneses uniflora** (L.) Gray One-flowered Pyrola. Seeds linear; testa forming a long white network around the embryo.

6 **Monotropa hypopithys** L. Pinesap. Seeds linear, testa forming a long white network around the embryo.

7 **Monotropa uniflora** L. Indian Pipe. Seeds similar to the previous species.

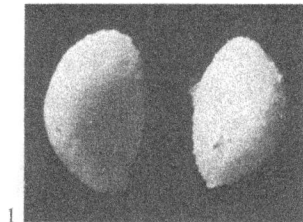
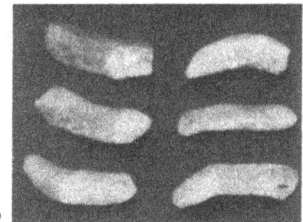
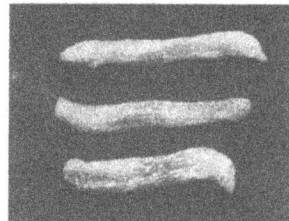
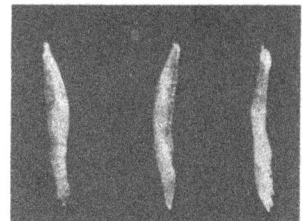
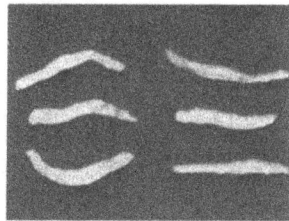
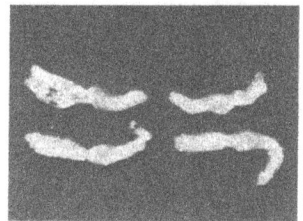
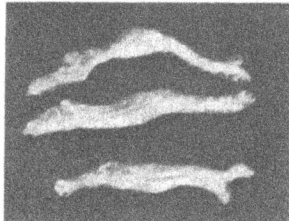

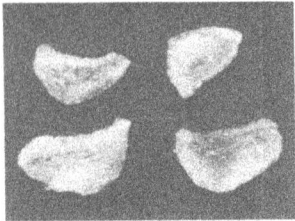

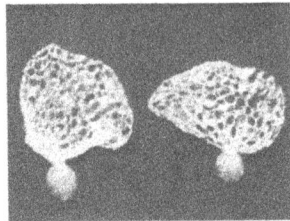

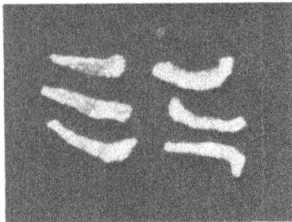

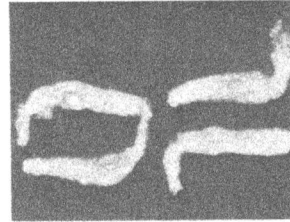

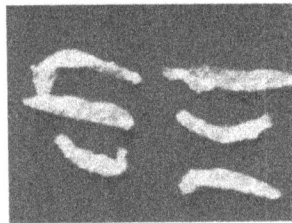

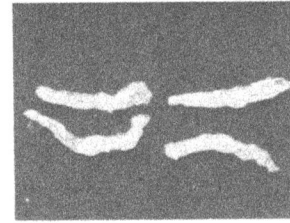

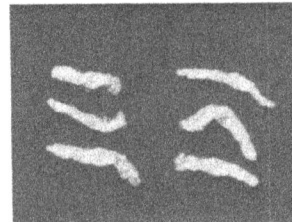

1 **Phyllodoce caerulea** (L.) Bab. (x30) Seeds 0.5 x 0.3 x 0.1 mm, irregular in form; surface reticulate or alveolate.

2 **Pterospora andromedea** Nutt. Giant Bird's-nest. Testa forming a white fan-shaped network at the seed apex, body longitudinally nerved and reticulate.

Pyrola spp. The seeds of this genus have the testa forming a loose, elongated network around the embryo. Except for minor differences in shape or size, the seeds of the species are not distinguishable from one another.

3 **Pyrola asarifolia** Michx. Pink Pyrola.

4 **Pyrola elliptica** Nutt. Shinleaf.

5 **Pyrola grandiflora** Radius. Arctic Pyrola.

6 **Pyrola minor** L.

7 **Pyrola rotundifolia** L. Wild Lily-of-the-valley.

1 **Pyrola secunda** L. One-sided Pyrola.

2 **Pyrola virens** Schweigg. Green Pyrola.

3 **Rhododendron canadense** (L.) Torr. (x15) Rhodora. Seeds 2.0 x 0.7 x 0.2 mm, elliptic 2–3 in l.s., elliptic 10–11 in c.s.; body surrounded by a lacerated, membranous wing notched at the apex and the base; surface longitudinally striate, brown.

4 **Vaccinium angustifolium** Ait. (x11) Low Sweet Blueberry. Seeds 1.4 x 0.8 x 0.5 mm, elliptic 3–4 in l.s., obliquely elliptic 8–9 in c.s.; surface reticulate and brown.

5 **Vaccinium caespitosum** Michx. (x22) Dwarf Blueberry. Seeds 0.7 x 0.5 x 0.3 mm, obliquely elliptic 4–5 to obovate 48–49 in l.s., obliquely elliptic 8–9 in c.s.; surface brown, reticulate, and slightly glossy.

6 **Vaccinium corymbosum** L. (x14) Highbush Blueberry. Seeds 1.4 x 1.2 x 1.1 mm, obliquely elliptic 5–6 in l.s., elliptic 6–7 in c.s.; surface areolate and brown.

7 **Vaccinium macrocarpon** Ait. (x10) Large Cranberry. Seeds 1.8 x 1.1 x 1.0 mm, elliptic 3–4 in l.s., elliptic 6–7 in c.s.; surface finely areolate, light brown, and somewhat glossy.

8 **Vaccinium myrtilloides** Michx. (x14) Seeds 1.0 x 0.8 x 0.5 mm, obliquely elliptic 4–5 in l.s., elliptic 8–9 in c.s.; surface reticulate and brown.

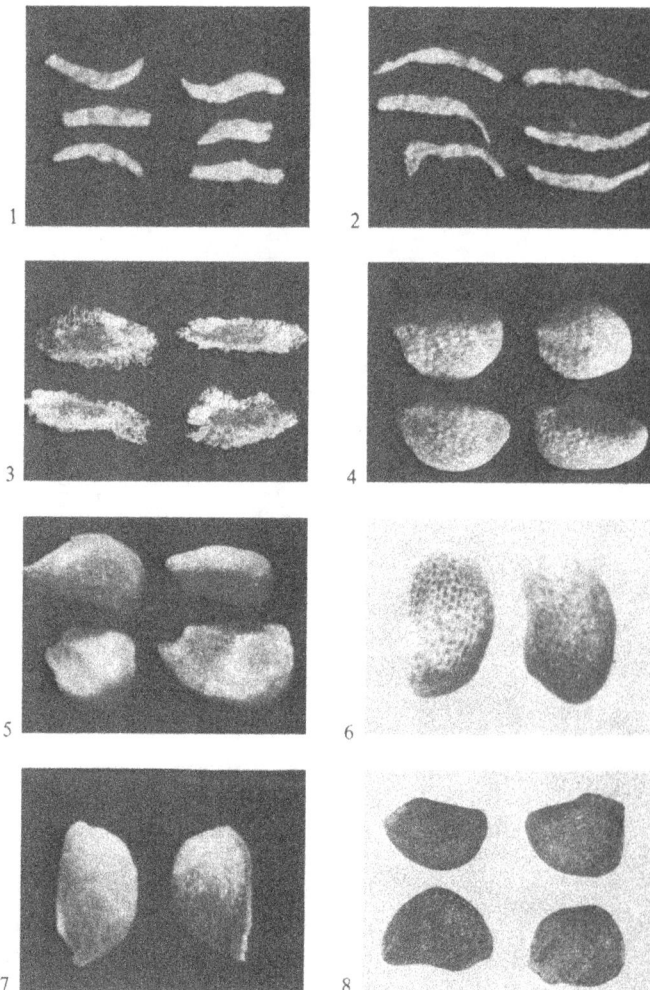

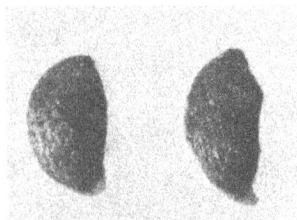

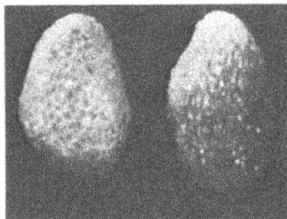

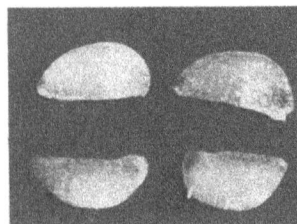

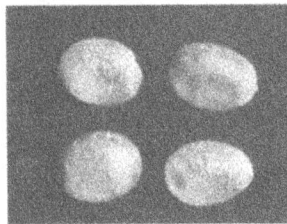

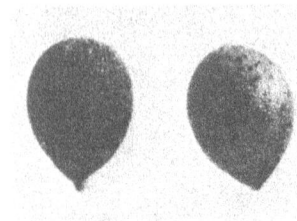

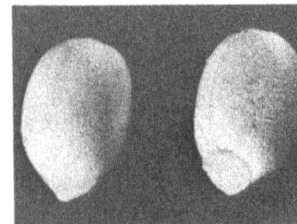

ERICACEAE

1 **Vaccinium oxycoccos** L. (x13) Small Cranberry. Seeds 1.5 x 0.9 x 0.8 mm, elliptic 3–4 in l.s., elliptic 6–7 in c.s.; surface reticulate, brown, and shiny.

2 **Vaccinium vacillans** Torr. (x15) Early Sweet Blueberry. Seeds 1.3 x 1.1 x 0.4 mm, obliquely elliptic 5–6 in l.s., elliptic 9–10 in c.s.; surface reticulate, slightly glossy, and brown.

3 **Vaccinium vitis-idaea** L. (x15) Mountain Cranberry. Seeds 1.0 x 0.6 x 0.5 mm, elliptic 3–4 in l.s., tending to be semicircular, elliptic 7 in c.s.; surface reticulate and dark brown.

ERIOCAULACEAE

4 **Eriocaulon septangulare** With. (x20) Pipewort. Seeds 0.7 x 0.5 x 0.5 mm, elliptic 4–5 in l.s., elliptic 6 in c.s.; surface obscurely rugulose, light brown.

EUPHORBIACEAE

The two obvious features of the seeds of this family are the prominent caruncle on most species and the conspicuous raphe line or ridge.

5 **Acalypha rhomboidea** Raf. (x12) Three-seeded Mercury. Seeds 1.5 x 1.1 x 1.1 mm, obovate 48–49 in l.s., elliptic 6 in c.s.; raphe ridge and caruncle evident, surface rugulose and areolate, light brown to black, often with a thin, gray, dark-dotted outer coating.

6 **Euphorbia corollata** L. (x8) Flowering Spurge. Seeds 2.3 x 1.7 x 1.7 mm, elliptic 4–5 in l.s., elliptic 6 in c.s.; caruncle comparatively small, surface with a thin outer coating which flakes off at maturity, finely puncticulate or reticulate and brown.

7 **Euphorbia cyparissias** L. (x11) Cypress Spurge. Seeds 1.8 x 1.3 x 1.3 mm, elliptic 4–5 in l.s., elliptic 6 in c.s.; raphe line not prominent but caruncle large; surface finely areolate, dull, gray or brown.

1 **Euphorbia dentata** Michx. (x7) Seeds 2.3 x 1.8 x 1.8 mm, obovate 48–49 in l.s., elliptic 6 in c.s. or inconspicuously 4-angled; surface rough, nodulose or rugose, finely papillose and areolate, black.

2 **Euphorbia esula** L. (x10) Leafy Spurge. Seeds 2.0 x 1.4 x 1.4 mm, elliptic 4–5 in l.s., elliptic 6 in c.s.; raphe obvious, caruncle rather small and flat; surface faintly areolate, brown or mottled with gray.

3 **Euphorbia glyptosperma** Engelm. (x20) Seeds 1.2 x 0.7 x 0.7 mm, obovate 47–48 in l.s., oblong 18 in c.s.; surfaces transversely ridged and faintly areolate, light gray; outer surfaces rounded, and inner surfaces somewhat concave.

4 **Euphorbia helioscopia** L. (x10) Sun Spurge. Seeds 2.0 x 1.5 x 1.5 mm, obovate 48–49 in l.s., elliptic 6 in c.s.; caruncle prominent and flat, surface coarsely reticulate or alveolate and obscurely areolate, light brown.

5 **Euphorbia maculata** L. (x17) Spotted Spurge. Seeds 1.2 x 0.9 x 0.9 mm, oblong 16–17 to obovate 48–49 in l.s., oblong 18 in c.s.; raphe evident on one of the angles, caruncle small; surface with rounded, transverse ridges, finely areolate, black with a grayish bloom.

6 **Euphorbia peplus** L. (x12) Petty Spurge. Seeds 1.5 x 0.8 x 0.8 mm, oblong 15–16 in l.s., irregularly elliptic 6 or 5- to 6-angled in c.s.; raphe line on one angle, caruncle prominent and conical; outer faces with 2–4 black pits, inner faces with 2 black pits or a single black furrow; surface finely reticulate, white or gray on the ridges.

7 **Euphorbia platyphylla** L. (x10) Broad-leaved Spurge. Seeds 2.0 x 1.6 x 1.1 mm, elliptic 4–5 in l.s., elliptic 7–8 in c.s.; surface brown and obscurely areolate, slightly glossy, caruncle small.

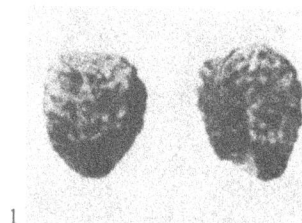

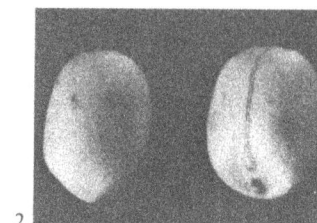

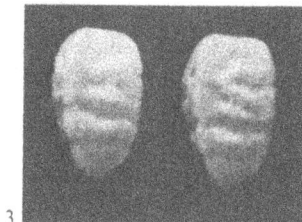

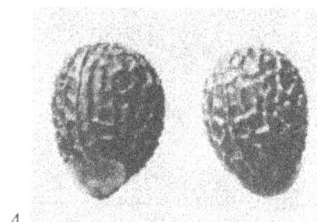

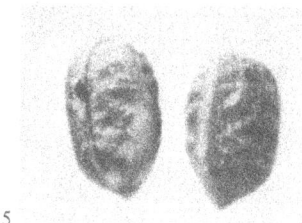

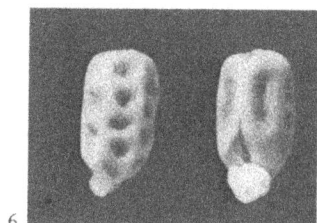

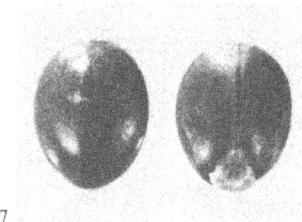

110

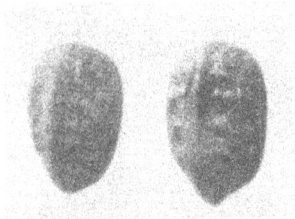

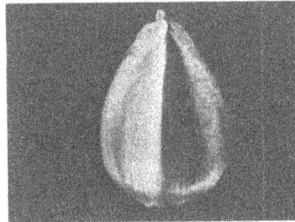

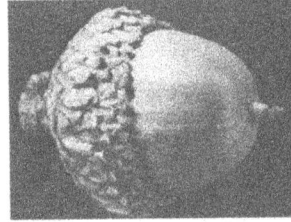

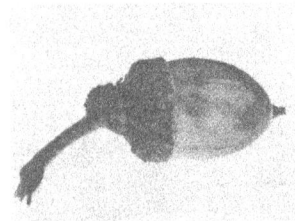

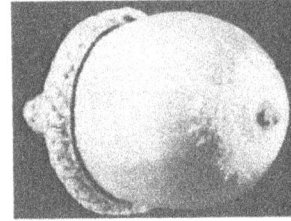

EUPHORBIACEAE

1 **Euphorbia vermiculata** Raf. (x16) Seeds 1.2 x 0.7 x 0.7 mm, obovate 47–48 in l.s., oblong 18 in c.s.; raphe on one of the angles, caruncle small, surfaces with inconspicuous, rounded transverse ridges, inner faces slightly concave and dorsal surfaces slightly convex, finely areolate, and gray to black.

2 **Ricinus communis** L. (x1.2) Castor Bean. Seeds 17 x 14 x 8 mm, elliptic 4–5 in l.s., elliptic 8–9 in c.s.; raphe prominent on one face and the caruncle quite large; surface mottled brown and gray and inconspicuously areolate.

FAGACEAE

3 **Fagus grandifolia** Ehrh. (x1.6) Beech. Fruit a nut and usually in pairs surrounded by a bur-like involucre; nut 15 x 10 x 6 mm, ovate 39 in l.s., pointed at the apex, triangular 80–81 in c.s.; sides slightly concave, smooth, and brown.

4 **Quercus alba** L. (x2) White Oak. Acorn about 20 mm long, cup covering about one-third of the nut, pubescent on the inner surface, scales thick and puberulent.

5 **Quercus bicolor** Willd. (x0.8) Swamp White Oak. Acorns usually in pairs and on long, pubescent stalks 4–10 cm long; acorn about 25 mm long with about one-third of its length in the cup, cup scales with loose tips forming a fringe.

6 **Quercus borealis** Michx. f. var. **maxima** (Marsh.) Ashe (x1.5) Red Oak. Acorns about 25 mm long, cup covering about one-quarter of the nut, scales flat and thin, cup pubescent on the inner surface and puberulent on the outer with brownish hairs.

7 **Quercus macrocarpa** Michx. (x1) Mossy-cup Oak. Acorn about 25 mm long, cup deep, covering up to two-thirds of the nut, pubescent on the inner surface, scales thick, pubescent, the upper ones tapering and forming a terminal fringe around the rim of the cup.

FAGACEAE

1 **Quercus muhlenbergii** Engelm. (x2) Yellow or Chestnut Oak. Acorn about 15 mm long, cup covering one-third to one-half of the nut, scales small, thin, and finely puberulent.

2 **Quercus palustris** Muenchh. (x1.5) Pin Oak. Acorns about 13 mm long, cup thin and covering only the base of the nut, scales appressed, thin, and puberulent.

3 **Quercus prinoides** Willd. (x1) Chinquapin Oak. Acorns about 17 mm long, cup covering about one half of the nut, scales numerous, small, tuberculate or roughened on the backs, finely puberulent.

4 **Quercus velutina** Lam. (x1.6) Black Oak. Acorns about 15 mm long, frequently soft pubescent, cup deep, enclosing one-half or more of the nut, lower scales appressed, upper scales loose and the scarious tips forming a fringe.

FUMARIACEAE

5 **Adlumia fungosa** (Ait.) Greene (x10) Climbing Fumitory. Seeds 1.1 x 1.3 x 0.8 mm, elliptic 6–7 in l.s., slightly reniform, elliptic 8–9 in c.s.; surface obscurely areolate, black, and glossy.

6 **Corydalis aurea** Willd. (x8.5) Golden Corydalis. Seeds 1.5 x 1.8 x 1.0 mm, elliptic 7 in l.s., slightly reniform, elliptic 8–9 in c.s.; surface obscurely colliculose and with a spongy caruncle, black and glossy.

7 **Corydalis sempervirens** (L.) Pers. (x11) Pale Corydalis. Seeds 1.1 x 1.3 x 0.6 mm, elliptic 6–7 in l.s., slightly reniform and with a small, spongy caruncle, elliptic 9–10 in c.s.; surface colliculose or papillose, papillae in concentric rows.

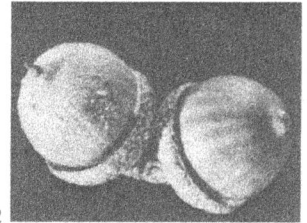

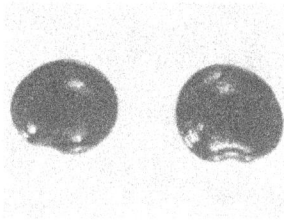
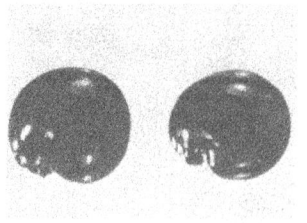
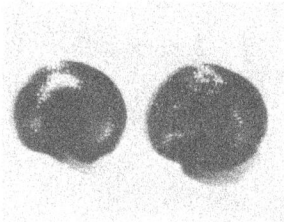

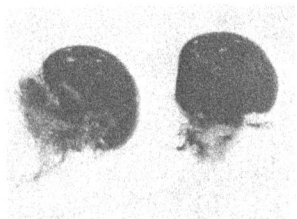
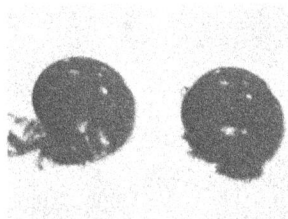

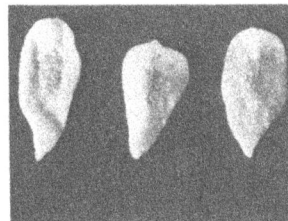
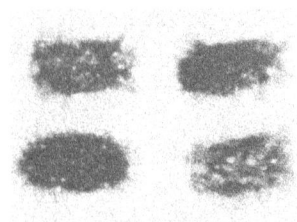
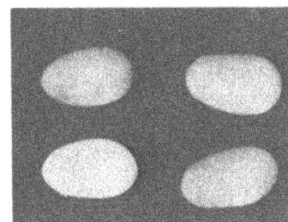
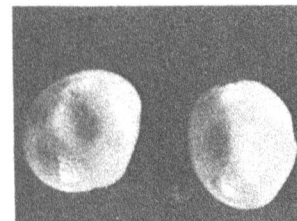
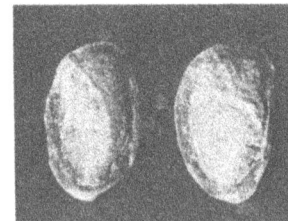

FUMARIACEAE

1 **Dicentra canadensis** (Goldie) Walp. (x7) Squirrel Corn. Seeds 1.9 x 2.4 x 1.4 mm, elliptic 7–8 in l.s., caruncle prominent and lacerate, elliptic 8–9 in c.s.; surface glossy, black, smooth or faintly colliculose.

2 **Dicentra cucullaria** (L.) Bernh. (x7) Dutchman's-breeches. Seeds 1.7 x 2.0 x 1.2 mm, elliptic 6–7 in l.s., elliptic 8–9 in c.s.; caruncle large and lacerate, surface faintly colliculose or smooth, black, and glossy.

3 **Fumaria officinalis** L. (x10) Common Fumitory. Fruit a single-celled pod 2.2 x 2.5 x 1.8 mm, elliptic 6–7 in l.s., slightly winged at the base, elliptic 7–8 in c.s.; surface rugulose or alveolate, green or brownish, pod indehiscent, style persistent.

GENTIANACEAE

4 **Gentiana andrewsii** Griseb. (x7.5) Bottle Gentian. Seeds 2.2 x 1.0 x 0.2 mm, obovate 46–47 in l.s., oblong 22–23 in c.s. with a white, rather spongy wing; surface glossy and longitudinally striate or finely ridged.

5 **Gentiana procera** Holm. (x25) Fringed Gentian. Seeds 0.6 x 0.4 x 0.4 mm, irregular in form or obliquely oblong 16–17 in l.s., irregularly angular in c.s.; surface with numerous, elongated, inflated papillae, brown.

6 **Helenia deflexa** (Sm.) Griseb. (x9) Spurred Gentian. Seeds 1.5 x 0.8 x 0.8 mm, elliptic 3–4 in l.s., elliptic 6 in c.s.; surface faintly rugulose or areolate, sometimes appearing smooth, brown.

7 **Menyanthes trifoliata** L. (x7) Buckbean. Seeds 2.4 x 2.2 x 1.4 mm, obliquely elliptic 5–6 in l.s., obliquely elliptic 8–9 in c.s.; surface smooth to faintly rugulose, brown, and slightly glossy.

8 **Swertia caroliniensis** (Walt.) Kuntze (x3) American Columbo. Seeds 7.0 x 4.2 x 0.7 mm, elliptic 3–4 in l.s., oblong 23 in c.s. with a narrow marginal wing; surface rugulose and finely areolate, dark brown.

1 **Erodium cicutarium** (L.) L'Her. (x2.7) Alfileria. Body of the fruit 4.5 x 1.0
1.0 mm, obtriangular 85–86 to obovate 45–46 in l.s. terminating in a long,
twisted awn, elliptic 6 in c.s.; stalk of the awn about 6.5 mm long, the awn
at right angles to the stalk and about 13 mm long; body with 2 lateral ridges
and covered with stiff hairs, base sharp-pointed and with a tuft of light-
coloured hairs.

2 **Geranium bicknellii** Britt. (x10) Seeds 2.2 x 1.4 x 1.2 mm, elliptic 3–4 in
l.s., elliptic 6–7 in c.s.; surface brown and reticulate.

3 **Geranium maculatum** L. (x8) Spotted Geranium. Seeds 2.6 x 1.7 x 1.5 mm,
elliptic 3–4 in l.s., elliptic 6–7 in c.s.; surface reticulate and brown.

4 **Geranium pusillum** L. (x11) Seeds 1.6 x 1.0 x 0.7 mm, elliptic 3–4 in l.s.,
elliptic 7–8 in c.s.; surface smooth and brown to black.

5 **Geranium robertianum** L. (x10) Herb-Robert. Seeds 2.0 x 1.3 x 1.3 mm,
elliptic 3–4 in l.s., elliptic 6 in c.s.; surface faintly rugulose or reticulate and
dark brown.

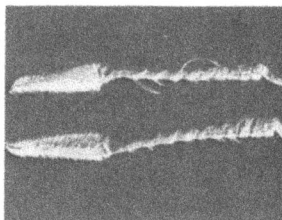

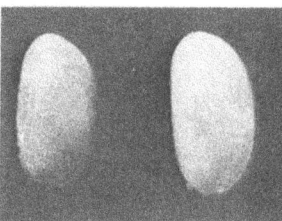

GRAMINEAE

The grass grain or caryopsis is usually enclosed in two opposite bracts. The
palea is the innermost and the outer one is the lemma. These, along with
the ovary, stamens, and lodicules, form the grass floret. In many species the
caryopsis is free from the lemma and palea, but in many other species it remains enclosed in these bracts. Two other bracts, the upper and lower glumes,
are usually present, and along with the floret, or group of florets, form the
spikelet.

6 **Agropyron dasystachyum** (Hook.) Scribn. (x4) Thickspike Wheatgrass. Spikelets several-flowered, disarticulating above the glumes; lemma densely pubescent or sometimes becoming nearly glabrous, awns absent, lemmas about 7.5
mm long and mucronate; palea about as long as the lemma and the apex
rounded.

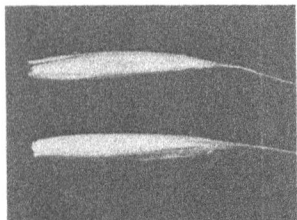

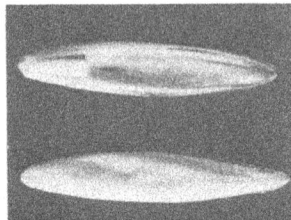

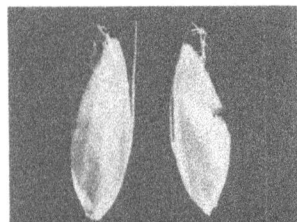

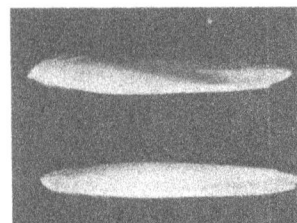

1 **Agropyron repens** (L.) Beauv. (x3) Quackgrass. Spikelets several-flowered, disarticulating above the glumes, glumes 5–7-nerved and short-awned; floret about 10 mm long, lemma inconspicuously 5-nerved at the rounded apex and terminating in a long or short awn; palea about as long as the lemma, apex rounded and the keel scabrous.

2 **Agropyron trachycaulum** (Link) Malte (x5) Slender Wheatgrass. Spikelets with few florets, disarticulating above the glumes, floret about 7 mm long, glumes rather broad, 5–7 nerved; lemmas awnless or short-awned.

3 **Alopecurus aequalis** Sobol. (x15) Short-awned Foxtail. Spikelets 1-flowered, disarticulating below the glumes; floret about 2.0 mm long, laterally compressed; glumes united at the base and with straight, stiff hairs on the keel; lemma 5-nerved, awned from near the base and the awn about as long as the floret; palea absent.

4 **Alopecurus pratensis** L. (x13) Meadow Foxtail. Spikelet as in the previous species, glumes with fine hairs on the keel and on the sides; floret about 4.3 mm long, awns of the lemmas longer than the floret.

5 **Ammophila breviligulata** Fern. (x5) American Beachgrass. Spikelets 1-flowered, disarticulating above the glumes, glumes of about equal length and longer than the lemma, slightly keeled and the keel scabrous, surface inconspicuously pubescent; floret about 7.0 mm long and with a tuft of fine hairs at the base, lemma and palea about the same length, inconspicuously nerved and blunt-tipped, palea with a short spine just behind the apex.

6 **Andropogon gerardi** Vitman (x2) Big Bluestem. Spikelets of two kinds, a sessile, perfect floret and a stalked, sterile or staminate floret; first glume of the sessile spikelet 7–10 mm long, concave on the back, the second glume very narrow; fertile lemma membranous and with a long, twisted, geniculate awn; second floret similar in structure but awnless and with or without anthers.

1 **Andropogon scoparius** Michx. (x3.5) Little Bluestem. Structure of the spikelets as in the previous species, sessile spikelet 6–8 mm long, very pubescent,
scabrous, and long-awned; sterile spikelet often reduced to a short awn.

2 **Anthoxanthum odoratum** L. (x6) Sweet Vernal Grass. Spikelet with one
terminal fertile floret and two sterile florets represented by the lemmas
only; spikelet about 3.3 mm long, disarticulating above the glumes, glumes
scabrous, second glume about twice as long as the first; sterile lemmas shorter
than the glumes and with a soft, golden pubescence, one sterile lemma with
a long, twisted, geniculate awn and the other with a shorter, straight awn,
fertile lemma shorter than the sterile ones and not awned, glossy brown;
palea enclosed in the lemma.

3 **Arrhenatherum elatius** (L.) Mert. & Koch (x3.7) Tall Oatgrass. Spikelet
2-flowered, the lower one staminate and the upper one perfect, florets disarticulating above the glumes, glumes papery, pubescent on the nerves, first
glume 1-nerved and smaller than the second which is 3-nerved; floret about
8.0 mm long and with a tuft of hairs at the base; fertile lemma scabrous and
rather shiny, staminate lemma with a long, geniculate awn from its base,
upper fertile lemma with a short straight awn from just below the apex.

4 **Avena fatua** L. (x2.2) Wild Oats. Spikelets usually with 3 florets disarticulating above the glumes; glumes membranous with 7–9 nerves; lemmas indurate, about 15 mm long and with long stiff brown hairs on the lower
half; awn 3–4 cm long, spirally twisted at the base and geniculate above
the florets; base of the floret with a hollow, circular scar surrounded by
stiff hairs.

5 **Beckmannia syzigachne** (Steud.) Fern. (x6) American Sloughgrass. Spikelets 1-flowered and closely overlapping each other in two rows on one side
of the rachis; spikelet about 3.0 mm long, disarticulating below the glumes,
laterally flattened; glumes oval to obovate, about equal in length, slightly
inflated and 3-nerved, apiculate and transversely wrinkled; lemma and palea
about equal, white and membranous, lemma 5-nerved and sharp-pointed.

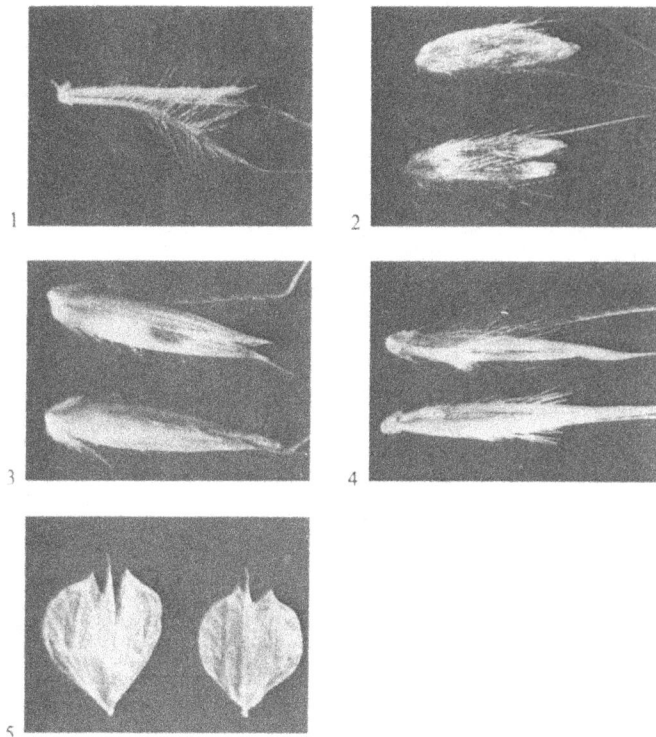

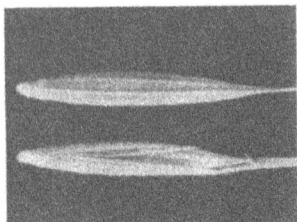
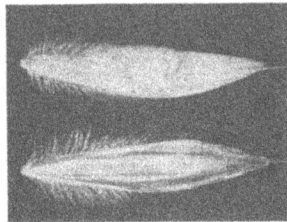
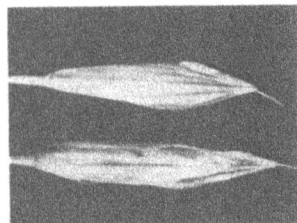

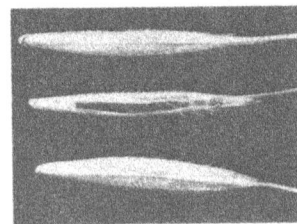

1 **Brachyelytrum erectum (Schreb.) Beauv.** (x3.3) Spikelets 1-flowered, disarticulating above the glumes, rachilla prolonged behind the palea to form a conspicuous bristle; glumes small or absent; lemma narrow, coriaceous, scabrous, 5-nerved, the nerves hispid, base forming an oblique callus and the apex prolonged into a long, straight, scabrous awn.

2 **Bromus ciliatus L.** (x3.5) Fringed Brome. Spikelets several-flowered, disarticulating above the glumes; glumes unequal in length, the first 1-nerved and the second 3-nerved; lemma 10–12 mm long, the lower half with an appressed pubescence near the margins, dorsal surface glabrous, awned at the bifid apex.

3 **Bromus inermis Leyes.** (x3.2) Smooth Brome. Spikelets several-flowered, disarticulating above the glumes; glumes unequal in length, the first 1-nerved and the second 3-nerved; floret about 10 mm long, lemma 3–5 nerved, glabrous or obscurely pubescent, awnless or with a very short awn from behind the bifid apex.

4 **Bromus secalinus L.** (x3.8) Chess. Spikelets of the same structure as the previous species; glumes rounded at the apex, first glume 3–5-nerved and the second 7-nerved, smooth; florets 6–8 mm long, lemma obscurely 7-nerved, rounded at the bifid apex, smooth or slightly roughened, margin inrolled, apex with a short awn or awnless; palea concave on the back, obtuse at the apex and the margins with conspicuous spine-like hairs.

5 **Bromus tectorum L.** (x3) Downy Brome. Spikelets as in *B. inermis*; glumes soft-pubescent; lemma and palea lanceolate and densely silky-pubescent, awns long and from between the two long teeth of the lemma.

6 **Calamagrostis canadensis (Michx.) Beauv.** (x8) Bluejoint. Spikelets 1-flowered, disarticulating above the glumes; glumes about equal in length, smooth or scabrous; florets about 2.0 mm long, lemma thin, membranous, 5-nerved with the middle nerve projected to form a hair-like awn, base of the lemma with numerous fine hairs; palea small and hyaline.

1 **Calamagrostis inexpansa** Gray (x4.5) Spikelets as in the previous species; glumes scabrous; florets about 3.8 mm long with a tuft of fine hairs at the base; lemma chartaceous and with thin hyaline margins, scabrous, awn from about the middle of the lemma and slightly longer.

2 **Calamovilfa longifolia** (Hook.) Scribn. (x4) Spikelets 1-flowered, disarticulating above the glumes; glumes chartaceous, 1-nerved, the outer glume shorter than the inner; florets about 4.5 mm long and with a tuft of white hairs at the base, lemmas chartaceous, 1-nerved, glabrous and awnless; palea about as long as the lemma and very similar.

3 **Cenchrus longispinus** (Hack.) Fern. (x2) Field Sandbur. Spikelets with a single fertile floret and a basal sterile floret, but several spikelets may aggregate and are surrounded by a bur-like, spiny involucre.

4 **Cinna arundinacea** L. (x9) Stout Woodreed. Spikelets 1-flowered, disarticulating below the glumes; glumes sub-equal, the first 1-nerved and the second 3-nerved, scabrous; floret about 3.5 mm long, lemma 3-nerved and with a minute awn from between the two teeth at the apex; palea small, 1-nerved or apparently so.

Cinna latifolia (Trev.) Griseb. (x9) Spikelets as in the previous species, about 3.0 mm long, awn of the lemma longer, up to 1.0 mm long; palea with 2 nerves very close together.

Dactylis glomerata L. (x8) Orchard Grass. Spikelets 3–6-flowered, disarticulating above the glumes, glumes unequal, keeled and with stiff hairs on the keel; florets about 5.0 mm long, lemma 5-nerved, keeled and ciliate on the keel, sharply pointed or awned at the apex; palea shorter than the lemma.

Danthonia spicata (L.) Beauv. (x6) Poverty Oatgrass. Spikelets several-flowered, disarticulating above the glumes, glumes sub-equal, papery; lemma about 4.0 mm long, finely pubescent except at the bifid apex, margins scarious, awn longer than the floret and geniculate; palea broad, flat, rounded at the apex, finely pubescent on the margins.

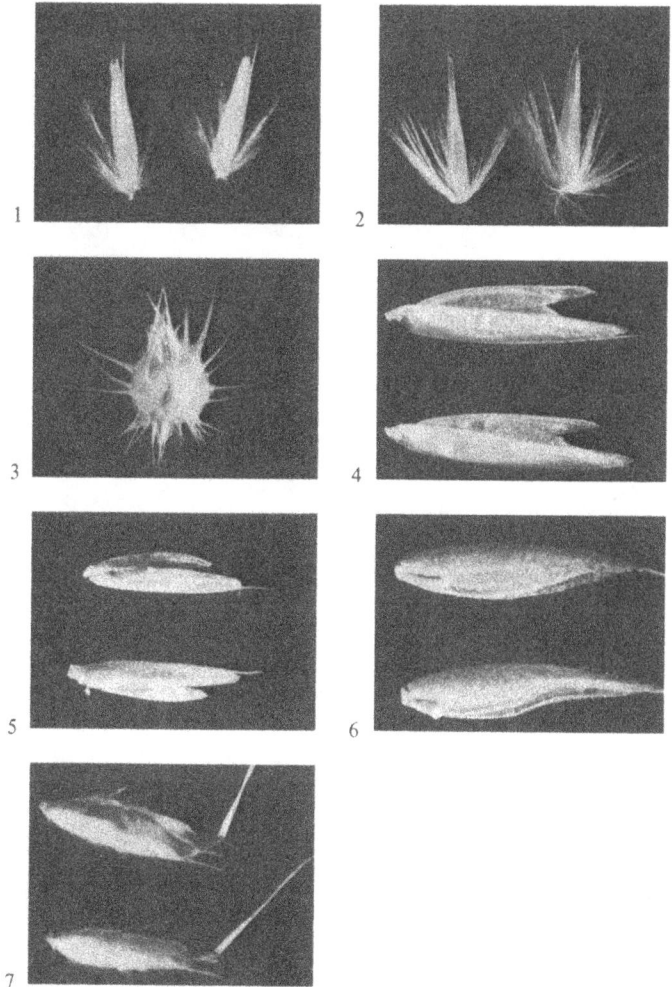

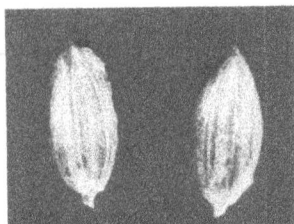
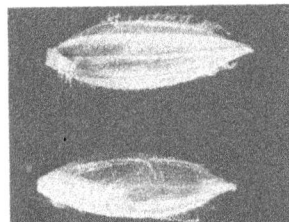
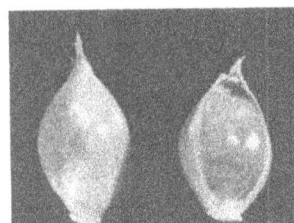

1 **Deschampsia caespitosa** (L.) Beauv. (x9) Tufted Hairgrass. Spikelets 2-flowered, disarticulating above the glumes, glumes about equal, membranous, shiny, pale or purple-tinged, first glume 1-nerved, the second 3-nerved; floret about 3.0 mm long, lemma thin, membranous, obscurely 5-nerved, bearded at the base, apex truncate and 2–3-toothed, awn from near the base hair-like, straight or bent and about as long as the lemma; palea thin and hyaline.

2 **Deschampsia flexuosa** (L.) Trin. (x8) Crinkled Hairgrass. Spikelets as in the previous species, glumes 1-nerved; florets about 3–8 mm long, lemma pubescent at the base, awn from near the base and longer than the floret and geniculate.

3 **Digitaria ischaemum** (Schreb.) Muhl. (x10) Smooth Crabgrass. Spikelets plano-convex about 2.0 mm long and with one fertile floret and a basal sterile floret, glumes very unequal, the first small or absent; sterile lemma membranous, purplish and with marginal, capitate hairs; fertile lemma brown to black, cartilaginous and with hyaline margins; palea completely enclosed by the lemma.

4 **Digitaria sanguinalis** (L.) Scop. (x9) Hairy Crabgrass. Spikelets similar to those of the previous species, about 3.0 mm long, second glume ciliate; veins of the sterile lemma purplish, conspicuous and pubescent; fertile lemma and palea smooth and greenish.

5 **Echinochloa crusgalli** (L.) Beauv. (x6.5) Barnyard Grass. Spikelets about 3.0 mm long and consisting of a terminal fertile floret and a basal sterile floret, spikelet oval in outline, plano-convex in cross section, glumes very unequal, first glume about one-half as long as the spikelet, the second about equal to the sterile lemma, sharply pointed or awned; sterile lemma purplish, long-awned, nerves with long stiff hairs, some of which have pustulate or swollen bases, spaces between the nerves with short, stiff hairs; fertile lemma indurate, smooth, and white; palea similar with the tip not enclosed by the lemma.

1 **Echinochloa muricata** (Beauv.) Fern. (x8) Spikelets similar to the previous species and distinguished from it chiefly by the greater number of pustulate-based hairs.

2 **Echinochloa walteri** (Pursh) Nash (x8) Spikelets as in *E. crusgalli*, about 4.0 mm long and much longer than broad; second glume long-awned, hairs of the glumes and sterile lemma without pustulate bases, sterile lemma long-awned.

3 **Eleusine indica** (L.) Gaertn. (x8) Goosegrass. Spikelets more than 1-flowered, florets disarticulating above the glumes, glumes unequal, 1-nerved, shorter than the first floret and keeled; floret about 3.0 mm long, lemma with three prominent nerves close together and forming a keel; palea shorter than the lemma and with a narrow, winged keel.

4 **Elymus canadensis** L. (x3.5) Canada Wild-rye. Spikelets 2 or more flowered, commonly clustered, florets disarticulating above the glumes, glumes unequal, narrow, indurate, 2–4-nerved, rough, and long-awned; lemma tough, hispid, prominently nerved, awn 2.5 cm or more long and divergent.

5 **Elymus virginicus** L. (x4.2) Virginia Wild-rye. Spikelets similar to the previous species, glumes nerveless at the base, bowed outward, scabrous; awns of the lemmas up to 1 cm long and straight.

6 **Eragrostis cilianensis** (All.) Link. (x10) Stinkgrass. Spikelets with several florets, disarticulating above the glumes and overlapping, glumes unequal, 1-nerved; lemmas 2-nerved, about 2.2 mm long, ovate, acute, the keel rough near the apex and with a few glands; palea shorter than the lemma, keel with short, straight hairs, rather persistent on the rachilla, grains reddish-brown.

Eragrostis pectinacea (Michx.) Nees. (x10) Spikelets similar to the previous species but much smaller, florets smaller, about 1.5 mm long, lemmas lacking glands on the keel.

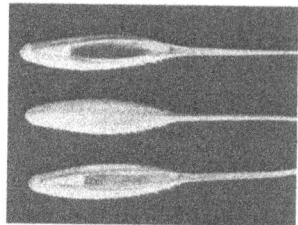
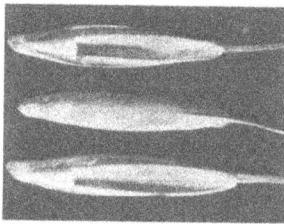

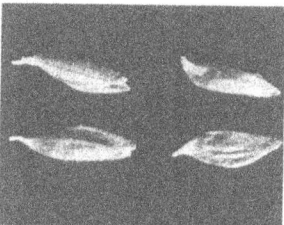

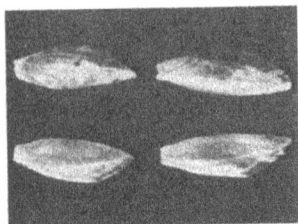
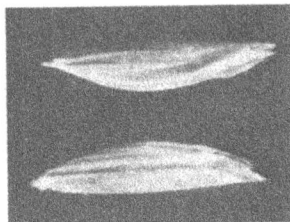

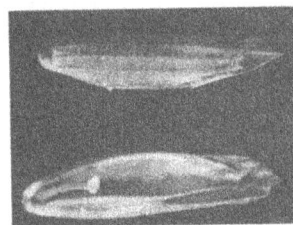
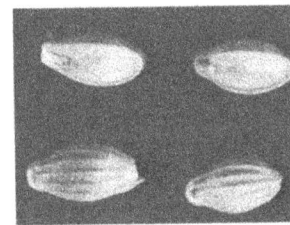
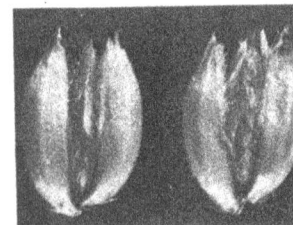

1 **Eragrostis poaeoides** Beauv. (x10) Spikelets similar to *E. cilianensis*, florets about 1.8 mm long, glands present on the keel of the lemma.

2 **Glyceria borealis** (Nash) Batchelder (x8.5) Northern Mannagrass. Spikelets with several florets disarticulating above the glumes, glumes unequal, obscurely nerved, short and rounded at the apex, margins thin and scarious; lemma about 3.5 mm long, broad, obtuse, scarious at the apex, 7-nerved, the nerves prominent and parallel and not converging at the apex, finely scabrous; palea thin and membranous.

3 **Glyceria canadensis** (Michx.) Trin. (x6.5) Rattlesnake Mannagrass. Spikelets short and broad and on flexuous pedicels, florets somewhat inflated; lemma about 3.5 mm long, 7-nerved but nerves obscure, apex membranous and purple-tinged, back of the palea concave.

4 **Glyceria grandis** S. Wats. (x6.5) American Mannagrass. Glumes small and whitish; lemmas about 2.5 mm long, purplish, strongly nerved; palea thin and papery.

5 **Glyceria septentrionalis** Hitchc. (x9) Eastern Mannagrass. Structure of the spikelets similar to those of *G. borealis* but 10–20 mm long, florets loosely overlapping; lemma green, white and hyaline at the apex, about 4.0 mm long, strongly veined and slightly roughened, veins not converging at the apex; palea membranous.

6 **Glyceria striata** (Lam.) Hitchc. (x7.5) Fowl Mannagrass. Spikelets as in *G. borealis*, 3–4 mm long, lemma about 2.0 mm long, nerves prominent; palea about as long as the lemma and with a bowed or arched keel.

7 **Hierochloe odorata** (L.) Beauv. (x5) Sweetgrass. Spikelets about 5.0 mm long with one fertile floret and two lateral sterile or staminate florets and disarticulating above the glumes, glumes equal, 3-nerved, thin and chartaceous, glabrous; lemma of sterile florets about 4.0 mm long, boat-shaped and with short, stiff hairs, margins quite hairy; fertile lemma indurate, pubescent only near the apex; palea 3-nerved and convex on the back.

GRAMINEAE

1 **Hordeum jubatum** L. (x4.3) Squirreltail Grass. Spikelets 1-flowered but usually 3 together at a node, middle floret sessile and perfect, the lateral ones short-stalked and reduced to awns only; glumes narrow and tapering to long awns; lemma broad, about 6 mm long, faintly 5-nerved and tapering to an awn; palea 2-nerved and awned.

2 **Hystrix patula** Moench. (x3.1) Bottlebrush. Spikelets in pairs at each node and spreading horizontally when mature, each spikelet with 2–4 florets, glumes modified to merely short awns; florets about 8.0 mm long, lemma rounded on the back, obscurely 5-nerved, tapering to a long awn, glabrous or pubescent; palea about as long as the lemma and awnless.

3 **Koeleria cristata** (L.) Pers. (x9) Junegrass. Spikelets 2–4-flowered, disarticulating above the glumes, glumes unequal in length, the first narrow, 1-nerved, the second broader and widest above the middle, 3–5-nerved, both glumes with a hispid pubescence; floret laterally compressed, about 3.0 mm long, lemma thin, chartaceous, shiny, obscurely 5-nerved and hispid, sometimes with a short awn or merely mucronate; palea narrow, 3-nerved.

4 **Leersia oryzoides** (L.) Swartz (x5) Rice Cutgrass. Spikelets 1-flowered, about 5.0 mm long, laterally compressed, glumes wanting; lemma chartaceous, 5-nerved but rather obscurely so, broad, boat-shaped, hispid on the keel and nerves; palea about as long as the lemma but narrow, 3-nerved, boat-shaped and hispid on the keel.

5 **Leersia virginica** Willd. (x9) Whitegrass. Spikelet as in the previous species but about 3.0 mm long, pubescence on the keel and nerves less hispid than in *L. oryzoides*.

6 **Milium effusum** L. (x11) Millet Grass. Spikelets 1-flowered and disarticulating above the glumes, glumes membranous, equal in length, obtuse at the apex, convex on the back and minutely scabrous; floret about 2.3 mm long, lemma indurate, obtuse at the apex, faintly nerved, shining and with inrolled margins closing over the palea.

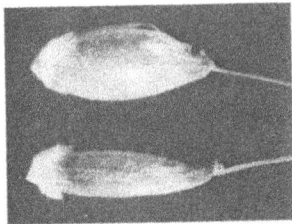
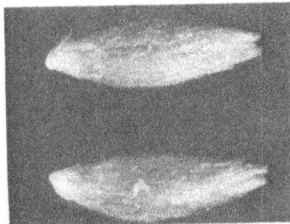

1 **Muhlenbergia frondosa** (Poir.) Fern. (x10) Spikelets 1-flowered, disarticulating above the glumes, glumes about equal in length, shorter than the lemma, acuminate, 1-nerved, margins scarious; floret about 2.5 mm long, lemma obscurely 3–5-nerved and acuminate but not awned, base with a tuft of silky hairs.

2 **Oryzopsis asperifolia** Michx. (x4.2) Ricegrass. Spikelets 1-flowered, disarticulating above the glumes, glumes about equal in length, 7-nerved, spiculose; floret about 6.0 mm long, lemma as long as the glumes and enclosing the palea, indurate, tapering to an awn about 10 mm long, pubescent on the dorsal surface and with a basal tuft of hairs.

3 **Oryzopsis pungens** (Torr.) Hitchc. (x8.5) Spikelets as in the previous species, glumes membranous, often tinged with purple on the margins; floret about 3.5 mm long, lemma indurate, appressed pubescent, enclosing the palea, awn short or often absent.

4 **Oryzopsis racemosa** (Smith) Ricker (x4.2) Spikelets as in *O. asperifolia*, glumes chartaceous; florets about 6 mm long, lemma indurate, black, pubescent, and tapering to an awn 1–2 cm long; palea enclosed by the lemma.

5 **Panicum capillare** L. (x10) Old Witchgrass. Spikelet with one terminal fertile floret and a sterile, basal floret, glumes unequal, the second glume about equal to the sterile lemma, sterile lemma thin and chartaceous; floret about 1.5 mm long, fertile lemma indurate, shiny, obtuse, 5-nerved, the margins inrolled and embracing the indurate palea.

6 **Panicum latifolium** L. (x8) General structure of the spikelet as in the previous species; floret about 2.8 mm long, glumes and sterile lemma pubescent; fertile lemma indurate, shiny, faintly areolate, margins inrolled around the palea.

1 **Panicum virgatum** L. (x9) Switchgrass. Structure of the spikelet as in
P. capillare, first glume conspicuous and clasping the floret, second glume
as long as the sterile lemma; floret about 3.3 mm long, fertile lemma indurate, glossy and embracing the palea.

2 **Phalaris arundinacea** L. (x6) Reed Canary Grass. Spikelets with one perfect floret and two lateral, sterile florets at its base, sterile florets represented by merely hairy scales, florets disarticulating above the glumes,
glumes equal, flattened or boat-shaped and longer than the lemma;
lemma about 4.0 mm long, indurate, faintly nerved, shiny, brown; palea
similar to the lemma, 2-nerved.

3 **Phleum pratense** L. (x10) Timothy. Spikelets 1-flowered, laterally compressed, disarticulating above the glumes, glumes strongly keeled with the
keels conspicuously ciliate, abruptly tapering at the apex to a short awn,
margins membranous; lemma shorter than the glumes, about 1.8 mm long,
hyaline, 3-5-nerved; palea as long as, but narrower than the lemma.

4 **Phragmites communis** Trin. (x1.6) Common Reed. Spikelets several-
flowered, florets disarticulating above the glumes, rachilla densely silky-
pubescent and the pubescence longer than the florets; glumes unequal,
3-nerved or the upper one sometimes 5-nerved; lemma 3-nerved, long-
acuminate, about 10 mm long; palea smaller and shorter than the lemma.

5 **Schizachne purpurascens** (Torr.) Swallen (x2.2) False Melic. Spikelets
several-flowered, disarticulating above the glumes, glumes unequal,
purplish, 3-5-nerved; lemma about 10 mm long, 7-nerved, pubescent
at the base, apex 2-toothed with an awn originating just below the teeth;
palea with the margins keeled and soft-pubescent.

6 **Setaria glauca** (L.) Beauv. (x7) Yellow Foxtail. Spikelets with a terminal
perfect floret and a basal sterile floret subtended by a cluster of 5-20
bristles, glumes unequal, yellowish at maturity, shorter than the spikelet,
3-5-nerved, sterile lemma membranous, yellowish and veined; fertile lemma
cartilaginous, finely transversely rugulose; fertile palea cartilaginous and flat.

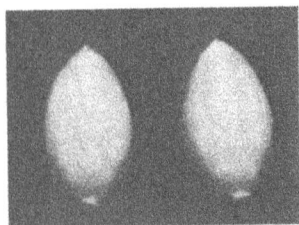
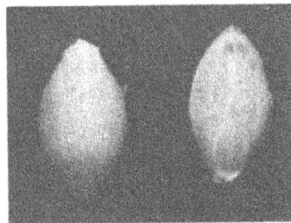

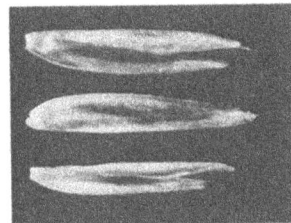
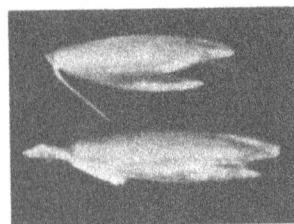

1 **Setaria verticillata** (L.) Beauv. (x10) Spikelets similar in structure to the previous species; florets subtended by only one bristle, fertile lemma white and cartilaginous, inconspicuously transversely rugulose; palea cartilaginous.

2 **Setaria viridis** (L.) Beauv. (x8) Green Foxtail. Spikelets similar to those of *S. glauca*, bristles 1–3 under each spikelet; fertile lemma cartilaginous, inconspicuously transversely rugulose or reticulate, brownish or mottled at maturity.

3 **Sorghastrum nutans** (L.) Nash (x4.5) Indian Grass. Spikelets in pairs, one sessile and perfect and the other represented by a hairy pedicel, glumes rather coriaceous, light brown, or yellowish, first glume densely pubescent with the margins inrolled over the second glume; floret about 4.5 mm long, lemma and palea thin and hyaline, lemma terminating in a long flexuous awn.

4 **Spartina pectinata** Link. (x3.5) Prairie Cordgrass. Spikelets 1-flowered, laterally compressed, disarticulating below the glumes, glumes unequal, nerved, flat and keeled, keels scabrous, taper-pointed or awned; floret 8–9 mm long, lemma folded and keeled, apex rounded or with 2 rounded teeth; palea flattened, keeled and 2-nerved.

5 **Sphenopholis intermedia** (Rydb.) Rydb. (x10) Slender Wedgegrass. Spikelets 2–3-flowered, disarticulating below the glumes, glumes unlike, the first 1-nerved, narrow and acute, the second obovate, obscurely 3–5-nerved, margins hyaline, apex apiculate; floret about 2.5 mm long, lemma chartaceous, indistinctly nerved, inconspicuously papillose, awnless or short-awned; palea thin and membranous, rachilla prolonged behind the palea in the upper floret.

6 **Sporobolus cryptandrus** (Torr.) Gray (x10) Sand Dropseed. Spikelets 1-flowered, disarticulating above the 1-nerved , unequal glumes; floret about 2.5 mm long, lemma membranous, 1-nerved, acute; palea similar to the lemma and as long or longer and acute.

1 **Stipa spartea** Trin. (x1.2) Porcupine Grass. Spikelets 1-flowered, disarticulating obliquely above the glumes and leaving a sharp-pointed, bearded callus on the base of the floret, glumes chartaceous. lanceolate, and acute; lemma cartilaginous, strongly inrolled and surrounding the palea, awn long (12–20 cm), twisted and twice geniculate, brown-pubescent on the lower half of the lemma.

2 **Zizania aquatica** L. (x1.7) Wild Rice. Spikelets unisexual with the staminate florets at the base of the inflorescence and the pistillate ones at the apex, glumes absent, fertile floret about 15 mm long, fertile lemma membranous, 3-nerved, tapering to a long, slender awn, hispid margins inrolled and clasping the palea; staminate lemma 5-nerved, chartaceous, taper-pointed or short-awned; palea 3-nerved; staminate spikelet soon deciduous, stamens six.

HALORAGACEAE

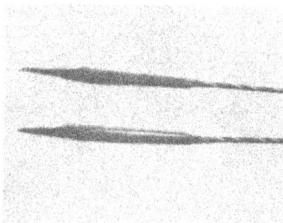
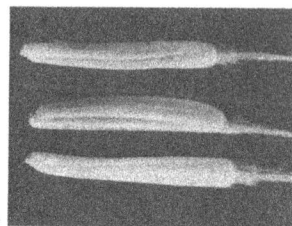
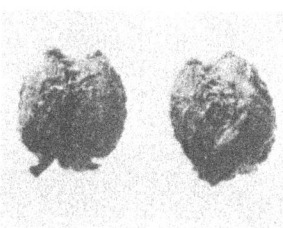
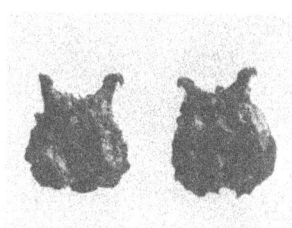
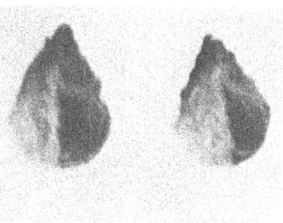

3 **Myriophyllum exalbescens** Fern. (x10) Fruit 2.3 x 2.0 x 2.0 mm, elliptic 5–6 in l.s., elliptic 6 or oblong 18 in c.s., 4-lobed; surface rugulose, papillose and slightly pubescent, style remnants small; fruit separating into 4 mericarps at maturity.

4 **Myriophyllum heterophyllum** Michx. (x8) Fruit 1.7 x 1.9 x 1.9 mm, elliptic 6–7 in l.s., oblong 18 in c.s., 4-lobed and 4-locular; surface ridged and rugose, persistent styles and stigmas conspicuous; fruit separating into 4 mericarps at maturity.

5 **Proserpinaca palustris** L. (x6) Mermaid-weed. Fruits nut-like, 3.3 x 2.5 x 2.5 mm, ovate 39–40 in l.s., triangular 78 in c.s.; surface hard, thick-walled, nutlets 3-celled and indehiscent; surface smooth or rugulose.

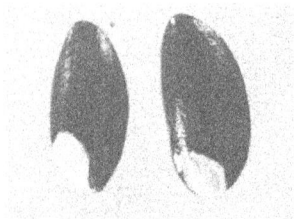

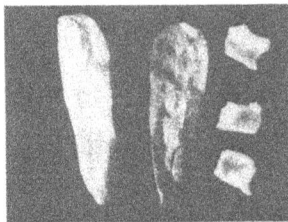

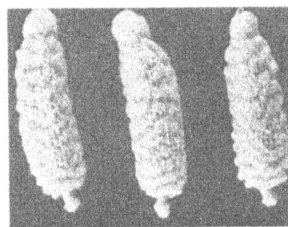

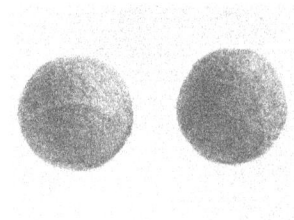

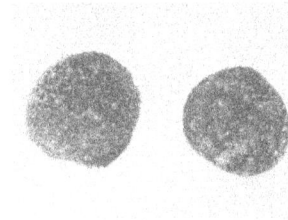

HAMAMELIDACEAE

1 **Hamamelis virginiana** L. (x3) Witch Hazel. Seeds 7.5 x 4.0 x 3.4 mm, elliptic 3–4 in l.s., elliptic 6–7 in c.s.; surface with a large basal scar, smooth, glossy, and black.

2 **Liquidambar styraciflua** L. (x3) Sweet Gum. Two forms of seeds occur in the fruit, the small, angular, sterile seeds are numerous, but the winged, fertile seeds are fewer; fertile seeds 8.0 x 2.5 x 0.8 mm, obovate 45–46 in l.s., oblong 22–23 in c.s.; wing membranous and sub-terminal; surface reticulate and often marked by oblong resin ducts.

HIPPURIDACEAE

3 **Hippuris vulgaris** L. (x10) Mare's-tail. Fruit hard and nut-like, 1-celled, 2.1 x 1.2 x 1.2 mm, elliptic 3–4 in l.s., elliptic 6 in c.s.; surface faintly longitudinally striate, obscurely areolate and brown to black.

HYDROCHARITACEAE

4 **Vallisneria americana** Michx. (x10) Eelgrass. Seeds 2.4 x 0.8 x 0.8 mm, oblong 14 in l.s., tapering slightly towards the apex which is rather bulbous, elliptic 6 in c.s.; surface with a thick, mucilaginous coating when wet; outer coating reticulate or alveolate with the areolae in longitudinal rows; remnant of the funiculus obvious, gray, and somewhat irridescent when dry.

HYDROPHYLLACEAE

5 **Hydrophyllum appendiculatum** Michx. (x4.5) Seeds 3.3 x 3.3 x 3.3 mm, elliptic 6 in l.s. and in c.s.; surface with a single slight ridge, finely reticulate and dark brown.

6 **Hydrophyllum canadense** L. (x6) Broad-leaved Waterleaf. Seeds 2.7 x 2.7 x 2.7 mm, elliptic 6 in l.s., and in c.s.; surface finely reticulate, slightly glossy and dark brown.

1 **Hydrophyllum virginianum** L. (x6.5) Virginia Waterleaf. Seeds 3.0 x 3.0 x 3.0 mm, elliptic 6 in l.s. and in c.s.; surface reticulate and light or dark brown.

HYPERICACEAE

2 **Hypericum boreale** (Britt.) Bickn. (x25) Seeds 0.6 x 0.2 x 0.2 mm, oblong 14 in l.s., elliptic 6 in c.s.; apiculate at both ends, surface faintly ribbed and reticulate.

3 **Hypericum canadense** L. (x30) Seeds 0.5 x 0.2 x 0.2 mm, oblong 14–15 in l.s., elliptic 6 in c.s.; apiculate at both ends, surface faintly striate or ribbed and reticulate.

4 **Hypericum ellipticum** Hook. (x30) Seeds 0.5 x 0.2 x 0.2 mm, oblong 14–15 in l.s., elliptic 6 in c.s.; apiculate at both ends, surface faintly ribbed and reticulate.

5 **Hypericum kalmianum** L. (x17) Shrubby St John's-wort. Seeds 1.0 x 0.4 x 0.4 mm, oblong 14–15 in l.s., elliptic 6 in c.s.; apiculate at both ends, raphe ridge evident, surface reticulate.

6 **Hypericum perforatum** L. (x18) Common St John's-wort. Seeds 1.1 x 0.5 x 0.5 mm, oblong 14–15 in l.s., elliptic 6 in c.s.; apiculate at the ends, raphe ridge or wing present, surface glossy, black, and reticulate.

7 **Hypericum punctatum** Lam. (x16) Seeds 0.9 x 0.3 x 0.3 mm, oblong 14 in l.s., elliptic 6 in c.s.; apiculate at the ends, surface with a raphe ridge and reticulate.

8 **Hypericum pyramidatum** Ait. (x10) Great St John's-wort. Seeds 1.4 x 0.4 x 0.4 mm, oblong 13–14 in l.s., elliptic 6 in c.s.; apiculate at both ends, surface with a prominent, longitudinal raphe ridge and reticulate.

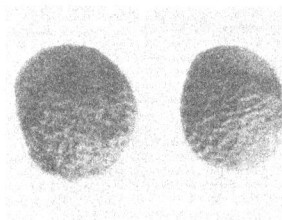
1

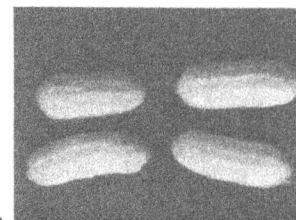
2

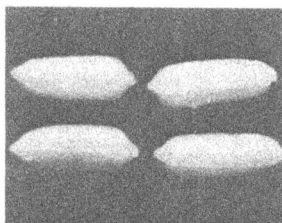
3

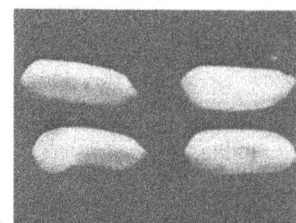
4

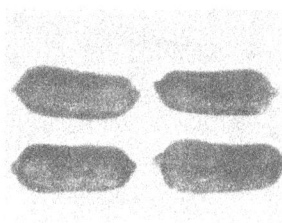
5

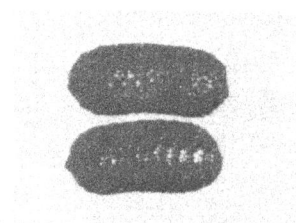
6

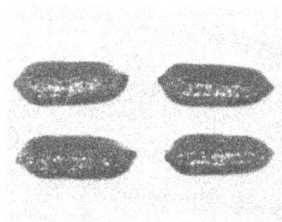
7

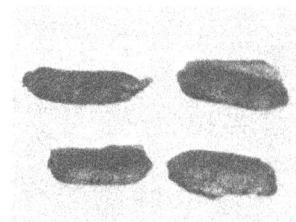
8

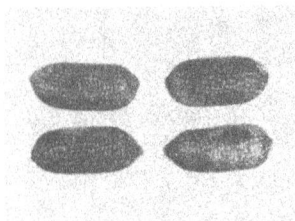

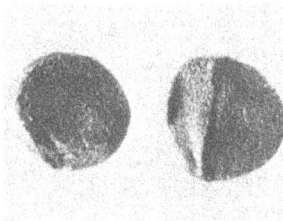

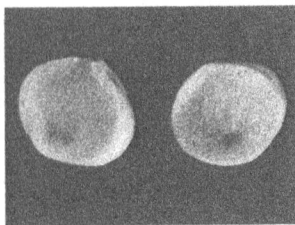

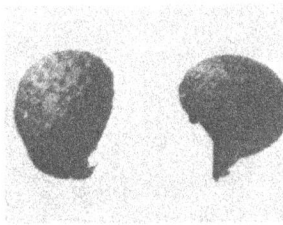

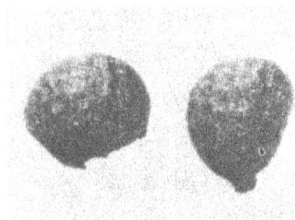

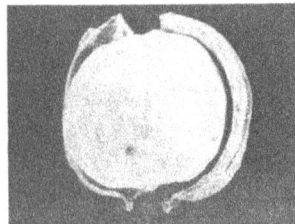

HYPERICACEAE

1 **Triadenum fraseri** (Spach.) Gl. (x13) Marsh St John's-wort. Seeds 1.1 x 0.5 x 0.5 mm, oblong 14–15 in l.s., elliptic 6 in c.s.; ends apiculate and the surface with a raphe ridge and reticulate.

IRIDACEAE

2 **Iris versicolor** L. (x2.5) Blue Flag. Seeds very irregular in form; surface rugulose, somewhat glossy and brown.

3 **Iris virginica** L. (x1.8) Seeds very irregular in form and dimensions; surface smooth, rather glossy, and brown.

4 **Sisyrinchium montanum** Greene (x12) Blue-eyed Grass. Seeds 1.2 x 1.2 x 1.2 mm, elliptic 6 to obovate 50 in l.s., elliptic 6 in c.s.; surface rugulose, dull, and black.

5 **Sisyrinchium mucronatum** Michx. (x16) Seeds 0.9 x 0.7 x 0.7 mm, elliptic 4–5 in l.s., elliptic 6 in c.s.; surface rugulose, dull, and black.

JUGLANDACEAE

The fruit of this family is a nut surrounded by a fleshy involucre or husk which becomes dry and hard at maturity. In *Carya* species the husk dehisces into 4 sections at maturity but in *Juglans* species it remains intact.

The nut shell is thin in *Carya* and thick in *Juglans* and the kernel is divided by septa extending inward from the shell.

6 **Carya cordiformis** (Wang.) K. Koch (x1) Bitternut Hickory. Fruit 2–3 cm long and with a thin husk; nut subglobose to elliptic, slightly angled in c.s., shell faintly ridged but otherwise smooth, kernel of the nut bitter.

7 **Carya ovalis** (Wang.) Sarg. (x1) Sweet Pignut Hickory. Fruit about 2 cm long with a thin husk; nut sub-globose to broadly elliptic, shell slightly ridged and nerved, kernel edible.

1 **Carya ovata** (Mill.) K. Koch (x0.8) Fruit about 4 cm long and with a thick husk; nut elliptical to obovate and slightly 4-angled in c.s.; shell smooth, kernel edible.

2 **Juglans cinerea** L. (x0.8) Butternut. Fruit elliptic, husk thin, glandular pubescent; nut elliptic, about 4 cm long and 2.5 cm in diameter, apex pointed, base rounded, surface rough with jagged ridges.

3 **Juglans nigra** L. (x1) Black Walnut. Fruit sub-globose with a thin, glandless, somewhat roughened husk; nut elliptic or ovate, about 2.5 cm long and 3.0 cm in diameter, surface strongly rugose.

JUNCACEAE

Juncus seeds are more or less fusiform and usually caudate or apiculate at one or both ends. Some species may have an aril-like coating.

4 **Juncus acuminatus** Michx. (x30) Seeds 0.5 x 0.3 x 0.3 mm, ovate 38–39 in l.s., elliptic 6 in c.s.; longitudinally striate and faintly areolate.

5 **Juncus balticus** Willd. (x25) Seeds 0.8 x 0.5 x 0.5 mm, ovate 38–39 in l.s., elliptic 6 in c.s.; surface with a whitish, membranous, silvery, aril-like coating, faintly longitudinally striate and areolate.

6 **Juncus brevicaudatus** (Engelm.) Fern. (x20) Seeds 0.9 x 0.2 x 0.2 mm, elliptic 1–2 in l.s., elliptic 6 in c.s.; body of the seed with a membranous coating prolonged at each end (caudate); surface faintly striate and areolate.

7 **Juncus bufonius** L. (x35) Toad-rush. Seeds 0.4 x 0.2 x 0.2 mm, elliptic 3 in l.s., elliptic 6 in c.s.; slightly caudate at the apex and base, surface striate and finely areolate.

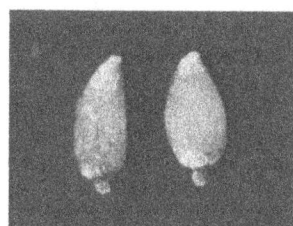

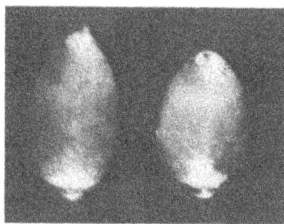

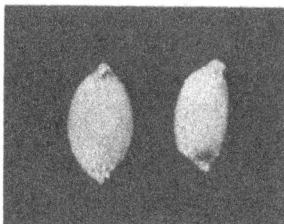

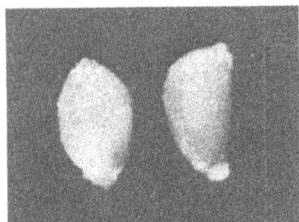
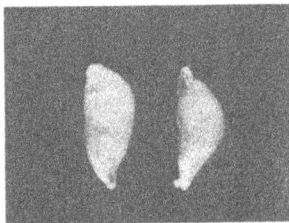
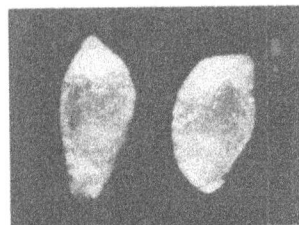
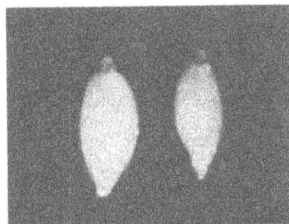
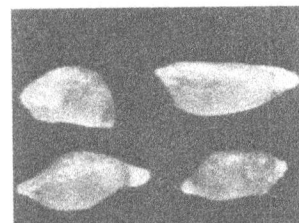
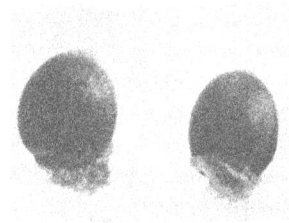
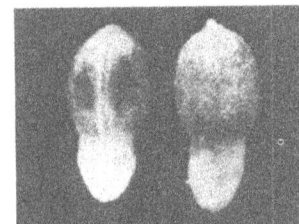
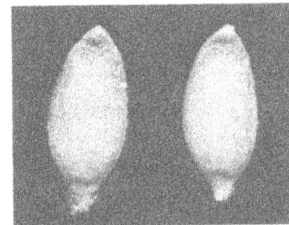

1 **Juncus dudleyi** Wieg. (x40) Seeds 0.4 x 0.25 x 0.25 mm, obliquely elliptic 3–4 in l.s., slightly caudate, elliptic 6 in c.s.; surface faintly striate and areolate.

2 **Juncus effusus** L. (x40) Soft Rush. Seeds 0.4 x 0.2 x 0.2 mm, elliptic 3 in l.s., short caudate at each end, elliptic 6 in c.s., surface finely striate and areolate.

3 **Juncus filiformis** L. (x40) Seeds 0.5 x 0.25 x 0.25 mm, elliptic 3 in l.s., elliptic 6 in c.s.; surface with a thin, membranous coating, apiculate at both ends, finely striate and areolate.

4 **Juncus nodosus** L. (x30) Seeds 0.5 x 0.2 x 0.2 mm, elliptic 2–3 in l.s., apiculate at each end, elliptic 6 in c.s.; surface longitudinally striate and areolate.

5 **Juncus tenuis** Willd. (x30) Trail-rush. Seeds 0.4 x 0.25 x 0.25 mm, obliquely elliptic 3–4 in l.s., apiculate at both ends, elliptic 6 in c.s.; surface finely striate and areolate.

6 **Luzula acuminata** Raf. (x13) Woodrush. Seeds 1.3 x 1.2 x 1.2 mm, elliptic 5–6 in l.s., elliptic 6 in c.s.; base with a prominent caruncle, surface areolate and brown.

7 **Luzula campestris** (L.) DC. (x20) Seeds 1.0 x 0.7 x 0.7 mm, elliptic 4–5 in l.s., elliptic 6 in c.s.; caruncle large and whitish (1.5 mm), surface reticulate and brown.

8 **Luzula parviflora** (Ehrh.) Desv. (x16) Seeds 1.2 x 0.6 x 0.5 mm, elliptic 3 in l.s., elliptic 7 in c.s.; caruncle prominent, surface areolate and brown.

1 **Scheuchzeria palustris** L. var. **americana** Fern. (x5.5) Seeds 4.0 x 2.2 x 2.0 mm, elliptic 3–4 in l.s., obliquely elliptic 6–7 in c.s.; surface with a prominent, longitudinal raphe ridge, slightly rugulose and black.

2 **Triglochin maritima** L. (x4.5) Arrowgrass. Fruit consisting of 6 united carpels cohering centrally, 4.5 x 2.2 x 2.2 mm, elliptic 2–3 in l.s., elliptic 6 in c.s. but coarsely longitudinally ribbed; individual carpels obtriangular in c.s. with sharp margins and U- or V-shaped on the dorsal surface, styles and stigmas persistent and reflexed.

3 **Triglochin palustris** L. (x4) Fruits consisting of 3 united carpels, 6.5 x 1.2 x 1.2 mm, elliptic 1–2 to obovate 45–46 in l.s., elliptic 6 in c.s.; carpels separating from the base upward upon maturity.

LABIATAE

The four nutlets or achenes of the fruit of the Labiatae develop from a 4-lobed ovary, each lobe completely separating into a single nutlet at maturity. Contact with the adjacent nutlets often makes the cross section of the nutlet obtriangular or obovate if the dorsal surface is considered the apex of the cross section. The hilum may be basal or more or less lateral. Many species have separated or united depressions or markings at the inner base and these may be characteristic for a species.

4 **Agastache foeniculum** (Pursh) Kuntze (x13) Blue Giant Hyssop. Nutlets 1.5 x 0.9 x 0.6 mm, elliptic 3–4 in l.s., obtriangular 92 in c.s. with the dorsal surface rounded; basal depressions conspicuous, remnant of the funiculus obvious, surface grayish-pubescent near the apex, smooth and brown.

5 **Agastache nepetoides** (L.) Kuntze (x13) Yellow Giant Hyssop. Nutlets 1.6 x 0.8 x 0.6 mm, elliptic 3 in l.s., obtriangular 91–92 in c.s.; basal depressions prominent, upper half of the nutlet densely pubescent, surface smooth and black.

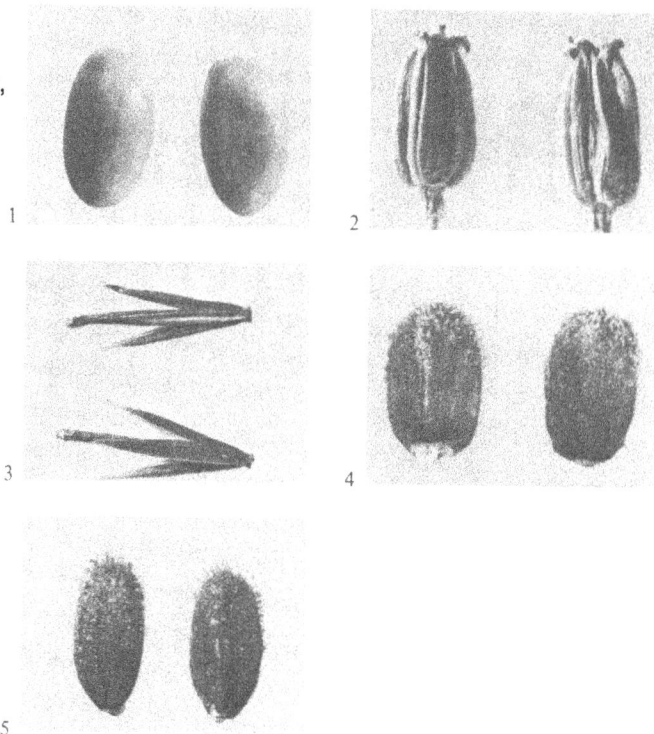

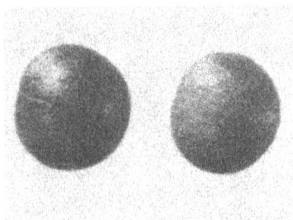

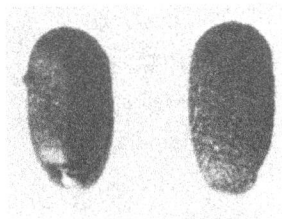
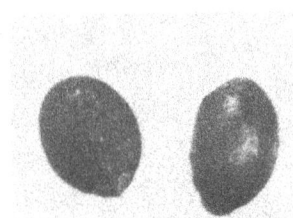

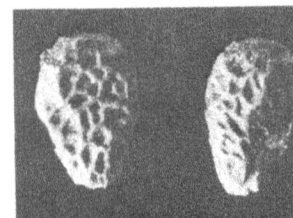
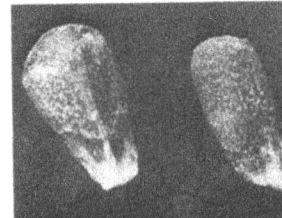

1 **Collinsonia canadensis** L. (x6.5) Stoneroot. Nutlets 2.5 x 2.5 x 2.5 mm, elliptic 6 in l.s. and in c.s.; surface veined, brown and shiny.

2 **Dracocephalum parviflorum** Nutt. (x8.5) Dragonhead. Nutlets 2.2 x 1.4 x 1.0 mm, elliptic 3–4 in l.s., obtriangular 91–92 in c.s.; inner faces with a grayish, crescent-shaped depression at the base, surface smooth or faintly areolate, slightly ragged at the apex (erose), brown to black.

3 **Galeopsis tetrahit** L. (x6.5) Hemp Nettle. Nutlets 3.0 x 2.3 x 1.6 mm, obovate 48–49 in l.s., elliptic 7–8 to obtriangular 91–92 in c.s.; surface colliculose, mottled gray, brown and black.

4 **Glecoma hederacea** L. (x15) Creeping Charlie. Nutlets 1.5 x 0.9 x 0.7 mm, elliptic 3–4 in l.s., obtusely obtriangular 91–92 in c.s., dorsal surface rounded; surface brown and glandular dotted.

5 **Hedeoma pulegioides** (L.) Pers. (x20) American Pennyroyal. Nutlets 0.9 x 0.7 x 0.5 mm, elliptic 4–5 in l.s., obtriangular 91–92 in c.s.; basal depressions small and whitish, surface faintly areolate, brown to black.

6 **Hyssopus officinalis** L. (x9.5) Hyssop. Nutlets 2.1 x 1.0 x 0.7 mm, obovate 46–47 in l.s., rather abruptly contracted at the pointed base, obtriangular 91–92 in c.s.; basal depressions grayish, surface finely areolate, dull and brown.

7 **Isanthus brachiatus** (L.) BSP. (x8) False Pennyroyal. Nutlets 2.5 x 1.7 x 1.1 mm, obovate 48–49 in l.s., obtriangular 92–93 in c.s., angles and dorsal surface rounded; hilar scar almost central on the inner angle and large, surface coarsely reticulate or alveolate.

8 **Lamium album** L. (x9) Nutlets 2.5 x 1.6 x 1.1 mm, obovate 47–48 in l.s., obtriangular 91–92 in c.s.; base with a conspicuous caruncle; surface rugulose with small, whitish ridges and faintly reticulate.

1 **Lamium amplexicaule** L. (x13) Henbit. Nutlets 1.9 x 1.0 x 0.6 mm, obovate 47–48 in l.s., base with a prominent caruncle, obtriangular 92–93 in c.s.; basal depressions obvious, surface light brown mottled with white, rugulose with the ridges rounded, finely reticulate.

2 **Leonurus cardiaca** L. (x11) Motherwort. Nutlets 1.9 x 1.2 x 0.8 mm, obtriangular 87–88 in l.s., truncate at the apex, obtriangular 92 in c.s.; apex with a ridge and grayish pubescence, surface black, slightly glossy, and finely areolate.

3 **Lycopus americanus** Muhl. (x12) Bugleweed. Nutlets 1.2 x 0.8 x 0.4 mm, obovate 48 in l.s., obliquely obtriangular 93 in c.s.; margins with whitish, suberose ridges, surface brown and glandular-dotted.

4 **Lycopus uniflorus** Michx. (x10) Water Horehound. Nutlets 1.4 x 1.0 x 0.5 mm, obovate 48–49 in l.s., obtriangular 93 in c.s.; margins with corky wings, apex truncate and toothed, surface dull, brown to black, surface glandular-dotted.

5 **Marrubium vulgare** L. (x10) Common Horehound. Nutlets 1.7 x 1.1 x 0.8 mm, obovate 47–48 in l.s., obtriangular 91–92 in c.s.; surface finely reticulate, olive-green, or mottled, or all black.

6 **Melissa officinalis** L. (x13) Balm. Nutlets 1.8 x 0.9 x 0.7 mm, obovate 47 in l.s., obtriangular 91–92 in c.s.; surface faintly areolate and black, basal depressions prominent and white.

7 **Mentha arvensis** L. (x20) Canada Mint. Nutlets 0.7 x 0.6 x 0.5 mm, elliptic 5–6 in l.s., elliptic 7 in c.s.; triangular at the base, basal depression crescent-shaped, surface faintly areolate and brown.

8 **Mentha piperita** L. (x17) Peppermint. Nutlets 0.7 x 0.5 x 0.4 mm, elliptic 4–5 in l.s., elliptic 7–8 in c.s.; slightly triangular at the base, basal depressions prominent, surface brown and faintly areolate.

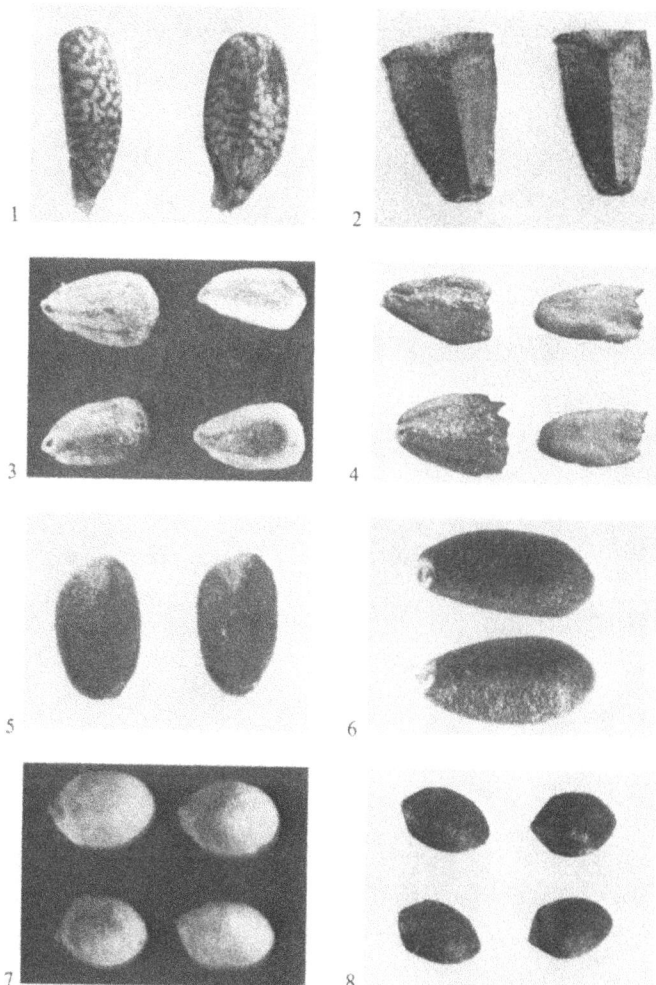

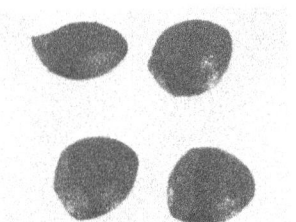

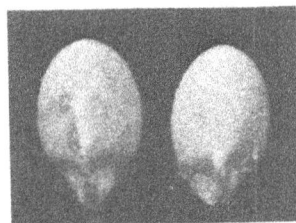

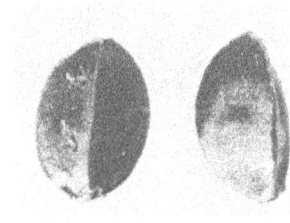

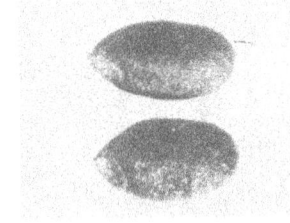

1 **Mentha spicata** L. (x15) Spearmint. Nutlets 0.8 x 0.5 x 0.4 mm, elliptic 3–4 in l.s., elliptic 7–8 in c.s.; slightly triangular at the base, basal depression large and lunate, surface faintly areolate, dull, and brown.

2 **Monarda didyma** L. (x10) Oswego Tea. Nutlets 1.5 x 1.0 x 0.7 mm, elliptic 4 in l.s., obliquely obtriangular 91–92 in c.s.; basal depression elliptical, hilum with a circular orifice, surface smooth and brown.

3 **Monarda fistulosa** L. (x11) Wild Bergamot. Nutlets 1.6 x 0.8 x 0.5 mm, elliptic 3 in l.s., obtriangular 92–93 in c.s.; basal depression large, hilum with a circular orifice, surface faintly areolate and brown.

4 **Nepeta cataria** L. (x10) Catnip. Nutlets 1.5 x 1.0 x 0.7 mm, elliptic 4 in l.s., obtriangular 91–92 in c.s.; base with two eye-like depressions, surface smooth, dull and brown to black.

5 **Origanum vulgare** L. (x27) Wild Marjoram. Nutlets 0.8 x 0.6 x 0.4 mm, elliptic 4–5 in l.s., abruptly contracted to a triangular base, elliptic 8 in c.s.; basal depressions prominent, surface smooth, dull, and brown.

6 **Physostegia virginiana** (L.) Benth. (x9) Obedient Plant. Nutlets 3.6 x 2.4 x 1.4 mm, elliptic 4 in l.s., obtriangular 92–93 in c.s. with the margins sharply angled or almost winged; surface faintly areolate, slightly glandular, dull, and brown.

7 **Prunella vulgaris** L. (x10) Heal-all. Nutlets 1.9 x 1.1 x 0.8 mm, elliptic 3–4 to obovate 47–48 in l.s., obtriangular 91–92 in c.s.; inner angle, dorsal surface, and margins with narrow ridges and sulcate, base with a prominent, white, pointed remains of the funiculus; surface brown, finely nerved, and areolate.

8 **Pycnanthemum virginianum** (L.) Durand & Jackson (x22) Mountain Mint. Nutlets 1.0 x 0.5 x 0.3 mm, elliptic to oblong 15 in l.s., obliquely obtriangular 92–93 in c.s.; basal depression obvious, surface finely areolate and black.

1 **Salvia nemorosa** L. (x10) Nutlets 1.8 x 1.5 x 1.0 mm, elliptic 5 in l.s., elliptic 8 to obliquely obtriangular 92 in c.s.; surface smooth or faintly nerved, dull, and brown.

2 **Salvia verticillata** L. (x12) Whorled Sage. Nutlets 1.7 x 1.3 x 1.0 mm, elliptic 4–5 in l.s., elliptic 7–8 to obtriangular 91–92 in c.s.; surface smooth or nerved, dull, and brown.

3 **Satureja acinos** (L.) Scheele (x11) Mother-of-thyme. Nutlets 1.3 x 0.7 x 0.5 mm, obovate 47–48 in l.s., obtriangular 91–92 in c.s.; basal depressions prominent and light-coloured, surface reticulate and brown.

4 **Satureja glabella** (Michx.) Briquet var. **angustifolia** (Torr.) Svenson (x13) Nutlets 0.9 x 0.5 x 0.4 mm, elliptic 3–4 in l.s., obtriangular 91–92 in c.s.; pointed at the base and with white basal depressions, surface reticulate, dull, and black.

5 **Satureja vulgaris** (L.) Fritch (x11) Basil. Nutlets 1.1 x 0.9 x 0.7 mm, elliptic 4–5 in l.s., obliquely obtriangular 91–92 to elliptic 7–8 in c.s.; basal depression tending to be lunate, surface brown.

6 **Scutellaria galericulata** L. (x10) Common Skullcap. Nutlets 1.1 x 1.3 x 1.1 mm, elliptic 6–7 in l.s. and in c.s.; surface verrucose and brown.

7 **Scutellaria lateriflora** L. (x14) Mad-dog Skullcap. Nutlets 0.7 x 1.4 x 1.1 mm, elliptic 9 in l.s., elliptic 7–8 in c.s., surface verrucose.

8 **Scutellaria parvula** Michx. (x14) Nutlets 0.7 x 1.2 x 1.2 mm, elliptic 8–9 in l.s., elliptic 6 in c.s.; surface strongly verrucose.

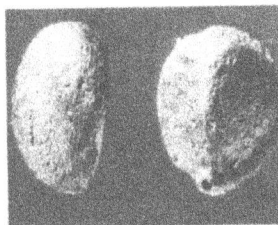
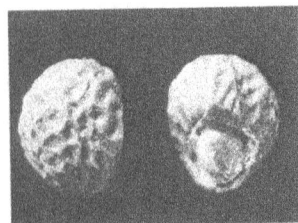

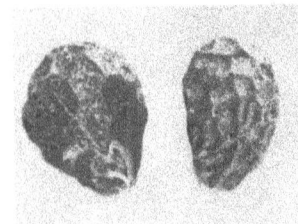

1 **Stachys germanica** L. (x10) Nutlets 2.0 x 1.5 x 1.1 mm, obovate 48–49 in l.s., obtriangular 91–92 in c.s. with the surfaces slightly convex; surface faintly areolate and striate, dull, and black.

2 **Stachys olympica** Poir. (x10) Lamb's-ears. Nutlets 2.1 x 1.6 x 1.1 mm, obovate 48–49 in l.s., elliptic 7–8 to obtriangular 91–92 in c.s., surfaces slightly convex; dorsal surface rugulose or minutely areolate and brown.

3 **Stachys palustris** L. (x10) Woundwort. Nutlets 1.9 x 1.5 x 1.0 mm, elliptic 4–5 in l.s., elliptic 8 or obtriangular 98 in c.s., surfaces slightly rounded; dorsal surface rugulose and inconspicuously areolate, black.

4 **Stachys tenuifolia** Willd. (x10) Hedge Nettle. Nutlets 2.2 x 1.9 x 1.3 mm, elliptic 5–6 in l.s., acute at the base, obtriangular 91–92 in c.s. with the angles and sides slightly rounded; surface minutely areolate, dull, and black.

5 **Teucrium canadense** L. (x10) Wood Sage. Nutlets 2.0 x 1.7 x 1.3 mm, obovate 49–50 in l.s., elliptic 7–8 in c.s.; hilar scar covering about one-third of the inner faces; surface coarsely alveolate, brown and with scattered white hairs and glandular dots.

6 **Thymus serpyllum** L. (x15) Wild Thyme. Nutlets 0.7 x 0.5 x 0.4 mm, elliptic 4–5 in l.s., elliptic 7–8 in c.s.; surface faintly areolate, light brown to black.

7 **Trichostema dichotomum** L. (x9) Nutlets 2.2 x 1.7 x 1.3 mm, obovate 48–49 in l.s., elliptic 7–8 to obtriangular 91–92 in c.s.; surface coarsely alveolate and obscurely areolate, attachment scar on the inner face large, gray to black.

1 **Lindera benzoin** (L.) Blume (x3) Spice Bush. Stones 7.0 x 5.8 x 5.8 mm, elliptic 4–5 in l.s., elliptic 6 in c.s.; surface smooth, dull, brown, or mottled gray and brown.

2 **Sassafras albidum** (Nutt.) Nees. (x4) Sassafras. Stones 5.0 x 5.0 x 5.0 mm, elliptic 6 in l.s. and in c.s.; surface with an obscure raphe ridge, smooth or slightly rugulose, almost black.

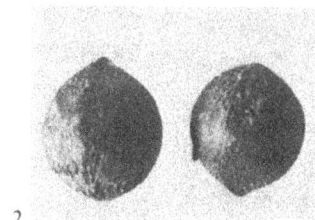

LEGUMINOSAE

The seeds of most of the Leguminosae, particularly in the sub-family Papilionoideae, are amphitropous and are usually depressed at the hilum, thus giving rise to a reniform outline on an otherwise elliptic seed. The position of the radicle is often indicated by a ridge near the hilum.

3 **Amphicarpa bracteata** (L.) Fern. (x6) Hog Peanut. Seeds 3.0 x 3.9 x 1.8 mm, elliptic 7–8 in l.s., elliptic 9–10 in c.s.; hilum large and white, surface smooth, dull, mottled light and dark brown.

4 **Apios americana** Medic. (x3) Groundnut. Seeds of this species are very rare either in the wild or on herbarium specimens. It has been found that within our area it is a triploid and, therefore, sterile. A specimen in the Herbarium of the Biosystematic Research Institute, Ottawa, Ontario, had seed pods and seeds but it was from Weeksville, North Carolina. The seeds of this specimen were 4.2 x 5.2 x 4.2 mm, and were oblong 19–20 in l.s., oblong 16–17 in c.s.; surface wrinkled and almost black.

5 **Astragalus canadensis** L. (x7.5) Milk Vetch. Seeds 2.0 x 2.5 x 0.8 mm, obliquely elliptic 7–8 in l.s., oblong 22–23 in c.s.; hilar notch near one end and asymmetrical, surface very finely areolate, olivaceous, mottled with black or purple.

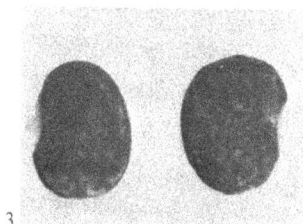

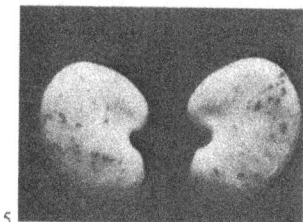

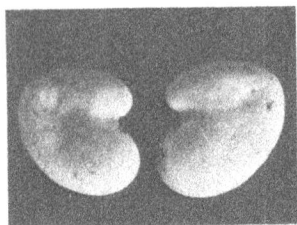

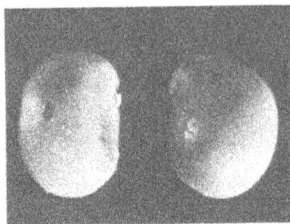

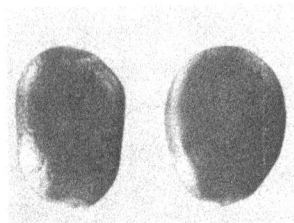

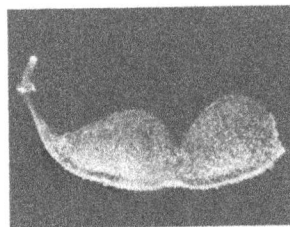

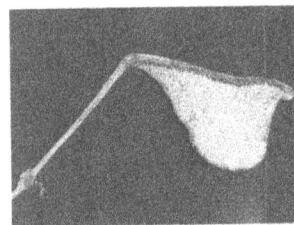

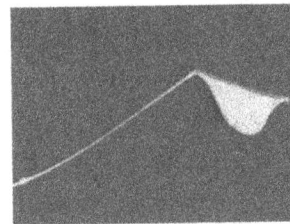

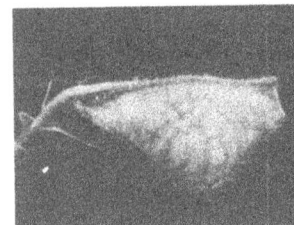

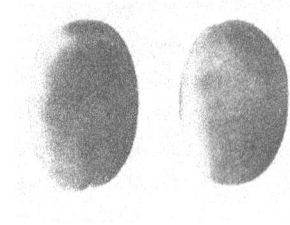

1 **Astragalus neglectus** (T.&G.) Sheld. (x7) Seeds 2.2 x 2.8 x 0.8 mm, obliquely elliptic 7–8 in l.s., oblong 22–23 in c.s.; hilar notch asymmetrical and near one end; surface finely areolate, brownish, mottled with black and purple.

2 **Baptisia tinctoria** (L.) R. Br. (x7.5) Wild Indigo. Seeds 2.0 x 2.6 x 1.6 mm, elliptic 7–8 in l.s., elliptic 8–9 in c.s.; surface puncticulate, dull, and olive-green.

3 **Cercis canadensis** L. (x3) Redbud. Seeds anatropous, 5.3 x 4.1 x 2.2 mm, obliquely obovate 48–49 in l.s., elliptic 8–9 in c.s.; surface smooth, dull, and brown.

4 **Desmodium canadense** (L.) DC. (x2.5) Tick-trefoil. Fruit a short-stalked loment of 3–5 segments; segments laterally flattened, rounded above and below, surface pubescent with uncinate, glandular hairs, reticulately veined.

5 **Desmodium glutinosum** (Muhl.) Wood (x2.3) Fruit a loment on a stipe about 8 mm, long, segments laterally flattened, nearly straight dorsally and asymmetrically rounded ventrally; surface with uncinate, non-glandular hairs, reticulately veined.

6 **Desmodium nudiflorum** (L.) DC. (x1.2) Loment with a stipe 10–20 mm long, segments laterally flattened, nearly straight dorsally and asymmetrically rounded below; surface with an uncinate pubescence and reticulately veined.

7 **Desmodium paniculatum** (L.) DC. (x4.5) Loment on a short stipe, segments laterally flattened, curved above and obtusely rounded below; surface with uncinate pubescence and reticulately veined.

8 **Gleditsia triacanthos** L. (x2.5) Honey Locust. Seeds anatropous, 10.0 x 6.8 x 3.5 mm, elliptic 4–5 in l.s., elliptic 8–9 in c.s.; surface smooth or slightly veined, dull, and dark brown.

1 **Glycyrrhiza lepidota** Pursh (x5) Wild Licorice. Pods with hooked spines and tardily dehiscent; seeds 2.2 x 2.7 x 1.9 mm, elliptic 7–8 in l.s., elliptic 7–8 in c.s.; surface smooth and brown.

2 **Gymnocladus dioica** (L.) K. Koch (x1.8) Kentucky Coffee Tree. Seeds anatropous, 15 x 15 x 8 mm, elliptic 6 in l.s., elliptic 8–9 in c.s.; surface smooth, brown to black at maturity; pods large and tardily dehiscent.

3 **Lathyrus latifolius** L. (x4.3) Everlasting Pea. Seeds 3.8 x 4.9 x 3.8 mm, oblong 19–20 in l.s., elliptic 7–8 in c.s.; hilum scar conspicuous and about one-half the length of the seed, surface rugose and dark brown.

4 **Lathyrus maritimus** (L.) Bigel. (x4.7) Beach Pea. Seeds 3.6 x 3.6 x 3.6 mm, elliptic 6 in l.s. and in c.s.; hilum scar about one-half the circumference of the seed, surface smooth and olive-green.

5 **Lathyrus ochroleucus** Hook. (x4.7) Seeds 3.2 x 3.2 x 3.2 mm, elliptic 6 in l.s. and in c.s.; hilum scar about one-quarter the seed circumference, surface smooth, olivaceous to brown.

6 **Lathyrus palustris** L. (x4.2) Vetchling. Seeds 3.3 x 3.3 x 3.3 mm, elliptic 6 in l.s. and in c.s.; surface mottled green and black.

7 **Lathyrus tuberosus** L. (x4.5) Tuberous Vetchling. Seeds 3.2 x 4.0 x 3.2 mm, elliptic 7–8 in l.s., elliptic 7–8 in c.s.; surface smooth, dull, and olive-green to brown.

8 **Lespedeza capitata** Michx. (x8) Seeds 1.4 x 2.2 x 0.9 mm, elliptic 8–9 in l.s., elliptic 9–10 in c.s.; surface smooth, dull, green to brown or black.

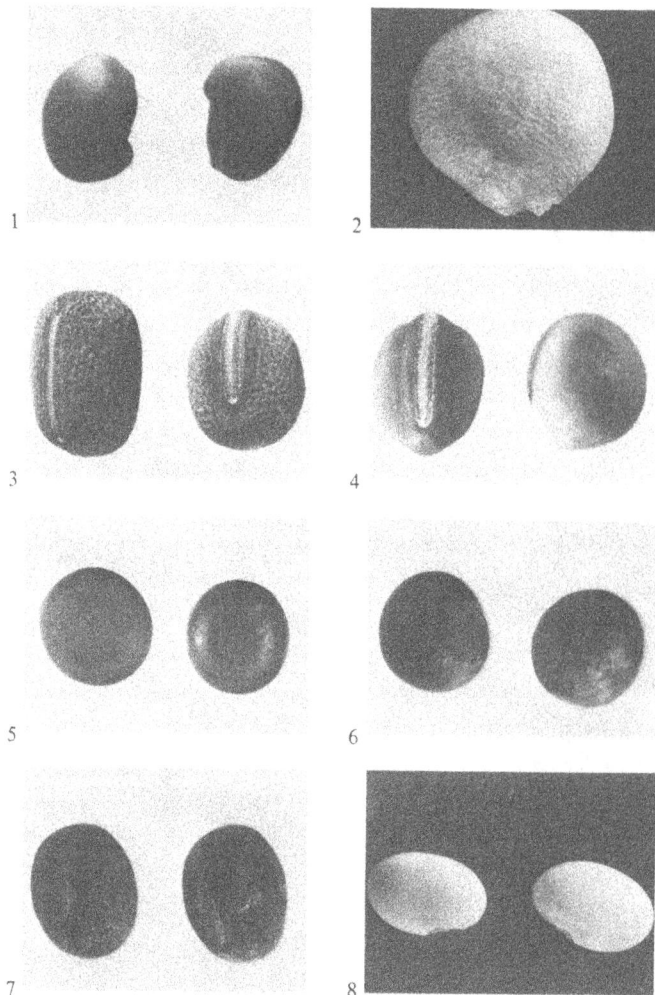

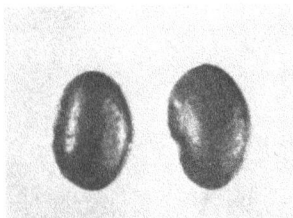
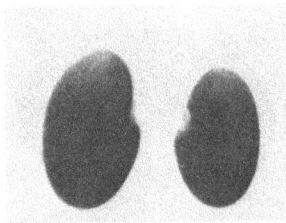

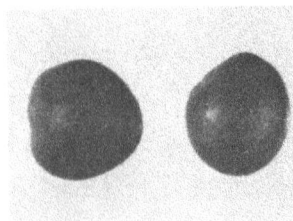
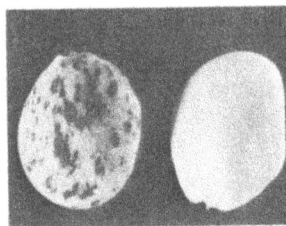

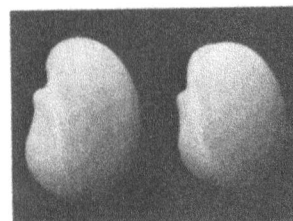
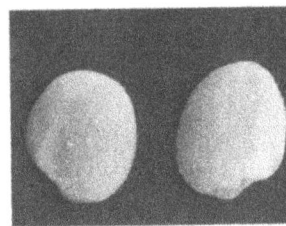

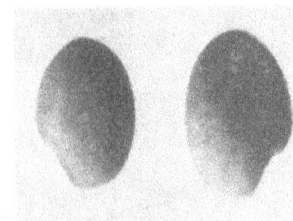
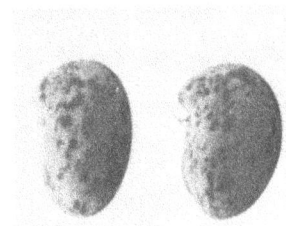

LEGUMINOSAE

1 **Lespedeza hirta** (L.) Hornem. (x6.3) Seeds 1.9 x 2.7 x 1.3 mm, elliptic 7–8 in l.s., elliptic 9–10 in c.s.; surface smooth and brown.

2 **Lespedeza intermedia** (Wats.) Britt. (x8) Seeds 1.5 x 2.5 x 0.9 mm, elliptic 8–9 in l.s., elliptic 9–10 in c.s.; surface smooth, dull and brown.

3 **Lotus corniculatus** L. (x10) Bird's-foot Trefoil. Seeds 1.1 x 1.4 x 0.9 mm, elliptic 7–8 in l.s., elliptic 8–9 in c.s.; surface smooth, brown or mottled with black.

4 **Lupinus perennis** L. (x5) Common Lupine. Seeds 4.2 x 3.4 x 2.2 mm, elliptic 4–5 in l.s., elliptic 8–9 in c.s.; surface,smooth, white or mottled with black.

5 **Medicago lupulina** L. (x13) Black Medick. Seeds contained in a more or less persistent pod almost circular in outline, about 2.0 mm in diameter and elliptic in c.s., surface pubescent and conspicuously veined; seeds 1.1 x 1.5 x 0.9 mm, elliptic 7–8 in l.s., elliptic 8–9 in c.s.; surface smooth and green to brown.

6 **Melilotus alba** Desr. (x10) White Sweet Clover. Seeds 1.8 x 1.5 x 1.0 mm, elliptic 5 in l.s., hilum sub-basal, elliptic 8 in c.s.; surface green to brown and smooth.

7 **Melilotus officinalis** (L.) Desr. (x12) Yellow Sweet Clover. Seeds 1.8 x 1.2 x 0.9 mm, elliptic 4 in l.s., hilum sub-basal, elliptic 7–8 in c.s.; surface smooth and green or mottled with brown.

8 **Robinia pseudoacacia** L. (x4.5) Black Locust. Seeds 3.0 x 4.3 x 1.7 mm, obliquely elliptic 7–8 in l.s., elliptic 9–10 in c.s.; surface smooth, dull, mottled green and black.

1 **Strophostyles helveola** (L.) Ell. (x3) Wild Bean. Seeds 4.1 x 7.0 x 4.1 mm, oblong 20–21 in l.s. and in c.s.; hilum long, white, margined with black, surface with a grayish, scurfy coating.

2 **Tephrosia virginiana** (L.) Pers. (x8) Goat's-rue. Seeds 2.7 x 3.5 x 1.5 mm, elliptic 7–8 in l.s., elliptic 9–10 in c.s.; surface smooth, mottled green, brown or black.

3 **Trifolium agrarium** L. (x13) Hop Clover. Seeds 1.1 x 0.9 x 0.6 mm, elliptic 4–5 in l.s., elliptic 8 in c.s.; surface smooth, dull, brown or straw-coloured.

4 **Trifolium arvense** L. (x10) Rabbit-foot Clover. Seeds 1.1 x 0.8 x 0.7 mm, elliptic 4–5 in l.s., elliptic 6–7 in c.s.; surface smooth and greenish to light brown.

5 **Vicia americana** Muhl. (x5.5) American Vetch. Seeds 2.6 x 2.6 x 2.6 mm, elliptic 6 in l.s. and in c.s.; surface smooth, brown or black.

6 **Vicia angustifolia** Reichard (x5.4) Common Vetch. Seeds 3.0 x 3.0 x 3.0 mm, elliptic 6 in l.s., and in c.s.; surface smooth, dark brown to black or mottled.

7 **Vicia cracca** L. (x6.5) Bird Vetch. Seeds 2.5 x 2.5 x 2.5 mm, elliptic 6 in l.s. and in c.s.; surface smooth, olive-green mottled with brown or black.

8 **Vicia tetrasperma** (L.) Moench. (x8) Four-seeded Vetch. Seeds 2.0 x 2.0 x 2.0 mm, elliptic 6 in l.s. and in c.s.; surface smooth, dull, mottled green and black.

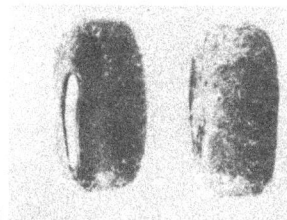

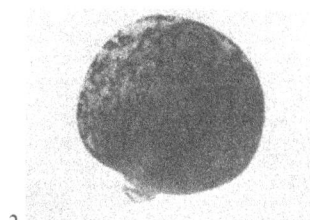

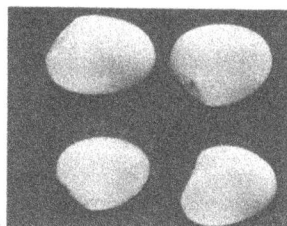

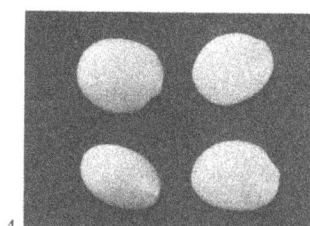

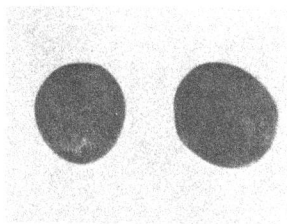

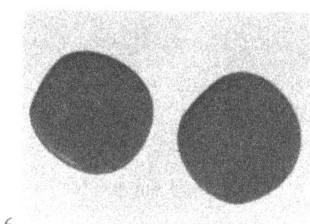

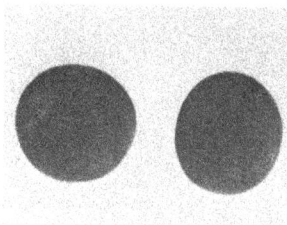

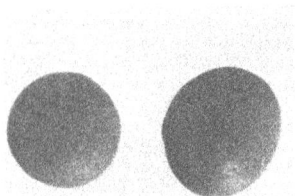

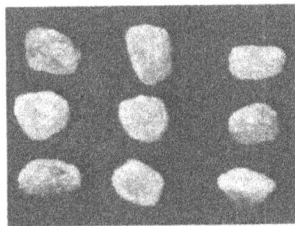

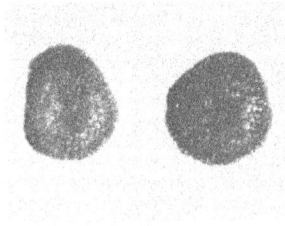

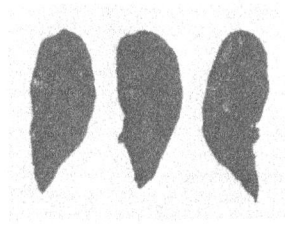

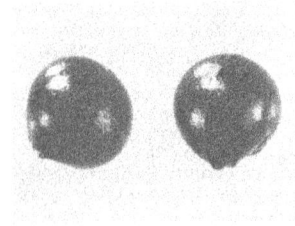

LEGUMINOSAE

1 **Vicia villosa** Roth (x4.5) Hairy Vetch. Seeds 3.4 x 3.4 x 3.4 mm, elliptic 6 in l.s. and in c.s.; surface smooth, dull, black or mottled.

LENTIBULARIACEAE

2 **Pinguicula vulgaris** L. (x25) Butterwort. Seeds 0.6 x 0.25 x 0.25 mm, elliptic 2–3 in l.s., elliptic 6 in c.s.; surface reticulate in more or less longitudinal lines, or scalariform, light brown in colour.

3 **Utricularia cornuta** Michx. (x30) Horned Bladderwort. Seeds 0.25 x 0.15 x 0.1 mm, irregular in form or elliptic 3–4 in l.s., elliptic 8 in c.s.; surface reticulate and light brown.

4 **Utricularia intermedia** Hayne (x20) Seeds 0.4 x 0.8 x 0.4 mm, irregular in shape or obliquely elliptic 9 in l.s., concave at the base and the apex, elliptic 9 in c.s.; surface reticulate and brown or black.

5 **Utricularia vulgaris** L. (x20) Common Bladderwort. Seeds 0.3 x 0.5 x 0.3 mm, irregular in shape or oblong 20–21 in l.s., hilum central on the lower flattened surface, 4–5-angled in c.s. with slightly winged angles; surface areolate and black.

LILIACEAE

6 **Allium schoenoprasum** L. var. **sibiricum** (L.) Hartm. (x8) Wild Chives. Seeds 2.8 x 1.2 x 0.8 mm, irregular in form or obliquely obovate 46–47 in l.s., irregularly angled in c.s.; surface ridged, reticulate, and black.

7 **Allium tricoccum** Ait. (x5.5) Wild Leek. Seeds 2.8 x 2.8 x 2.8 mm, elliptic 6 in l.s. and in c.s.; surface with a single ridge circling the seed, margin of the basal hilar cavity puckered, finely areolate and shiny, black.

1 **Asparagus officinalis** L. (x5) Asparagus. Seeds 2.5 x 3.6 x 3.0 mm, elliptic 7–8 in l.s., elliptic 7 in c.s., slightly angular or flattened at points of contact with other seeds; surface black, dull and finely areolate.

2 **Chamaelirium luteum** (L.) Gray (x4.5) Blazing Star. Seeds 4.0 x 1.4 x 0.6 mm, obliquely oblong 14–15 in l.s., obliquely oblong 21–22 in c.s.; central brown seed body surrounded by a white, spongy, reticulate wing.

3 **Clintonia borealis** (Ait.) Raf. (x6) Clintonia or Corn Lily. Seeds 3.7 x 2.1 x 1.6 mm, irregular in l.s. and in c.s. or obliquely elliptic 3–4 in l.s., irregularly angular in c.s. with the sides tending to be concave; surface areolate, glossy, and brown.

4 **Disporum lanuginosum** (Michx.) Nicholson (x4) Yellow Mandarin. Seeds 4.8 x 3.2 x 3.2 mm, elliptic 4 in l.s., elliptic 6 in c.s.; surface smooth, caruncle obvious, light brown or yellowish.

5 **Erythronium albidum** Nutt. (x5.5) White Dogtooth Violet. Seeds 3.4 x 1.5 x 1.8 mm, irregular or obliquely ovate 37–38 in l.s., elliptic 5 in c.s.; caruncle prominent, surface longitudinally rugose and transversely rugulose, brown.

6 **Erythronium americanum** Ker. (x6) Yellow Dogtooth Violet. Seeds 3.5 x 2.0 x 1.8 mm, surface features very similar to the previous species.

7 **Lilium canadense** L. (x3) Canada Lily. Seeds 7.6 x 6.3 x 0.3 mm, obovate 48–49 to obtriangular 88–89 in l.s., thin and wafer-like or oblong 24 in c.s.; seed body surrounded by a rather spongy wing, finely colliculose.

8 **Lilium philadelphicum** L. (x4) Philadelphia or Prairie Lily. Seeds 5.3 x 4.4 x 0.3 mm, in general appearance similar to the previous specious.

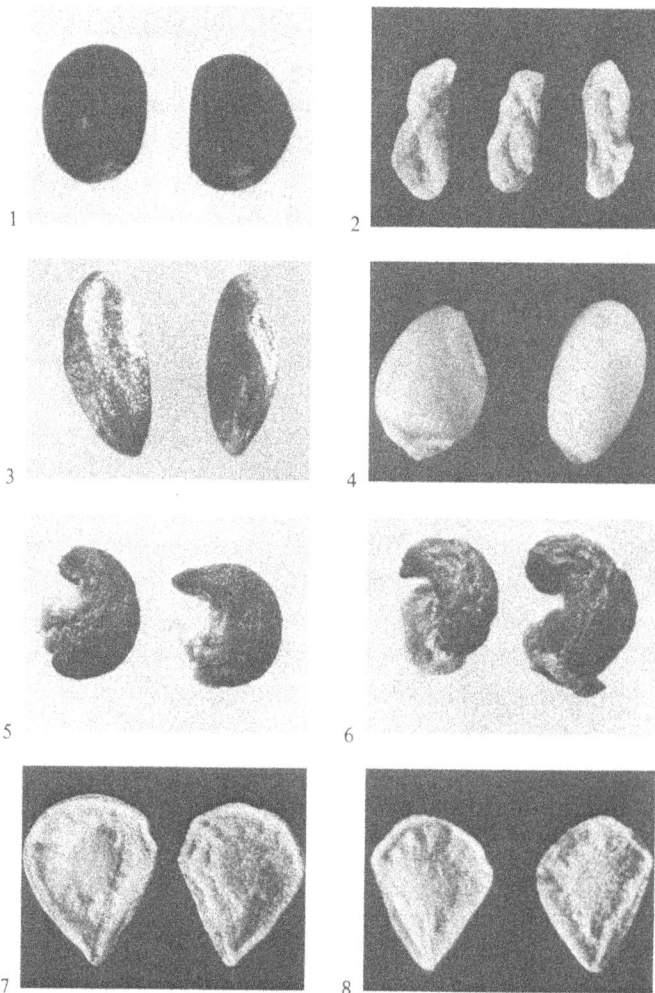

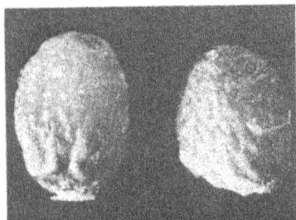
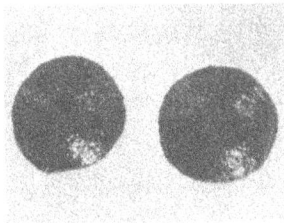
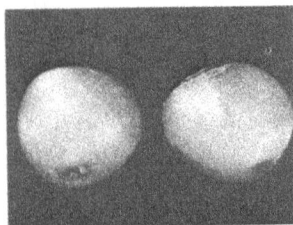
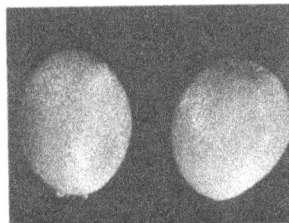
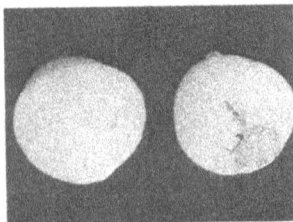
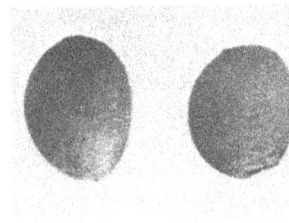
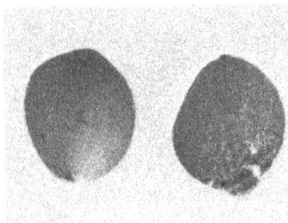
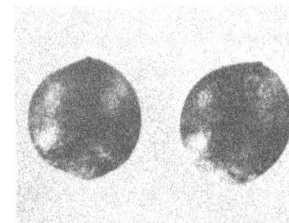

1 **Maianthemum canadense** Desf. (x9) False Lily-of-the-valley. Seeds 2.1 x 1.9 x 1.8 mm, elliptic 5–6 in l.s., elliptic 6–7 in c.s.; surface rugulose and areolate.

2 **Medeola virginiana** L. (x5) Indian Cucumber-root. Seeds 3.0 x 3.0 x 3.0 mm, elliptic 6 in l.s. and in c.s.; surface slightly flattened where other seeds have contacted, rugulose and finely areolate, brown.

3 **Polygonatum biflorum** (Walt.) Ell. (x5.5) Seeds 3.4 x 3.0 x 2.5 mm, elliptic 5–6 in l.s., elliptic 7 in c.s.; hilar scar prominent, surface slightly roughened or areolate.

4 **Polygonatum pubescens** (Willd.) Pursh (x6.5) Solomon's Seal. Seeds 2.8 x 2.4 x 2.0 mm, elliptic 5–6 in l.s., elliptic 7 in c.s.; surface faintly roughened or areolate and light brown.

5 **Smilacina racemosa** (L.) Desf. (x5) False Spikenard. Seeds 3.3 x 3.3 x 3.3 mm, elliptic 6 in l.s. and in c.s.; hilar scar prominent, surface rugulose and areolate.

6 **Smilacina stellata** (L.) Desf. (x6) Starry False Solomon's Seal. Seeds 3.2 x 3.2 x 3.2 mm, elliptic 6 in l.s. and in c.s.; surface faintly rugulose or areolate and brown.

7 **Smilacina trifolia** (L.) Desf. (x5.5) Three-leaved False Solomon's Seal. Seeds 3.0 x 2.5 x 2.0 mm, elliptic 5 in l.s., elliptic 7–8 in c.s.; surface faintly longitudinally striate and areolate, brown.

8 **Smilax herbacea** L. (x3.8) Carrion-flower. Seeds 4.2 x 4.0 x 3.6 mm, elliptic 5–6 in l.s., elliptic 6–7 in c.s., flattened where in contact with other seeds; surface faintly areolate and brown.

1 **Smilax hispida** Muhl. (x3.3) Bristly Greenbrier. Seeds 4.9 x 4.9 x 4.9 mm, elliptic 6 in l.s. and in c.s., slightly flattened where in contact with other seeds; surface faintly areolate and brown.

2 **Streptopus amplexifolius** (L.) DC. (x8) White Mandarin. Seeds 4.0 x 2.8 x 2.4 mm, elliptic 4–5 in l.s., elliptic 6–7 in c.s.; surface smooth and cream-coloured.

3 **Streptopus roseus** Michx. (x7.5) Twisted Stalk. Seeds 2.4 x 1.1 x 1.0 mm, obliquely ovate 37–38 in l.s., elliptic 6–7 in c.s.; surface longitudinally ribbed with the ribs transversely rugulose, white to cream-coloured.

4 **Tofieldia glutinosa** (Michx.) Pers. (x17) False Asphodel. Seeds 0.9 x 0.4 x 0.2 mm, obliquely elliptic 2–3 in l.s., elliptic 9 in c.s.; surface rugulose and brown.

5 **Trillium cernuum** L. (x7) Nodding Trillium. Seeds 1.6 x 2.2 x 1.5 mm, elliptic 7–8 in l.s., elliptic 7–8 in c.s.; caruncle large and prominent, surface longitudinally rugose and transversely rugulose, brown.

6 **Trillium erectum** L. (x7) Red Trillium. Seeds 1.7 x 2.4 x 1.7 mm, surface features similar to the above species.

7 **Trillium grandiflorum** (Michx.) Salisb. (x7) White Trillium. Seed features similar to those of *T. cernuum*, 1.9 x 3.0 x 1.8 mm.

8 **Trillium undulatum** Willd. (x7) Painted Trillium. Seeds 1.9 x 2.6 x 1.8 mm, surface features not different from *T. cernuum*.

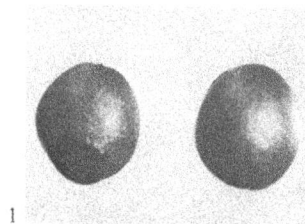
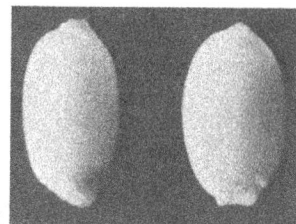
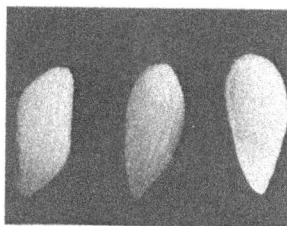
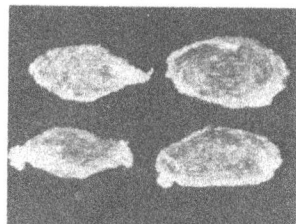
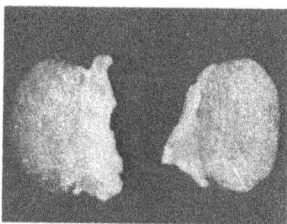
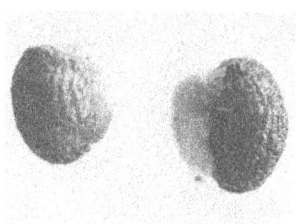
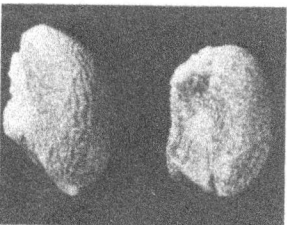
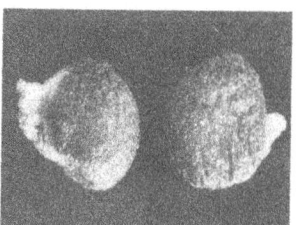

146

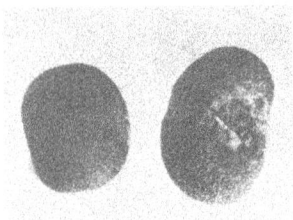
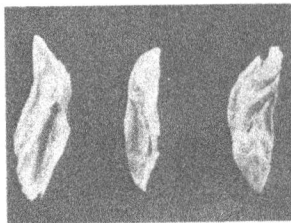
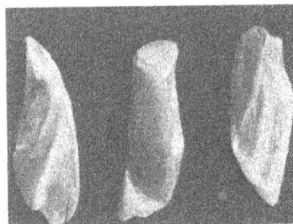
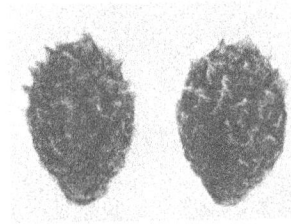

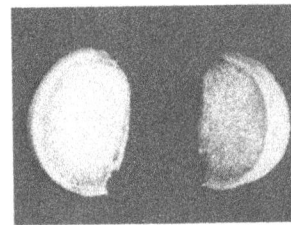
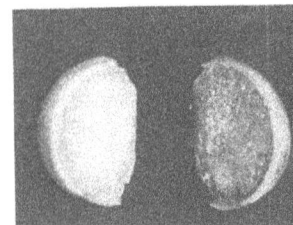

LILIACEAE

1 **Uvularia grandiflora** Sm. (x4.5) Bellwort. Seeds 3.1x 4.2 x 3.3 mm, elliptic 7–8 in l.s. and in c.s.; caruncle large, surface rugulose and light brown.

2 **Veratrum viride** Ait. (x3.3) False Hellebore. Seeds 6.0 x 2.0 x 0.4 mm, irregularly elliptic in outline, flattened in c.s. and the margins winged; surface rugulose and faintly reticulate, brown.

3 **Zygadenus glaucus** Nutt. (x5.5) White Camass. Seeds 4.5 x 1.5 x 0.7 mm, irregularly oblong in l.s., flattened and irregularly ridged in c.s.; surface finely striate, areolate, slightly glossy, and light brown.

LIMNANTHACEAE

4 **Floerkea proserpinacoides** Willd. (x6.5) Mermaid-weed. Fruit an achene, 3.1 x 2.1 x 1.7 mm, obovate 48–49 in l.s., elliptic 7–8 in c.s.; surface with numerous, flat, whitish, scaly projections, dull, and black.

LINACEAE

5 **Linum lewisii** Pursh (x7) Seeds 3.0 x 1.5 x 0.5 mm, elliptic 3 in l.s., elliptic 10 in c.s.; margin with a narrow, light-coloured wing; surface finely areolate, black, and shiny.

6 **Linum medium** (Planch.) Britt. (x12) Ovary globose and dividing into 10 semicircular sections appearing somewhat like the sections of an orange; sections 1.5 x 1.0 x 0.7 mm, obliquely elliptic 4 in l.s., obtriangular 88–89 in c.s.; surface gray or green and smooth; seeds closely invested by the ovary walls, semicircular in outline and obtriangular in c.s., surface faintly reticulate.

7 **Linum virginianum** L. (x12) Sections of the fruit 1.7 x 1.2 x 0.8 mm, not obviously different from the previous species.

1 **Lobelia cardinalis** L. (x30) Cardinal Flower. Seeds 0.9 x 0.4 x 0.4 mm, elliptic 2-3 to obovate 46–47 in l.s., elliptic 6 in c.s.; surface reticulate and brown.

2 **Lobelia dortmanna** L. (x30) Water Lobelia. Seeds 0.6 x 0.3 x 0.3 mm, obovate 47 in l.s., elliptic 6 in c.s.; surface reticulate and brown.

3 **Lobelia inflata** L. (x30) Indian Tobacco. Seeds 0.6 x 0.25 x 0.25 mm, elliptic 2-3 in l.s., elliptic 6 in c.s.; surface reticulate and light brown.

4 **Lobelia kalmii** L. (x35) Seeds 0.7 x 0.2 x 0.2 mm, elliptic 1–2 in l.s., elliptic 6 in c.s.; surface reticulate and brown.

5 **Lobelia siphilitica** L. (x35) Great Lobelia. Seeds 0.7 x 0.3 x 0.3 mm, obovate 46–47 in l.s., elliptic 6 in c.s.; surface reticulate and brown.

6 **Lobelia spicata** Lam. (x40) Spike Lobelia. Seeds 0.6 x 0.3 x 0.3 mm, obovate 47 in l.s., elliptic 6 in c.s.; surface reticulate and light brown.

LYTHRACEAE

7 **Decodon verticillatus** (L.) Ell. (x12) Water Willow. Seeds 1.6 x 1.5 x 1.1 mm, obtriangular 89–90 or obovate 49–50 in l.s., obliquely obtriangular 91–92 in c.s.; surface finely areolate, slightly glossy, and brown.

8 **Lythrum alatum** Pursh (x30) Winged Loosestrife. Seeds 0.5 x 0.25 x 0.25 mm, obliquely elliptic 3 in l.s., elliptic 6 in c.s.; surface smooth and light brown.

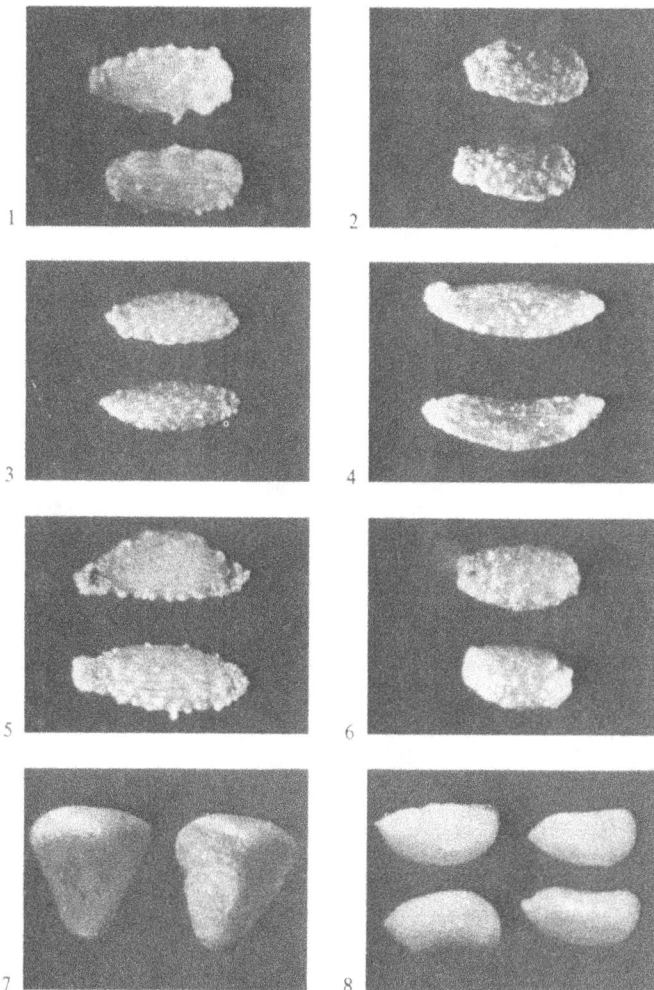

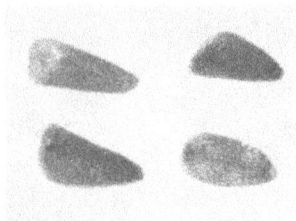
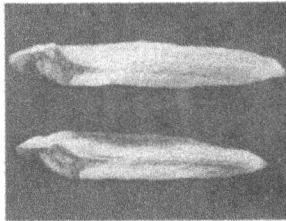
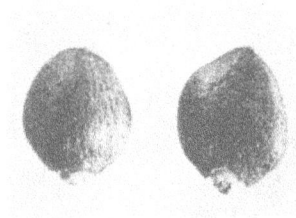

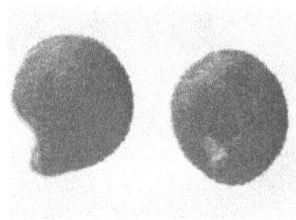
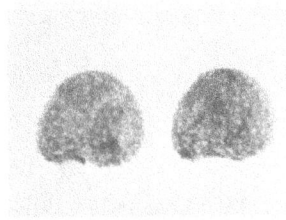

LYTHRACEAE

1 **Lythrum salicaria** L. (x13) Purple Loosestrife. Seeds 0.9 x 0.4 x 0.3 mm, obtriangular 86–87 in l.s., obliquely elliptic 7–8 in c.s., more or less plano-convex; surface obscurely striate, glossy, and light brown.

MAGNOLIACEAE

2 **Liriodendron tulipifera** L. (x1) Tulip Tree. Fruit a long, narrow samara about 3.3 cm long, wing elongate, midrib prominent, but the 3 or 4 longitudinal veins obscure; seed contained in the enlarged indehiscent base.

3 **Magnolia acuminata** L. (x2.5) Cucumber Tree. Seeds body covered with a red aril; seed 7.0 x 6.0 x 3.4 mm, ovate 40–41 in l.s., elliptic 8–9 in c.s.; surface dull and black, faintly rugulose.

MALVACEAE

The embryo is amphitropous or curved and the usually elliptic seeds frequently have a reniform appearance. In our species the raphe is obvious and often has an external remnant.

4 **Abutilon theophrasti** Medic. (x3.8) Velvet Leaf. Seeds 3.3 x 3.0 x 1.6 mm, obliquely elliptic 5–6 in l.s., oblong 20–21 in c.s. with the sides slightly concave; surface scaly and setose, finely areolate, dull, and black.

5 **Hibiscus palustris** L. (x5.5) Swamp Hollyhock. Seeds 3.0 x 2.6 x 2.6 mm, elliptic 5–6 to obovate 49–50 in l.s., elliptic 6 in c.s.; surface with light brown papillae, finely areolate.

6 **Hibiscus trionum** L. (x6.8) Flower-of-an-hour. Seeds 1.8 x 2.2 x 1.4 mm, obliquely elliptic 7–8 in l.s., oblong 20–21 in c.s. with the sides slightly concave; surface finely areolate and with light brown papillae, black.

1 **Malva moschata** L. (x9) Musk Mallow. Seeds 1.7 x 1.7 x 1.0 mm, obliquely elliptic 6 in l.s., oblong 20–21 in c.s. with the surfaces slightly concave; surface glabrous or slightly pubescent particularly around the hilum, obscurely areolate and gray.

2 **Malva neglecta** Wallr. (x10) Round-leaved Mallow. Seeds 1.4 x 1.4 x 0.7 mm, obliquely elliptic 6 in l.s. with a deep hilar notch, oblong 21 in c.s. with the sides slightly concave; surface finely areolate, slightly pubescent around the hilum, dark brown or grayish.

3 **Malva verticillata** L. (x10) Whorled Mallow. Seeds 2.0 x 2.0 x 1.3 mm, elliptic 6 in l.s., obovate 47–48 in c.s. with the sides slightly concave; surface smooth or faintly areolate.

4 **Sida spinosa** L. (x6) Prickly Mallow. Fruit irregular in outline, angular in c.s., apical opening forming two beaks and ciliate on the margins; seeds 2.0 x 1.5 x 1.5 mm, ovate 39–40 in l.s., obtriangular 90 in c.s. with the dorsal surface rounded; surface puberulent particularly around the hilum, faintly areolate, brown or black.

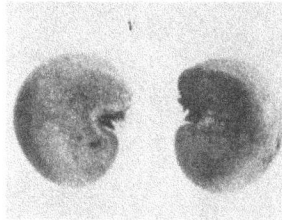

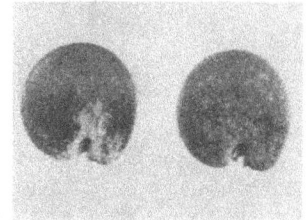

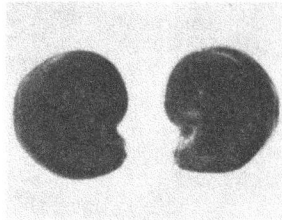

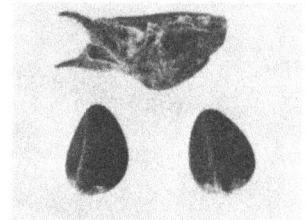

MARTYNIACEAE

5 **Proboscidea louisianica** (Mill.) Woot. & Standl. (x3.2) Unicorn Plant. Seeds 10.0 x 6.3 x 3.3 mm, obovate 47–48 in l.s., oblong 20–21 in c.s.; surface nodular and alveolate, black.

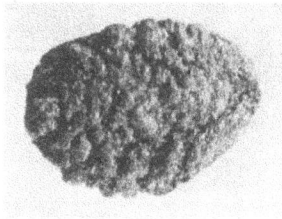

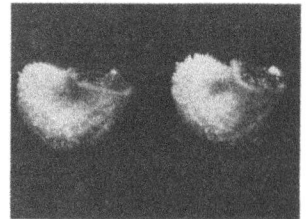

MELASTOMATACEAE

6 **Rhexia virginica** L. (x25) Deer Grass. Seeds 0.6 x 0.5 x 0.3 mm, very irregular in form, rather snail-shell in shape; surface ridged and verrucose, base with a reddish remains of the funiculus.

150

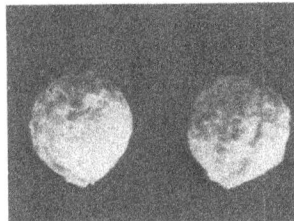
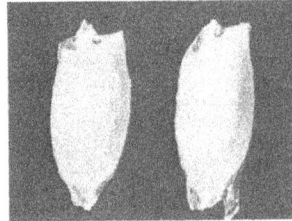

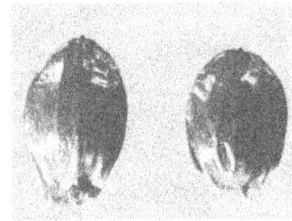

MENISPERMACEAE

1 **Menispermum canadense** L. (x3.5) Moonseed. Seeds 7.0 x 7.0 x 2.5 mm, elliptic 6 in l.s. with a deep hilar notch, oblong 21–22 in c.s. with the sides concave; embryo coiled to form almost a complete circle and this is indicated on the seed coat by the concentric ridges, ridges transversely rugose.

MORACEAE

2 **Humulus japonicus** Sieb. & Zucc. (x4.8) Japanese Hops. Achenes 3.3 x 3.3 x 3.3 mm, elliptic 6 in l.s. and in c.s.; outer coating light brown, dull, faintly veined, inner coating black.

3 **Humulus lupulus** L. (x6) Hops. Achenes 2.5 x 2.2 x 2.0 mm, elliptic 5–6 in l.s., elliptic 6–7 in c.s., margin with a distinct ridge; outer coating with glandular dots, brown to black.

4 **Maclura pomifera** (Raf.) Schneid. (x3) Osage Orange. Achenes buried in a tough, globose fusion of the calyx and receptacle; achenes 7.7 x 3.9 x 2.4 mm, elliptic 2–3 in l.s., 2-lobed at the apex, elliptic 8–9 in c.s.; margin prominently ridged or winged, surface white and slightly rugulose.

5 **Morus alba** L. (x6) White Mulberry. Berry-like fruit a cluster of several small drupe-like fruits on a common, enlarged receptacle (a multiple fruit); achenes 2.5 x 1.3 x 1.3 mm, elliptic 3–4 to ovate 38–39 in l.s., obliquely elliptic 6 in c.s.; surface smooth and light brown.

MYRICACEAE

6 **Myrica asplenifolia** L. (x4.5) Sweet Fern. Nutlets 4.0 x 3.1 x 2.3 mm, elliptic 4–5 in l.s., elliptic 7–8 in c.s.; basal area puckered and ridged, surface smooth or longitudinally ribbed, glossy, olive-green to brown.

1 **Myrica gale** L. (x7.5) Sweet Gale. Fruit a small nut enclosed by closely adhering, greenish bractlets with slightly inflated wings and resinous-dotted surfaces; nutlets 2.0 x 1.4 x 1.2 mm, elliptic 4-5 in l.s., elliptic 6-7 in c.s.; surface smooth and green.

2 **Myrica pensylvanica** Loisel. (x5) Bayberry. Fruits spherical nutlets 3.6 x 3.2 x 3.2 mm, elliptic 5-6 in l.s., elliptic 6 in c.s.; surface verrucose and the black protuberences mingled with white, waxy particles and hairs.

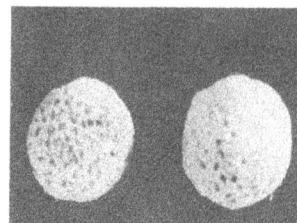

NAJADACEAE

3 **Najas flexilis** (Willd.) R.&S. (x8) Seeds 2.8 x 1.0 x 1.0 mm, elliptic 2-3 in l.s., elliptic 6 in c.s.; surface obscurely areolate, glossy and brownish.

4 **Potamogeton epihydrus** Raf. (x7) Achenes 2.6 x 2.1 x 0.8 mm, obliquely elliptic 4-5 in l.s., oblong 21-22 in c.s.; outer margin slightly winged, beak terminal at the apex of the inner margin.

5 **Potamogeton foliosus** Raf. (x7) Achenes 2.0 x 1.9 x 1.0 mm, elliptic 5-6 in l.s., elliptic 8-9 in c.s.; sides slightly flattened, margins narrowly winged, beak terminal.

6 **Potamogeton gramineus** L. (x9) Achenes 2.3 x 1.8 x 1.1 mm, obovate 48-49 in l.s., elliptic 8-9 in c.s.; beak short, terminal and recurved, margin keeled, surface rugulose or pitted.

7 **Potamogeton illinoensis** Morong (x6) Achenes 3.0 x 2.1 x 1.5 mm, obovate 48-49 in l.s., oblong 19-20 in c.s.; dorsal surface with three ridges or wings with the middle one the most prominent, surface rugose, beak short, erect or recurved.

8 **Potamogeton natans** L. (x4.5) Achenes 3.5 x 2.9 x 2.0 mm, obovate 48-49 in l.s., elliptic 7-8 in c.s. with the surfaces slightly flattened; beak short, broad and blunt, dorsal surface 3-ridged with the middle one the most prominent, surface rugose.

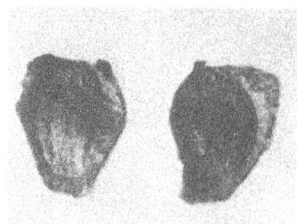

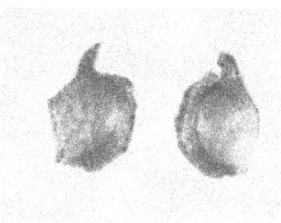

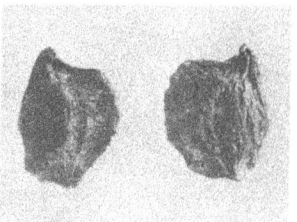

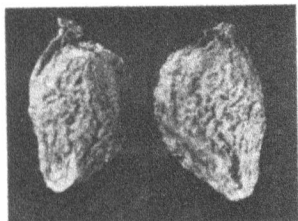
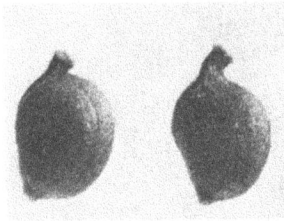
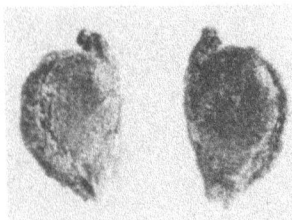
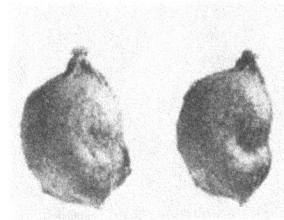
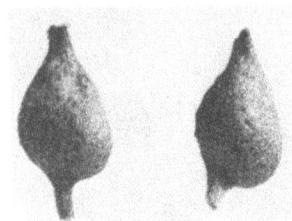
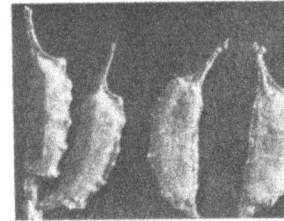
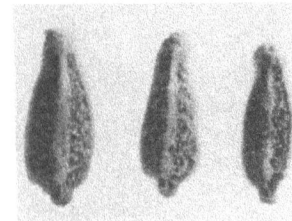

NAJADACEAE

1 **Potamogeton pectinatus** L. (x6.5) Achenes 3.2 x 2.4 x 1.7 mm, obliquely obovate 48–49 in l.s., elliptic 7–8 in c.s. with the sides slightly flattened; dorsal margin slightly winged, beak slender and incurved.

2 **Potamogeton pusillus** L. (x8) Achenes 2.1 x 1.6 x 1.0 mm, obovate 48–49 in l.s., obovate 47–48 in c.s.; dorsal surface broadly ridged, surfaces slightly roughened and reticulate.

3 **Potamogeton richardsonii** (Benn.) Rydb. (x8) Achenes 2.7 x 2.0 x 1.0 mm, obovate 48–49 in l.s., oblong 21 in c.s.; dorsal surface rounded, not keeled, lateral surfaces rugose, beak prominent.

4 **Potamogeton zosteriformis** Fern. (x5.3) Achenes 3.4 x 2.7 x 1.7 mm, elliptic 4–5 in l.s., elliptic 8–9 in c.s.; back of the achenes with a narrow, slightly toothed wing, base of the achenes broad and truncate, surface rugulose.

5 **Ruppia maritima** L. (x9) Ditch-grass. Fruits 2.2 x 1.5 x 1.1 mm, obliquely ovate 39–40 in l.s., elliptic 7–8 in c.s.; style persistent, short and stout, surface smooth, green, brown and dark-dotted or black.

6 **Zannichellia palustris** L. (x6) Horned Pondweed. Fruits a cluster of usually several oblong, short-stalked nutlets, 2.5 x 1.0 x 0.4 mm, oblong 14–15 in l.s., elliptic 9–10 in c.s.; margin on one side with a narrow, toothed wing, styles persistent, recurved, surface often faintly longitudinally striate and areolate.

NYCTAGINACEAE

7 **Oxybaphus nyctagineus** (Michx.) Sweet (x5) Umbrellawort. Achenes 4.0 x 1.7 x 1.7 mm, obovate 46–47 in l.s., elliptic 6 in c.s.; surface with 4 or 5 prominent whitish ribs, muricate, strigose with stiff, jointed hairs, brown.

NYMPHAEACEAE

1 **Brasenia schreberi** Gmel. (x6) Water Shield. Seeds 3.3 x 2.6 x 2.6 mm, elliptic 4–5 in l.s., elliptic 6 in c.s.; surface colliculose, brown, and dark-dotted.

2 **Nuphar microphyllum** (Pers.) Fern. (x10) Small-leaved Yellow Water Lily. Seeds 2.2 x 1.6 x 0.9 mm, obovate 48–49 in l.s., elliptic 8–9 in c.s.; surface with a narrow, dark ridge or wing, smooth, light brown or green.

3 **Nuphar variegatum** Engelm. (x4) Yellow Water Lily. Seeds 5.0 x 4.0 x 4.0 mm, obovate 48–49 in l.s., elliptic 6 in c.s.; surface areolate and greenish.

4 **Nymphaea odorata** Ait. (x8) Fragrant Water Lily. Seeds with an aril; 2.3 x 1.5 x 1.5 mm, elliptic 3–4 in l.s., elliptic 6 in c.s.; surface with a single longitudinal ridge, smooth and brown.

5 **Nymphaea tuberosa** Paine (x5) Tuberous Water Lily. Seeds with an aril; 4.0 x 3.2 x 3.2 mm, obovate 48–49 in l.s., elliptic 6 in c.s.; surface with a single longitudinal ridge, faintly areolate, glossy, and light brown.

OLEACEAE

6 **Fraxinus americana** L. (x1) White Ash. Samaras about 3.7 cm long, the body about 1.5 cm long; wing starting from near the apex of the fruit body, wing long and elliptic.

7 **Fraxinus nigra** Marsh. (x1) Black Ash. Samaras about 3.5 cm long, fruit body about 1.8 cm long; wing decurrent one-half or more the length of the fruit body, wing tip emarginate.

8 **Fraxinus pennsylvanica** Marsh. (x1) Red Ash. Samaras 4–7 cm long, fruit body about 2 cm long; wing decurrent to about one-half the length of the body, wing tending to oblong.

154

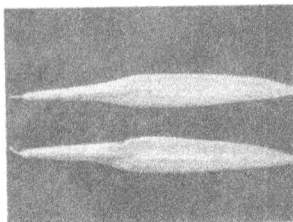
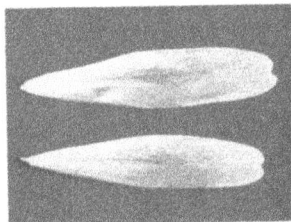
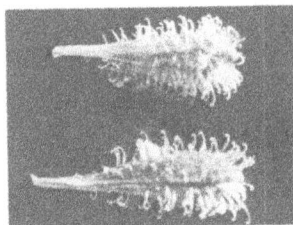

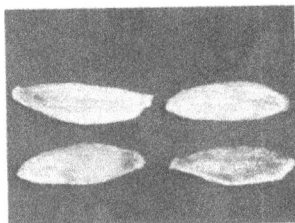
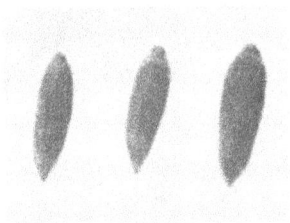
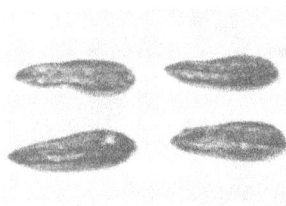
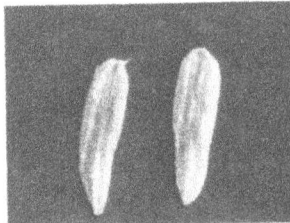

OLEACEAE

1 **Fraxinus pennsylvanica** Marsh. var. **subintegerrima** (Vahl.) Fern. (x1) Green Ash. Samaras 3–4 cm long, body about 1.5 cm long; wing decurrent to about one-half the length of the body and narrowly obovate.

2 **Fraxinus quadrangulata** Michx. (x1) Blue Ash. Samaras 2.5–5 cm long, body about 2.5 cm long; wing decurrent to near the base of the body, wing oblong and apex emarginate.

ONAGRACEAE

3 **Circaea alpina** L. (x10) Enchanter's Nightshade. Fruit 2.0 x 0.7 x 0.7 mm, obovate 46–47 in l.s., elliptic 6 in c.s.; surface longitudinally ribbed, covered with white, uncinate bristles, brown when mature.

4 **Circaea quadrisulcata** (Maxim.) Franch. & Sav. var. **canadensis** (L.) Hara (x7) Fruits 3.5 x 2.1 x 1.5 mm, obovate 47–48 in l.s., elliptic 7–8 in c.s.; surface with about 10 ribs, covered with white, uncinate bristles, brown.

5 **Epilobium angustifolium** L. (x18) Fireweed. Seeds 1.0 x 0.3 x 0.15 mm, elliptic 1–2 in l.s., elliptic 9 in c.s.; surface smooth or faintly longitudinally striate, terminating with a white coma, brown.

6 **Epilobium ciliatum** Raf. (x20) Seeds 1.0 x 0.3 x 0.2 mm, obovate 45–46 in l.s. with a short beak, elliptic 8 in c.s.; surface with longitudinal rows of papillae, apex with a whitish coma.

7 **Epilobium coloratum** Biehler (x14) Seeds 1.2 x 0.3 x 0.3 mm, obovate 45–46 in l.s., elliptic 6 in c.s.; surface papillose, light to dark brown, coma brownish.

8 **Epilobium glandulosum** Lehm. (x15) Seeds 1.3 x 0.3 x 0.2 mm, obovate 45–46 in l.s., elliptic 8 in c.s., surface papillose, light to dark brown, coma white.

1 **Epilobium hirsutum** L. (x22) Willow Herb. Seeds 1.0 x 0.5 x 0.4 mm, obovate 47 in l.s., elliptic 7–8 in c.s.; surface with longitudinal rows of papillae, brown, coma nearly white.

2 **Epilobium latifolium** L. (x10) River Beauty. Seeds 1.7 x 0.5 x 0.4 mm, elliptic 1–2 to obovate 45–46 in l.s., elliptic 7–8 in c.s.; surface smooth or rugulose, light brown, coma tawny in colour.

3 **Ludwigia palustris** (L.) Ell. var. **americana** (DC.) Fern. & Grisc. (x20) Water Purslane. Seeds 0.7 x 0.3 x 0.3 mm, oblong 14–15 in l.s., elliptic 6 in c.s.; surface with an obvious raphe ridge, obscurely longitudinally and transversely rugulose, white to brown and somewhat glossy.

4 **Oenothera biennis** L. (x8) Evening Primrose. Seeds 1.5 x 0.8 x 0.6 mm, irregularly angular, the angles sharp and prominent; surface rugose, brown and mottled with darker dots.

5 **Oenothera parviflora** L. (x9) Seeds 1.9 x 1.2 x 1.0 mm, irregularly angled with the angles somewhat winged; surface brown, longitudinally ridged and faintly reticulate.

6 **Oenothera perennis** L. (x18) Sun Drops. Seeds 0.9 x 0.4 x 0.3 mm, irregularly angled, the angles not sharp; surface finely reticulate and brown.

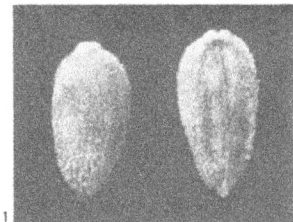

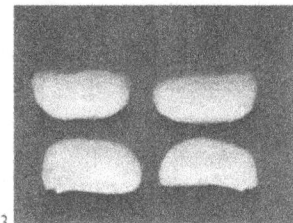
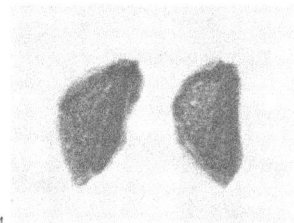
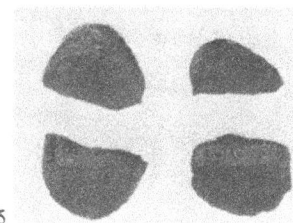
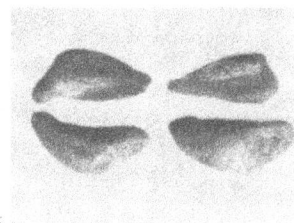

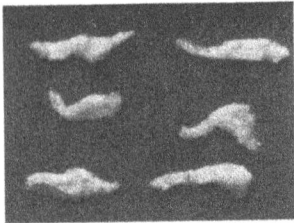

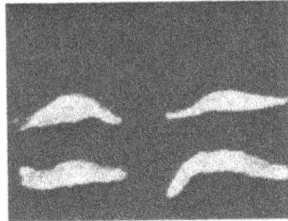

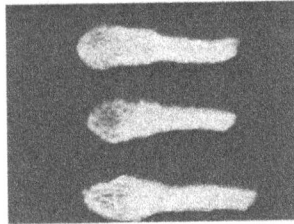

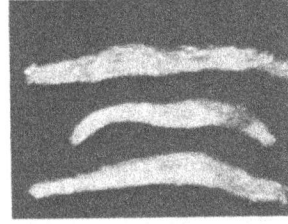

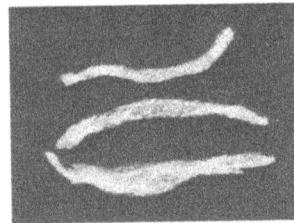

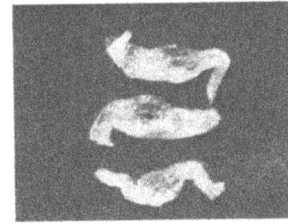

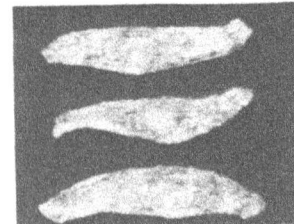

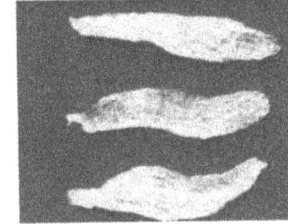

The seeds of orchids are extremely variable in shape and size. In general, they are linear or oblong in form. They may vary from 0.25 to 1.2 mm in length and from 0.090 to 0.27 mm in width. The seed consists of a small, spherical embryo suspended within a membranous, frequently transparent and mesh-like coat. They differ from the seeds with which we are most familiar in that they have no cotyledons and no endosperm, and the embryo consists of comparatively few, scarcely differentiated, isodiametric cells. However, one can recognize two regions, a posterior end of large cells which give rise to the root, and an anterior end of small and more dense cells which eventually develop into the growing point of the aërial shoot.

Represented here are:

1 **Calopogon pulchellus** (Sw.) R. Br. Grass Pink.

2 **Calypso bulbosa** (L.) Oakes. Calypso.

3 **Corallorhiza maculata** Raf. Spotted Coralroot.

4 **Corallorhiza striata** Lindl. Striped Coralroot.

5 **Cypripedium acaule** Ait. Stemless Lady's-slipper.

6 **Cypripedium arietinum** R. Br. Ram's-head Lady's-slipper.

7 **Cypripedium calceolus** L. Yellow Lady's-slipper.

8 **Cypripedium reginae** Walt. Showy Lady's-slipper.

1 **Epipactis helleborine** (L.) Crantz. Helleborine.

2 **Goodyera oblongifolia** Raf. Giant Rattlesnake Plantain.

3 **Goodyera pubescens** (Willd.) R. Br. Downy Rattlesnake Plantain.

4 **Goodyera tesselata** Lodd.

5 **Habenaria clavellata** (Michx.) Spreng. Club-spur Orchis.

6 **Habenaria dilatata** (Pursh) Hook. White Bog Orchis.

7 **Habenaria hyperborea** (L.) R. Br. Northern Green Orchis.

8 **Habenaria obtusata** (Pursh.) Richards. Blunt-leaf Orchis.

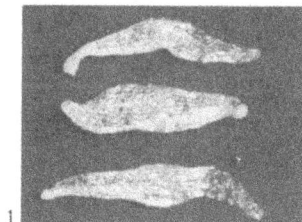
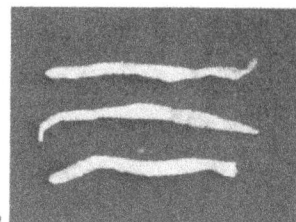
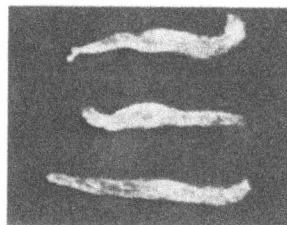
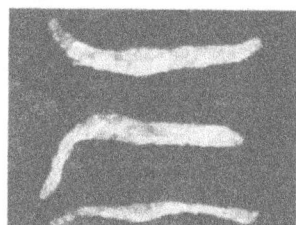
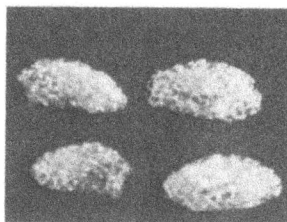
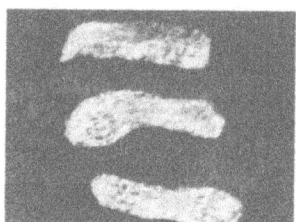
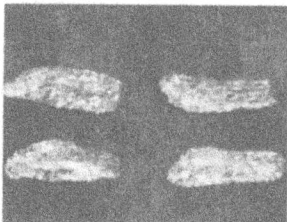
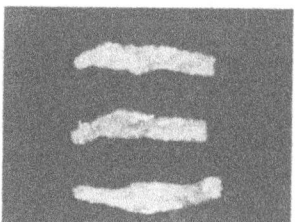

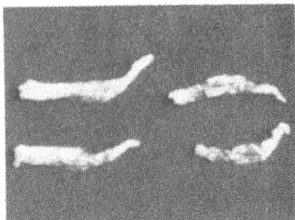

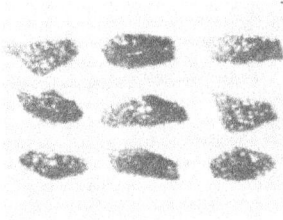

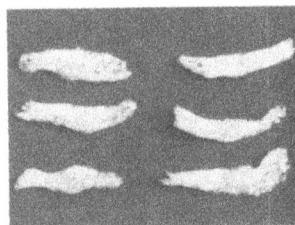

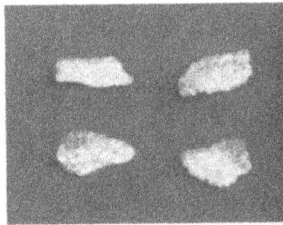

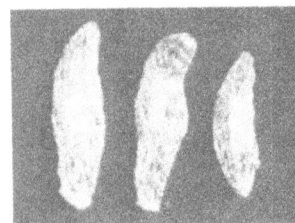

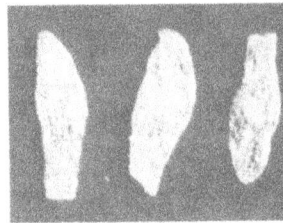

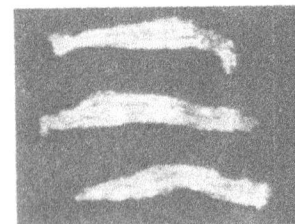

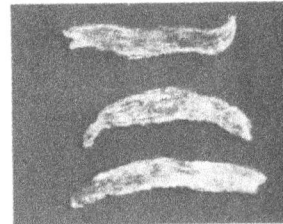

ORCHIDACEAE

1 **Habenaria orbiculata** (Pursh) Torr. Round-leaved Orchis.

2 **Habenaria psycodes** (L.) Spreng. Small Purple-fringed Orchis.

3 **Habenaria viridis** (L.) R. Br. var. **bracteata** (Muhl.) Gray. Bracted Green Orchis.

4 **Liparis loeselii** (L.) Rich. Loesel's Twayblade.

5 **Listera convallarioides** (Sw.) Torr. Broad-lipped Twayblade.

6 **Listera cordata** (L.) R. Br. Heart-leaf Twayblade.

7 **Orchis spectabilis** L. Showy Orchis.

8 **Pogonia ophioglossoides** (L.) Ker. Rose Pogonia.

ORCHIDACEAE

1 **Spiranthes cernua** (L.) Rich. Nodding Ladies'-tresses.

2 **Spiranthes gracilis** (Bigel.) Beck. Slender Ladies'-tresses.

OROBANCHACEAE

3 **Conopholis americana** (L.) Wallr. (x14) Squawroot. Seeds 1.0 x 0.7 x 0.5 mm, very irregularly angular in form with the angles rounded; surface finely reticulate, glossy and brown.

4 **Epifagus virginiana** (L.) Bart. (x25) Beechdrops. Seeds 0.5 x 0.1 x 0.1 mm, irregular or elliptic 1–2 in l.s., elliptic 6 in c.s.; surface longitudinally striate and reticulate.

5 **Orobanche uniflora** L. (x25) One-flowered Cancer-root. Seeds 0.4 x 0.2 x 0.2 mm, irregular or obliquely elliptic 3 in l.s., obliquely elliptic 6 in c.s. tending to be angular; surface coarsely reticulate and black.

OXALIDACEAE

6 **Oxalis acetosella** L. (x9) Common Wood Sorrel. Seeds 2.6 x 1.5 x 0.6 mm, obovate 47–48 in l.s., elliptic 9–10 in c.s.; surface longitudinally ribbed, slightly nodulose between the ribs, obscurely areolate and dark brown.

7 **Oxalis stricta** L. (x17) Seeds 1.3 x 1.0 x 0.4 mm, obovate 48–49 in l.s., elliptic 9–10 in c.s.; slightly oblique at the base, surface covered with a silvery, aril-like coating, transversely rugose and brown, margin narrowly ridged and sulcate.

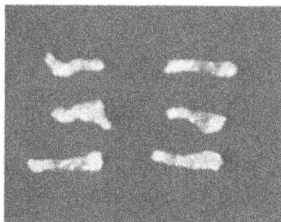
1

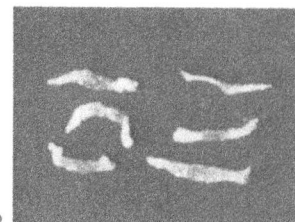
2

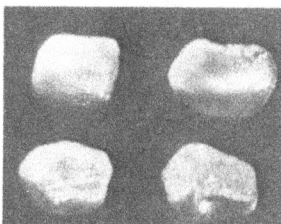
3

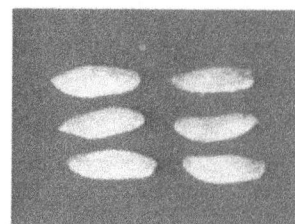
4

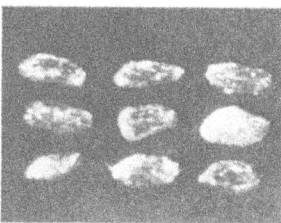
5

6

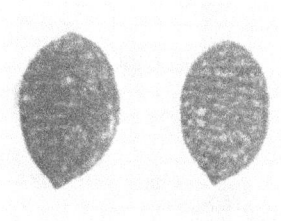
7

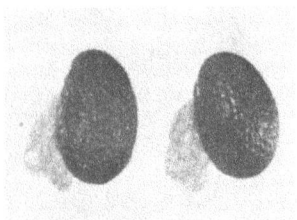

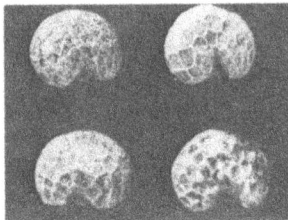

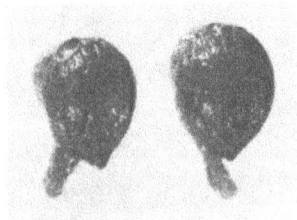

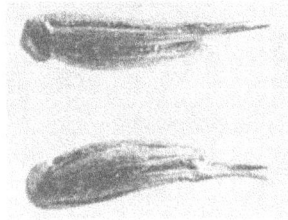

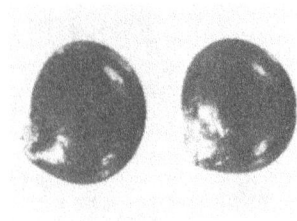

PAPAVERACEAE

1 **Chelidonium majus** L. (x20) Greater Celandine. Seeds 0.8 x 1.3 x 0.8 mm, obliquely elliptic 8–9 in l.s., elliptic 8–9 in c.s.; hilum with a prominent caruncle, surface reticulate, glossy and black.

2 **Papaver somniferum** L. (x10) Opium Poppy. Seeds 1.0 x 1.3 x 0.7 mm, elliptic 7–8 in l.s., slightly reniform, elliptic 8–9 in c.s.; surface coarsely reticulate or alveolate and colliculose.

3 **Sanguinaria canadensis** L. (x5.5) Bloodroot. Seeds 3.2 x 2.5 x 2.5 mm, elliptic 4–5 in l.s., elliptic 6 in c.s.; hilum with a prominent caruncle, surface rugulose and faintly reticulate, brown and glossy.

PHRYMACEAE

4 **Phryma leptostachya** L. (x6) Lopseed. Fruit inverted and enclosed in the persistent calyx, the lobes of which are hooked at the apex; fruit 4.2 x 1.4 x 1.1 mm, ovate 37 in l.s., elliptic 7–8 in c.s.; surface longitudinally nerved and striate.

PHYTOLACCACEAE

5 **Phytolacca americana** L. (x6.5) Pokeweed. Seeds 2.5 x 2.9 x 1.5 mm, elliptic 6–7 in l.s., embryo curved and seed tending to be reniform, elliptic 8–9 in c.s.; surface smooth or obscurely areolate, glossy and black, remnant of the funiculus obvious.

PINACEAE

6 **Abies balsamea** (L.) Mill. (x2.2) Balsam Fir. Seeds about 1.2 cm long, wing broad and obovate, gray.

1 **Larix laricina** (DuRoi) K. Koch (x3) Tamarack or Larch. Seeds about 7.5 mm long, wing broadly ovate, light brown or mottled.

2 **Picea abies** (L.) Karst. (x1.5) Norway Spruce. Seeds about 1.7 cm long, wing obovate, surface slightly scabrous and dark brown.

3 **Picea glauca** (Moench.) Voss (x2.5) White Spruce. Seeds about 1.0 cm long, wing broadly oblong or obovate, surface slightly roughened and dark brown.

4 **Picea mariana** (Mill.) BSP. (x3.5) Black Spruce. Seeds about 7 mm long, wing obovate, dark brown to black and slightly roughened.

5 **Picea rubens** Sarg. (x3) Red Spruce. Seeds about 9 mm long, wing broadly obovate, surface dark brown and slightly roughened.

6 **Pinus banksiana** Lamb. (x2) Jack Pine. Seeds about 1.3 cm long, wing oblong to semielliptic, slightly roughened and black.

7 **Pinus resinosa** Ait. (x2) Red Pine. Seeds about 1.5 cm long, wings obliquely ovate, surface slightly roughened and brown or mottled.

8 **Pinus rigida** Mill. (x2) Pitch Pine. Seeds about 1.6 cm long, wing more or less oblong, surface black and slightly glossy.

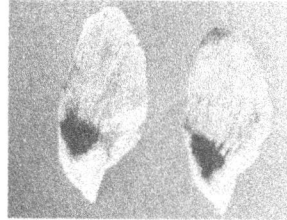
1

2

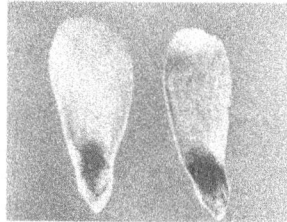
3

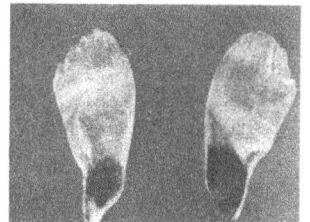
4

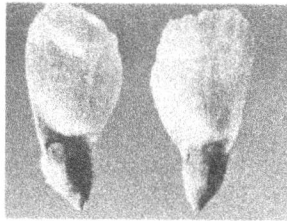
5

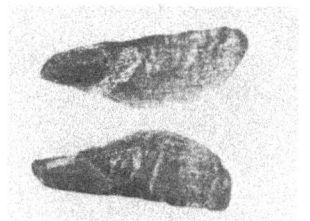
6

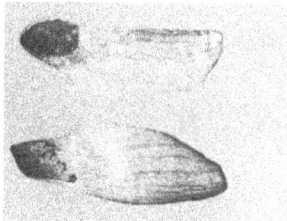
7

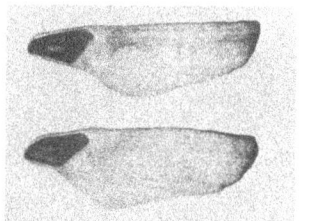
8

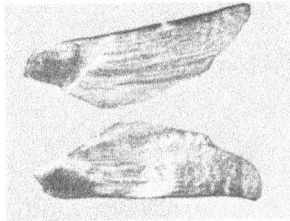

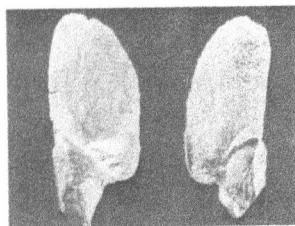

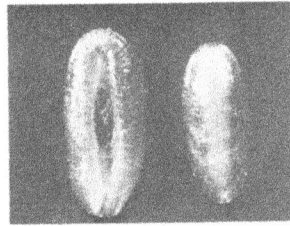

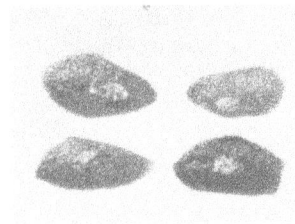

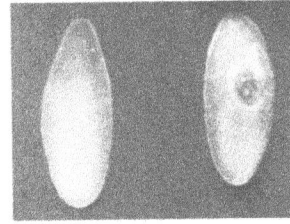

PINACEAE

1 **Pinus strobus** L. (x1.5) White Pine. Seeds about 2 cm long, wing oblong, slightly roughened, brown or mottled.

2 **Pinus sylvestris** L. (x1.5) Scotch Pine. Seeds about 2 cm long, wing obliquely ovate, surface slightly roughened, brown or mottled.

3 **Tsuga canadensis** (L.) Carr. (x3.5) Hemlock. Seeds about 7.7 mm long, wing obliquely ovate, surface roughened and dark brown.

PLANTAGINACEAE

Plantain seeds are peltate; therefore, the longitudinal axis is very short while the width in some species is very much greater. The hilum may be superficial or buried in the sulcus on the adaxial surface.

4 **Plantago lanceolata** L. (x10) Ribgrass. Seeds 0.6 x 2.1 x 1.1 mm, elliptic 10–11 in l.s., elliptic 8–9 in c.s.; ventral surface with a longitudinal sulcus in which is the hilum, surface smooth or faintly colliculose, glossy and brown with a median light stripe.

5 **Plantago major** L. (x15) Common Plantain. Seeds 0.4 x 1.0 x 0.6 mm, irregular in shape or obliquely elliptic 9–10 in l.s., obliquely elliptic 8–9 in c.s.; surface rugulose or reticulate, light to dark brown, appearing striate around the hilum.

6 **Plantago maritima** L. (x10) Seaside Plantain. Seeds 0.5 x 2.6 x 1.0 mm, elliptic 10–11 in l.s., elliptic 9–10 in c.s. with a narrow marginal wing; ventral surface with a longitudinal sulcus, upper surface smooth and light or dark brown.

7 **Plantago media** L. (x10) Seeds 0.3 x 1.8 x 0.9 mm, irregular in l.s. and in c.s. or obliquely elliptic 11 in l.s., elliptic 9 in c.s.; hilum in the concave adaxial surface, surface slightly rugulose or areolate, brown.

PLANTAGINACEAE

1 **Plantago psyllium** L. (x9) Seeds 0.8 x 2.6 x 1.3 mm, elliptic 10–11 in l.s., elliptic 9 in c.s., slightly constricted in the middle; hilum in the adaxial sulcus, surface with a light brown stripe the length of the seed, smooth or obscurely areolate, dark brown, and glossy.

2 **Plantago rugelii** Decne. (x9) Seeds 0.4 x 1.7 x 0.9 mm, irregular in form or obliquely elliptic 10–11 in l.s., obliquely elliptic 8–9 in c.s.; hilum in a concavity on the adaxial surface, upper surface faintly areolate or rugulose and black.

PLATANACEAE

3 **Platanus occidentalis** L. (x2.3) Sycamore. Fruit a globose head of obtriangular archenes each subtended by numerous brown hairs about as long as the achene; achenes 10.0 x 1.5 x 1.0 mm, obtriangular 84–85 in l.s., elliptic 8 in c.s.; surface slightly pubescent below the middle and densely so near the apex, apex with branching hairs or scales.

PLUMBAGINACEAE

4 **Limonium nashii** Small (x7) Sea Lavender. Seeds 3.5 x 0.8 x 0.5 mm, elliptic 1–2 in l.s., elliptic 8–9 in c.s.; surface dark brown, glossy, and faintly striate.

POLEMONIACEAE

5 **Collomia linearis** Nutt. (x6.5) Tiny Trumpet. Seeds 2.6 x 1.2 x 0.6 mm, obovate 46–47 in l.s., elliptic 9 in c.s.; one face with a longitudinal sulcus; surface rugulose and light brown; when moistened the seed coats form many mucilaginous threads.

1

2

3

4

5

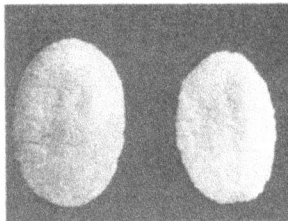
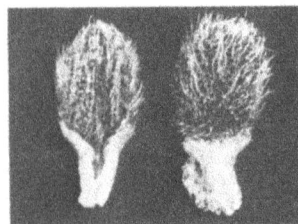
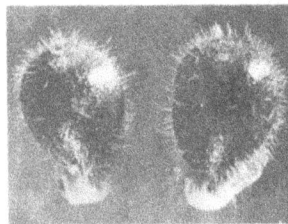
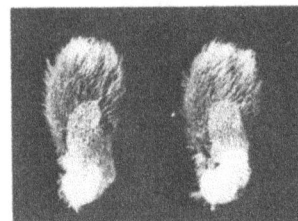
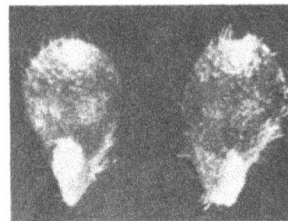
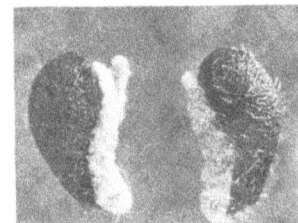

POLEMONIACEAE

1 **Phlox divaricata** L. (x8.5) Wild Phlox. Seeds 2.9 x 1.7 x 1.1 mm, obovate 47–48 in l.s., elliptic 8–9 in c.s.; one face with a longitudinal sulcus, surface rugulose, green, or light brown.

2 **Phlox subulata** L. (x10) Moss Pink. Seeds 2.1 x 1.4 x 0.7 mm, elliptic 4 in l.s., elliptic 9 in c.s.; one face with a longitudinal sulcus, surface light green or brown and rugulose.

POLYGALACEAE

The seeds of our species have a conspicuous, white, loosely cellular caruncle or aril which may be almost entire or 2–3-lobed. The surface is usually pubescent and clearly reticulate.

3 **Polygala incarnata** L. (x16) Seeds 1.5 x 1.0 x 1.0 mm, obovate 46–47 in l.s., elliptic 6 in c.s. caruncle small, white, spongy, and inconspicuously lobed, surface black, densely white pubescent, hairs with pustulate bases.

4 **Polygala paucifolia** Willd. (x9) Fringed Polygala. Seeds 2.7 x 2.0 x 1.6 mm, obovate 48–49 in l.s., elliptic 7–8 in c.s.; caruncle 2–3-lobed, surface black and with a dense, white pubescence, reticulate.

5 **Polygala polygama** Walt. (x11) Seeds 2.0 x 1.2 x 1.2 mm, obovate 47–48 in l.s., elliptic 6 in c.s.; caruncle lobed, surface reticulate, densely white-pubescent, hairs with pustulate bases.

6 **Polygala sanguinea** L. (x15) Seeds 1.4 x 0.8 x 0.8 mm, obovate 47–48 in l.s., elliptic 6 in c.s.; caruncle 2-lobed, surface rugulose, black but with a white pubescence.

7 **Polygala senega** L. (x8.5) Seneca Snakeroot. Seeds 2.8 x 1.3 x 1.3 mm, obliquely obovate 46–47 in l.s., elliptic 6 in c.s.; caruncle lobed, surface white-pubescent and reticulate.

POLYGALACEAE

Polygala verticillata L. (x15) Whorled Polygala. Seeds 1.4 x 0.6 x 0.6 mm, obliquely obovate 46–47 in l.s., elliptic 6 in c.s.; caruncle 2-lobed, surface black and finely pubescent, rugulose or reticulate.

POLYGONACEAE

Oxyria digyna (L.) Hill (x3.7) Mountain Sorrel. Fruits 4.0 x 4.8 x 0.5 mm, elliptic 7 in l.s., very thin in cross section, broadly winged, the wing retuse at the apex and cordate at the base, reticulately veined and reddish tinged.

The achenes of *Polygonum* species are lenticular or trigonous in c.s. They are usually enclosed in a persistent calyx which may be winged or wingless; the characteristics of the calyx are often useful in identification of the species.

Polygonum achoreum Blake (x10) Erect Knotweed. Achenes 2.4 x 1.5 x 1.1 mm, ovate 38–39 in l.s., obliquely triangular 79–80 in c.s.; surface dull, light brown and finely areolate.

Polygonum aviculare L. (x10) Prostrate Knotweed. Achenes 2.5 x 1.5 x 1.3 mm, ovate 39–40 in l.s., triangular 78–79 in c.s.; surface dull, dark brown, faintly rugulose, conspicuously longitudinally striate and areolate.

Polygonum cilinode Michx. (x8.5) Achenes 3.0 x 2.0 x 1.7 mm, elliptic 4 in l.s., triangular 78–79 in c.s.; surface smooth, glossy, and black.

Polygonum coccineum Muhl. (x7.5) Water Smartweed. Achenes 2.8 x 2.5 x 1.4 mm, obovate 49–50 in l.s., apiculate, elliptic 8–9 in c.s.; surface faintly areolate, dark brown.

Polygonum convolvulus L. (x5) Black Bindweed. Calyx narrowly winged; achenes 3.1 x 2.1 x 1.9 mm, elliptic 4–5 in l.s., pointed at the apex and base, triangular 78–79 in c.s. with the sides slightly concave; surface finely striate, dull, and black.

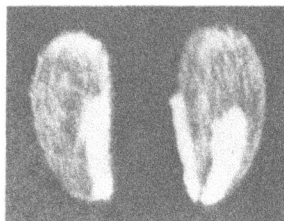

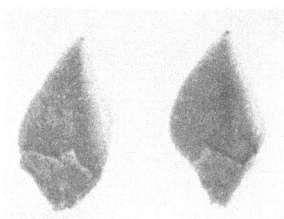

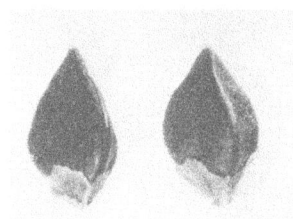

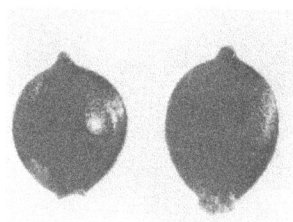

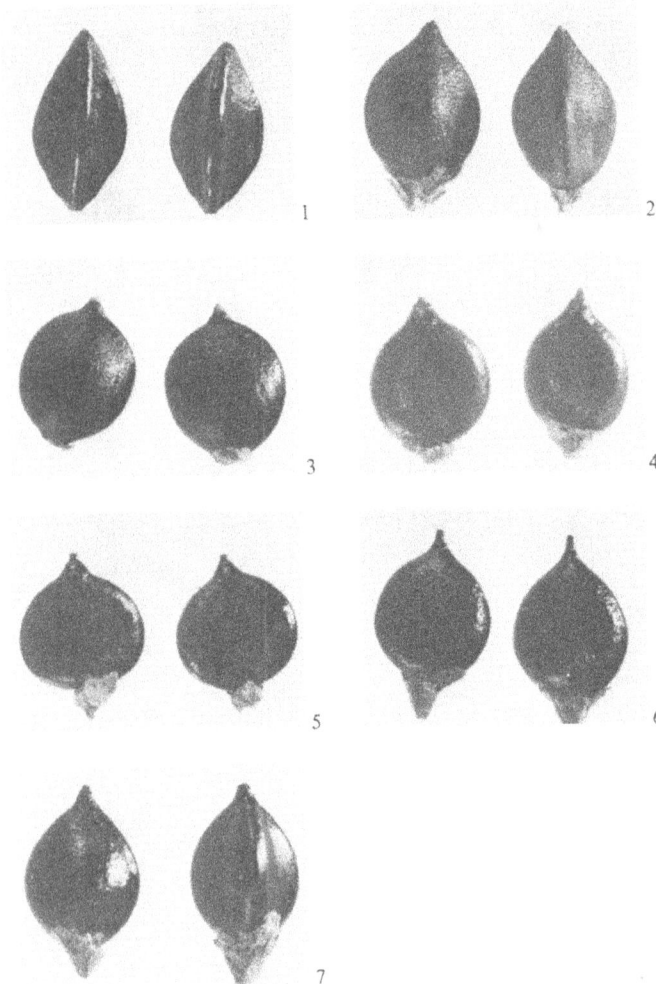

1 **Polygonum douglasii** Greene (x7) Achenes 3.1 x 1.7 x 1.7 mm, elliptic 3–4 in l.s., triangular 78 in c.s., with the faces slightly concave; surface faintly areolate, black, and glossy.

2 **Polygonum hydropiper** L. (x10) Water Pepper. Calyx with glandular, punctate dots; achenes 2.3 x 1.8 x 1.0 mm, ovate 39–40 in l.s., elliptic 8–9 to triangular 80–81 in c.s.; surface striate, puncticulate, dull and dark brown to black.

3 **Polygonum lapathifolium** L. (x10) Dock-leaved Smartweed. Achenes 2.0 x 1.7 x 0.6 mm, elliptic 5–6 to ovate 40–41 in l.s., apiculate, oblong 21–22 in c.s. with the margins rounded and the faces often concave; surface smooth, shiny, and brown.

4 **Polygonum natans** Eat. (x7) Water Smartweed. Achenes 2.7 x 2.1 x 1.3 mm, elliptic 4–5 to obovate 48–49 in l.s., apiculate, elliptic 8–9 in c.s.; surface faintly areolate, slightly glossy or dull, brown to black.

5 **Polygonum pensylvanicum** L. (x6) Pinkweed. Achenes 3.0 x 3.0 x 1.0 mm, elliptic 6 in l.s., apiculate, oblong 22 in c.s. with the faces slightly concave, margins rounded; surface black and puncticulate.

6 **Polygonum persicaria** L. (x9) Lady's Thumb. Achenes 2.2 x 1.8 x 0.8 mm, ovate 39–40 in l.s., apiculate, elliptic 9–10 in c.s. or occasionally triangular; surface puncticulate, black, and glossy.

7 **Polygonum punctatum** Ell. (x8) Water Smartweed. Calyx lobes glandular-punctate; achenes 2.4 x 1.7 x 1.2 mm, ovate 39–40 in l.s., apiculate, elliptic 8–9 or triangular 79–80 in c.s.; surface obscurely areolate, brown or black, and glossy.

Polygonum sagittatum L. (x8) Arrow-leaved Tear-thumb. Achenes 2.6 x 1.9 x 1.5 mm, ovate 39–40 in l.s., apiculate, triangular 79–80 in c.s. with the faces slightly concave; surface brown or black and slightly glossy.

Polygonum scandens L. (x6) Climbing False Buckwheat. Calyx lobes broadly winged and pubescent; achenes 3.8 x 2.2 x 2.0 mm, elliptic 3–4 in l.s., triangular 78–79 in c.s., with the faces slightly concave; surface smooth, glossy, and black.

Polygonum tenue Michx. (x7) Achenes 2.6 x 1.7 x 1.5 mm, elliptic 3–4 in l.s., triangular 78–79 in c.s., with the faces slightly concave; surface puncticulate, black, and shiny.

Polygonum virginianum L. (x6) Jumpseed. Achenes 3.3 x 2.4 x 1.6 mm, ovate 39–40 in l.s., elliptic 8 in c.s.; the two, long, deflexed styles often persistent, divided to the base and hooked at the apex, about 3 mm long; surface brown to black and glossy.

The achenes of the genus *Rumex* are enclosed in the persistent calyx. The inner 3 sepals are winged except in *R. acetosella*. They are reticulately veined and often have a swelling or tubercle at the base on the midvein. The achenes are all triangular in cross section.

Rumex acetosella L. (x10) Sheep Sorrel. Achenes 1.1 x 0.8 x 0.8 mm, elliptic 4–5 to ovate 39–40 in l.s., apiculate, triangular 78 in c.s. with the faces slightly concave; surface obscurely areolate, slightly glossy, and brown.

Rumex crispus L. (x10) Curled Dock. Achenes 2.0 x 1.2 x 1.0 mm, ovate 38–39 in l.s., acute at the apex, triangular 79 in c.s. with the faces slightly concave, margins distinctly ridged; surface puncticulate, brown and glossy; sepals tuberculate and reticulate.

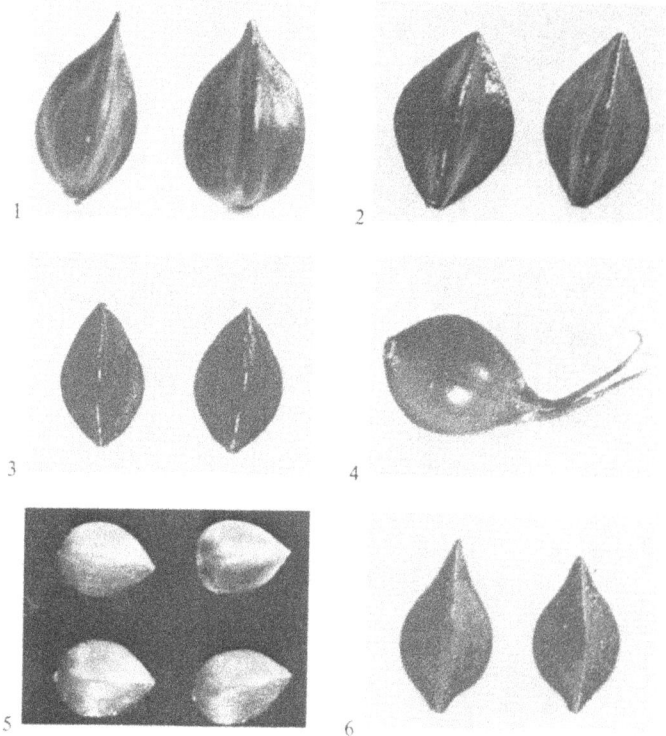

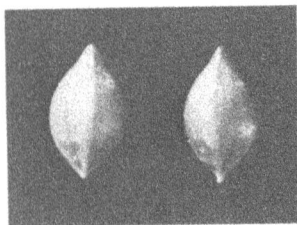

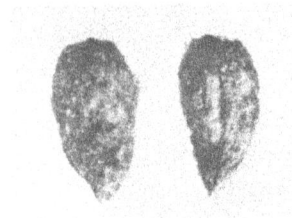
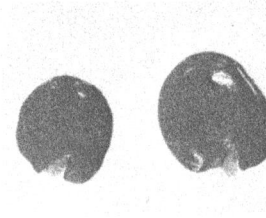

POLYGONACEAE

1 **Rumex maritimus** L. (x15) Golden Dock. Wings of the sepals spiny, each of the three inner lobes with a tubercle; achenes 1.2 x 0.7 x 0.6 mm, elliptic 3–4 in l.s., triangular 78–79 in c.s.; surface smooth, dull, and straw-coloured.

2 **Rumex mexicanus** Meissn. (x11) Calyx lobes winged and reticulate, each lobe with a prominent tubercle; achenes 1.9 x 1.1 x 1.1 mm, ovate 38–39 in l.s., sharp-pointed at the apex, triangular 78 in c.s., angles sharp and ridged; surface faintly pumcticulate, brown or black.

3 **Rumex obtusifolius** L. (x12) Broad-leaved Dock. Calyx margins toothed, surfaces reticulately veined, only one lobe with a large tubercle; achenes 2.0 x 1.2 x 1.1 mm, ovate 38–39 in l.s., acute at the apex, triangular 78–79 in c.s., angles acute; surface faintly puncticulate and brown.

4 **Rumex orbiculatus** Gray (x6.5) Calyx lobes winged and each lobe with an elliptical tubercle; achenes 3.4 x 1.6 x 1.5 mm, ovate 37–38 in l.s., acute at the apex, triangular 78–79 in c.s.; surface smooth, shiny, and brown.

PONTEDERIACEAE

5 **Pontederia cordata** L. (x6) Pickerel Weed. Seeds enclosed in the indehiscent ovary and the withered remains of the perianth; seeds 3.5 x 2.0 x 2.0 mm, ovate 38–39 in l.s., elliptic 6 in c.s.; surface irregularly longitudinally ridged and brownish.

PORTULACACEAE

6 **Claytonia caroliniana** Michx. (x7.5) Seeds 2.0 x 2.0 x 0.9 mm, elliptic 6 in l.s., embryo coiled and the seeds tending to be reniform in outline, elliptic 9–10 in c.s.; hilum with an obvious caruncle, surface faintly colliculose or papillose, black and glossy.

1 **Claytonia virginica** L. (7.5) Spring Beauty. Seeds 1.7 x 1.7 x 0.9 mm, in structure and surface characters very similar to the previous species.

2 **Portulaca oleracea** L. (x20) Common Purslane. Seeds 0.8 x 0.7 x 0.4 mm, obliquely elliptic 5–6 in l.s., embryo coiled within the seed coats and the seed tending to be reniform, elliptic 8–9 in c.s. with the faces slightly flattened or concave; base with an obvious caruncle, surface with 5–6 concentric rows of sculptured papillae, black and somewhat glossy.

PRIMULACEAE

The embryos of the seeds of this family are amphitropous and the seeds are borne on a free central placenta. They all are peltate or dome-shaped with the elongated hilum on the adaxial surface.

3 **Anagallis arvensis** L. (x12) Scarlet Pimpernel. Seeds 0.8 x 1.2 x 1.0 mm, very irregular in form but tending to be dome-shaped; margin narrowly winged; surface areolate, brown, or black with brownish scales.

4 **Glaux maritima** L. (x12) Sea Milkwort. Seeds 0.7 x 1.4 x 0.9 mm, irregular in shape but tending to be dome-shaped; surface areolate and brown.

5 **Lysimachia ciliata** L. (x9) Fringed Loosestrife. Seeds 1.0 x 1.7 x 1.2 mm, irregular in shape but rather dome-shaped in l.s., obliquely elliptic 7–8 in c.s., margin acute; surface reticulate and dark brown.

6 **Lysimachia quadriflora** Sims (x15) Seeds 0.8 x 1.3 x 0.9 mm, irregular in form but rather dome-shaped in l.s., angular at the base, obtusely elliptic 7–8 in c.s.; surface dark brown and faintly areolate.

7 **Lysimachia quadrifolia** L. (x10) Whorled Loosestrife. Seeds 1.1 x 1.9 x 1.5 mm, irregularly peltate and rather dome-shaped in l.s., obliquely elliptic 7–8 in c.s.; surface dull, brown and with minute scales or verrucose.

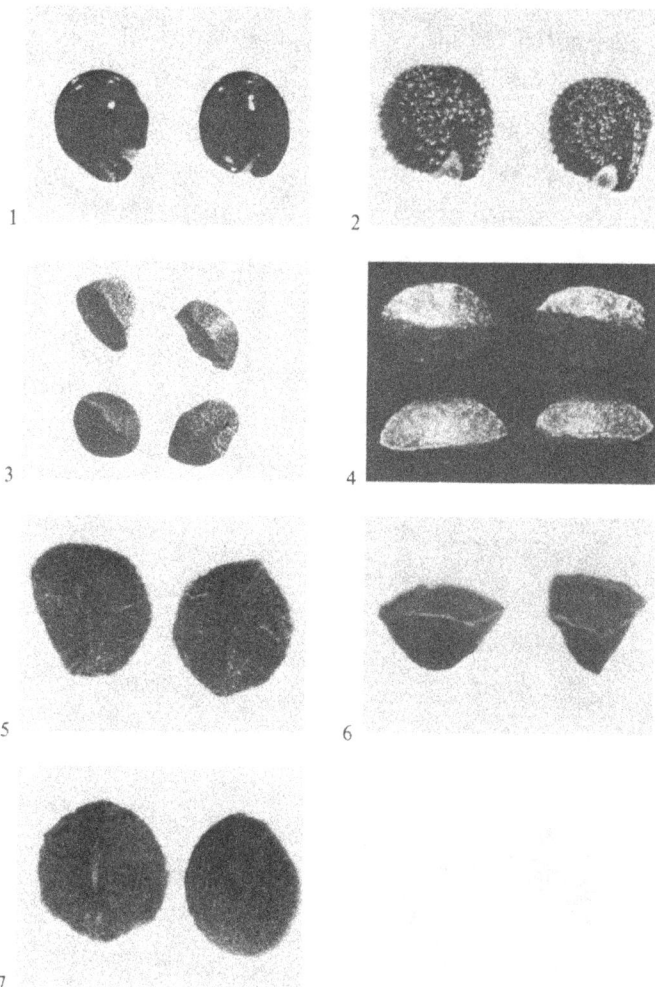

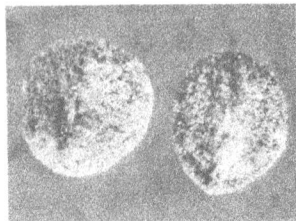
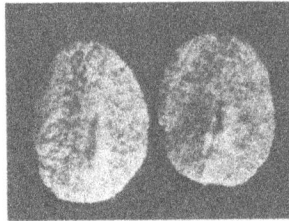
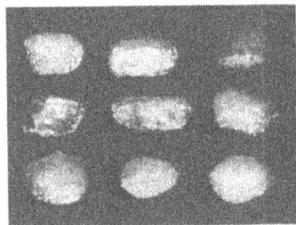

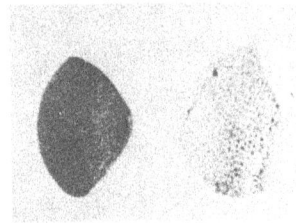

PRIMULACEAE

1 **Lysimachia terrestris** (L.) BSP. (x10) Swamp Loosestrife. Seeds 0.9 x 1.5 x 1.5 mm, irregular in l.s. and in c.s., surface dull, brown, and scaly or verrucose.

2 **Lysimachia thyrsiflora** L. (x15) Tufted Loosestrife. Seeds 0.7 x 1.3 x 1.4 mm, irregular in l.s., tending to be elliptic 7–8 in c.s.; surface grayish, rough, and areolate.

3 **Primula mistassinica** Michx. (x15) Bird's-eye Primrose. Seeds 0.2 x 0.6 x 0.4 mm, irregularly angular in outline and in c.s.; surface dark brown and finely reticulate.

4 **Samolus floribundus** HBK. (x25) Water Pimpernel. Seeds 0.3 x 0.3 x 0.3 mm, irregularly angular in l.s. and in c.s.; surface dull, black with a yellowish bloom, faintly areolate.

5 **Trientalis borealis** Raf. (x13) Star Flower. Seeds enclosed in an outer, white, reticulate coating, 0.7 x 1.3 x 1.1 mm, irregular in form or obovate 52–53 in l.s., elliptic 6–7 in c.s.; surface black, finely reticulate.

RANUNCULACEAE

6 **Actaea alba** (L.) Mill. (x6.5) White Baneberry. Seeds 3.0 x 1.7 x 1.7 mm, irregular in l.s. and c.s. or obliquely obovate 47–48 in l.s., obliquely obtriangular 90 to oblong 18 in c.s., angles strongly ridged; dorsal surface rounded and the lateral surfaces flat, dark brown, dull, and slightly areolate.

7 **Actaea rubra** (Ait.) Willd. (x5.5) Red Baneberry. Seeds not obviously different in features from those of the previous species, 3.5 x 2.5 x 1.8 mm.

1 **Anemone canadensis** L. (x5) Canada Anemone. Achenes 3.8 x 4.5 x 0.7 mm, elliptic 6–7 in l.s., terminating with a long, slender style, strongly flattened in c.s., margin with a broad wing; surface gray to brown and slightly pubescent.

2 **Anemone cylindrica** Gray (x6.5) Long-headed Anemone. Achenes 3.1 x 2.0 x 1.0 mm, obovate 47–48 in l.s., terminating in a rather short, stout style recurved at the apex, elliptic 9 in c.s.; surface dark but with a dense, gray, woolly pubescence.

3 **Anemone multifida** Poir. (x7) Cut-leaved Anemone. Achenes 3.0 x 1.6 x 1.0 mm, obovate 47–48 in l.s., elliptic 8–9 in c.s.; surface brown but with a dense, white, woolly pubescence.

4 **Anemone quinquefolia** L. (x4.2) Wood Anemone. Achenes 4.0 x 1.5 x 1.0 mm, elliptic 2–3 in l.s. with a long, stout style, elliptic 8 in c.s.; surface brown and with a short, spreading, silvery pubescence.

5 **Anemone virginiana** L. (x6) Thimbleweed. Achenes 3.0 x 2.0 x 0.8 mm, obovate 48 in l.s., tapering abruptly to a short style, elliptic 9–10 in c.s.; surface with a white, woolly pubescence.

6 **Anemonella thalictroides** (L.) Spach (x7) Rue Anemone. Achenes 3.8 x 1.4 x 1.4 mm, elliptic 2–3 in l.s., elliptic 6 in c.s. with prominent longitudinal ridges; surface smooth, green to brown.

7 **Aquilegia canadensis** L. (x10) Canada Columbine. Seeds 2.0 x 1.0 x 1.0 mm, irregular in form or obliquely elliptic 4 in l.s., obliquely elliptic 6 in c.s. with a prominent ridge; surface black and slightly shiny, faintly puncticulate.

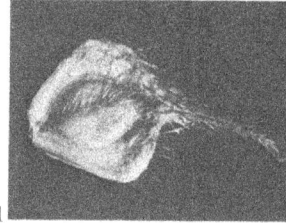

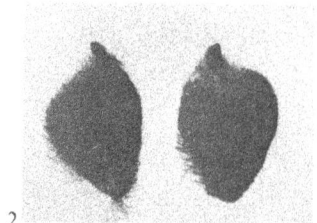

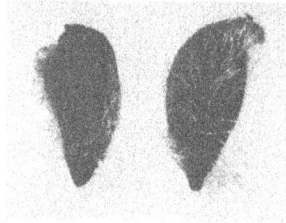

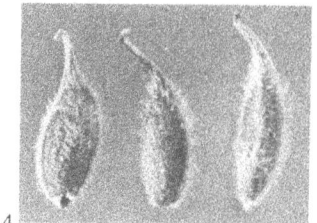

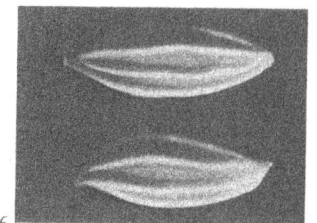

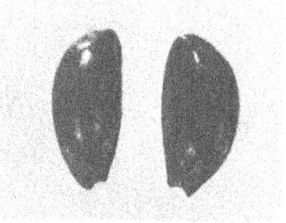

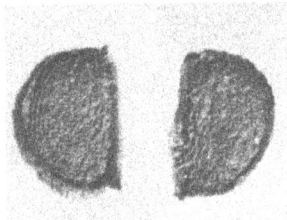

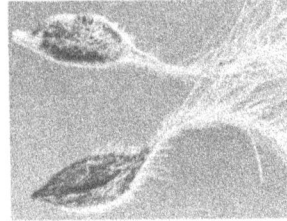
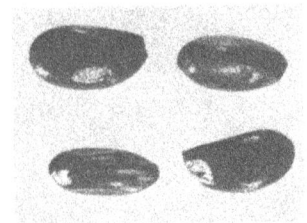
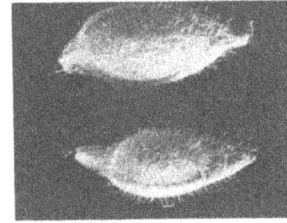
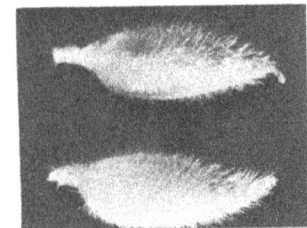
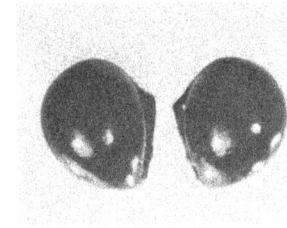

RANUNCULACEAE

1 **Caltha palustris** L. (x10) Marsh Marigold. Seeds 2.1 x 1.0 x 1.0 mm, obliquely obovate 46–47 in l.s., pointed at the base, elliptic 6 in c.s.; apex rather spongy, surface irregularly ribbed and finely reticulate.

2 **Cimicifuga racemosa** (L.) Nutt. (x9) Bugbane. Seeds 2.2 x 1.6 x 0.7 mm, irregular in form or obliquely elliptic 4–5 in l.s., obliquely obtriangular 86–87 in c.s. with the angles slightly winged; surface brown and rugulose or areolate.

3 **Clematis verticillaris** DC. (x4) Purple Virgin's-bower. Achenes 5.0 x 3.0 x 1.1 mm, obovate 47–48 in l.s., tapering to a long, plumose style, elliptic 9–10 in c.s.; surface brown and hirsute.

4 **Clematis virginiana** L. (x4) Virgin's-bower. Achenes 4.0 x 1.8 x 0.8 mm, obovate 46–47 in l.s., terminating in a long, plumose style, elliptic 9–10 in c.s.; surface brown and densely white-pubescent.

5 **Coptis trifolia** (L.) Salisb. var. **groenlandica** (Oder) Fassett (x12) Goldthread. Seeds 1.2 x 0.6 x 0.5 mm, elliptic 3 to obovate 47 in l.s., elliptic 7 in c.s.; surface black, glossy, and inconspicuously areolate.

6 **Hepatica acutiloba** DC. (x6) Hepatica. Achenes 3.5 x 1.6 x 1.4 mm, ovate 37–38 in l.s., elliptic 6–7 in c.s.; margin with a broad rib, surface green to brown and hirsute.

7 **Hepatica americana** (DC.) Ker. (x6) Achenes 4.0 x 1.5 x 1.3 mm, surface features not different from those of the previous species.

8 **Hydrastis canadensis** L. (x5.5) Golden Seal. Seeds 2.7 x 3.6 x 2.7 mm, elliptic 7–8 or obovate 47–48 in l.s., obovate 48–49 in c.s.; hilar scar on a prominent ridge, surface black and glossy.

1 **Isopyrum biternatum** (Raf.) T.&G. (x6) Seeds 2.5 x 1.5 x 1.3 mm, obovate 47–48 in l.s., elliptic 6–7 in c.s.; margin ridged, surface smooth or faintly areolate and glandular pubescent.

The achenes of *Ranunculus* are laterally compressed or turgid. The margins are usually conspicuous and often ridged. The apex tapers to the persistent style which varies greatly in size and shape and is often characteristic for the species.

2 **Ranunculus abortivus** L. (x10) Wood Buttercup. Achenes 1.3 x 1.1 x 0.6 mm, obliquely obovate 49–50 in l.s., elliptic 8–9 in c.s.; margin with a narrow, corky ridge, beak short and acentric, surface smooth, or slightly roughened.

3 **Ranunculus acris** L. (x6) Tall Buttercup. Achenes 2.5 x 2.0 x 0.8 mm, obliquely obovate 48–49 in l.s., elliptic 9–10 in c.s.; margin with a thick ridge, beak acentric, short, thick-based and slightly curved, surface smooth, or slightly roughened.

4 **Ranunculus aquatilis** L. (x10) White Water Crowfoot. Achenes 1.3 x 1.1 x 0.8 mm, elliptic 5–6 in l.s., elliptic 7–8 in c.s.; margin acute, beak short, surface roughly transversely ridged or rugose, glabrous or slightly pubescent near the dorsal ridge.

5 **Ranunculus bulbosus** L. (x7) Bulbous Buttercup. Achenes 3.0 x 2.5 x 0.7 mm, elliptic 5 to obovate 49 in l.s. with a short, stout, slightly recurved beak, elliptic 10–11 in c.s.; margin with a thick, slightly corky keel, surface areolate.

6 **Ranunculus fascicularis** Muhl. (x8) Achenes 2.7 x 2.5 x 1.1 mm, elliptic 5–6 in l.s., style long, slender and curved or straight, elliptic 9–10 in c.s. with a corky keel; surface slightly rugulose and areolate.

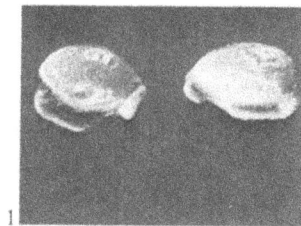

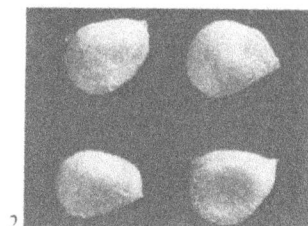

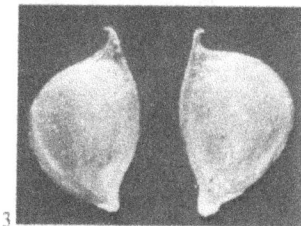

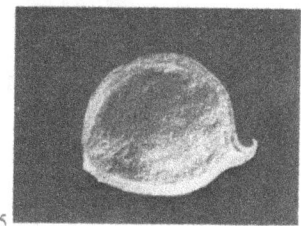

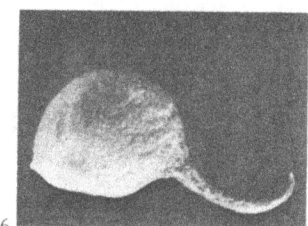

174

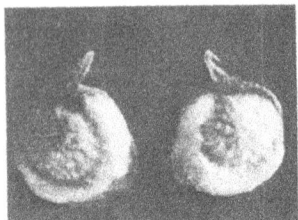
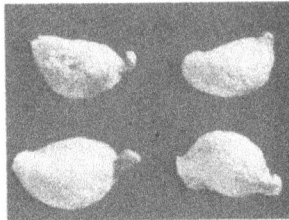

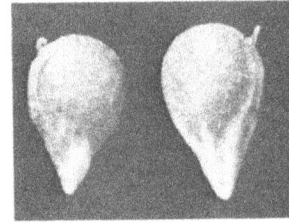
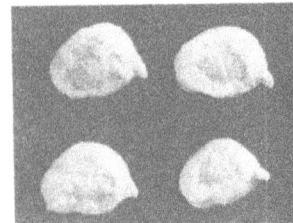

1 **Ranunculus flabellaris** Raf. (x7) Yellow Water Crowfoot. Achenes 1.3 x 1.2 x 1.0 mm, elliptic 6 in c.s., beak short, broad, flat and straight or slightly curved, elliptic 7–8 in c.s. with thick, spongy margins; surface smooth or slightly rugulose.

2 **Ranunculus flammula** L. (x12) Spearwort. Achenes 1.3 x 0.9 x 0.7 mm, elliptic 4–5 to obovate 48–49 in l.s., elliptic 7–8 in c.s.; beak short, stout, and straight, margin with a corky ridge, surface rugulose.

3 **Ranunculus pensylvanicus** L. f. (x10) Bristly Crowfoot. Achenes 2.5 x 1.7 x 0.6 mm, obliquely elliptic 4–5 to obovate 48–49 in l.s., elliptic 9–10 in c.s.; beak stout at the base and straight, margin with a thin, corky keel, surface finely areolate.

4 **Ranunculus recurvatus** Poir. (x10) Hooked Buttercup. Achenes 2.0 x 1.6 x 0.6 mm, obliquely elliptic 4–5 to obliquely obovate 48–49 in l.s., elliptic 9–10 in c.s.; beak long, thick-based, slender-tipped and distinctly recurved at the apex, margin slightly keeled and corky, surface faintly reticulate.

5 **Ranunculus repens** L. (x7) Creeping Buttercup. Achenes 2.4 x 2.0 x 0.9 mm, obliquely obovate 49 in l.s., with a stout, recurved beak, elliptic 9–10 in c.s.; margin thick and corky, surface glabrous, smooth or faintly areolate.

6 **Ranunculus rhomboideus** Goldie (x8.5) Prairie Buttercup. Achenes with a short, broad stalk; 2.4 x 1.5 x 1.0 mm, obliquely obovate 47–48 in l.s., elliptic 8 in c.s.; beak short and abruptly curved, margin not conspicuously ridged, surface faintly areolate, green or brown.

7 **Ranunculus sceleratus** L. (x10) Cursed Crowfoot. Achenes 1.1 x 0.7 x 0.5 mm, obliquely obovate 47–48 in l.s., elliptic 7–8 in c.s.; beak inconspicuous, margin thick and corky, keel obscure, surface with minute transverse ridges, faintly areolate on the margins.

1 **Ranunculus septentrionalis** Poir. (x6) Swamp Buttercup. Achenes 3.0 x 2.4 x 1.1 mm, obliquely obovate 48–49 in l.s., elliptic 9–10 in c.s.; beak long, flat, slightly curved or straight, margin with a broad, corky ridge, surface rugulose or faintly areolate.

2 **Thalictrum dasycarpum** Fisch. & Avé-Lall (x5.5) Achenes 3.7 x 1.6 x 1.6 mm, elliptic 2–3 in l.s., rounded at the base or with a very short stipe, elliptic 6 in c.s.; style long; surface coarsely longitudinally ridged, papillose, slightly glandular-pubescent, green to black.

3 **Thalictrum dioicum** L. (x6.5) Early Meadow Rue. Achenes 3.7 x 1.8 x 1.8 mm, elliptic 2–3 in l.s. with a long stigma, elliptic 6 in c.s.; surface with coarse longitudinal ridges, smooth, brown to black with a whitish bloom in the sulci.

4 **Thalictrum polygamum** Muhl. (x5) Tall Meadow Rue. Achenes 4.0 x 1.7 x 1.2 mm, elliptic 2–3 in l.s., tapering at the base to a distinct stipe, style long and slender, elliptic 7–8 in c.s.; surface coarsely ridged, black, and faintly areolate.

5 **Thalictrum revolutum** DC. (x6) Purple Meadow Rue. Achenes 3.0 x 1.6 x 1.3 mm, elliptic 3–4 in l.s., elliptic 7–8 in c.s.; coarsely longitudinally ridged, slightly glandular-pubescent, faintly reticulate and black.

RHAMNACEAE

6 **Ceanothus americanus** L. (x8) New Jersey Tea. Seeds 2.5 x 2.1 x 1.5 mm, obovate 49–50 in l.s., elliptic 7–8 in c.s.; surface glossy and brown.

7 **Ceanothus ovatus** Desf. (x10) Seeds 2.2 x 1.8 x 1.4 mm, obovate 48–49 in l.s., elliptic 7–8 in c.s., slightly angular where in contact with other seeds; surface brown to black and glossy.

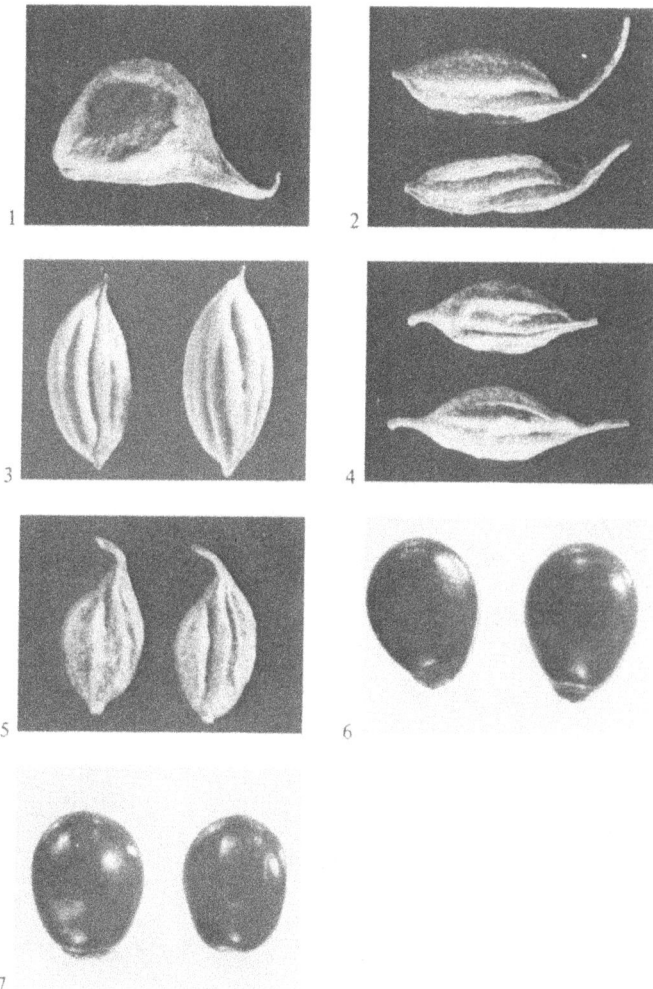

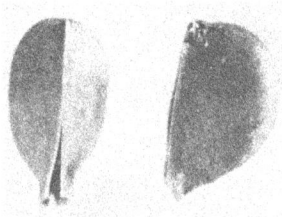

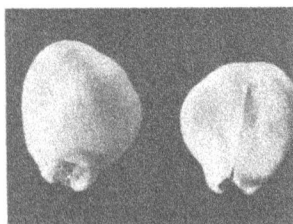
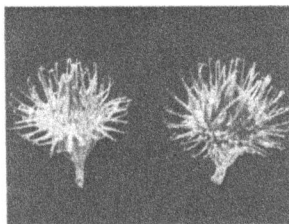

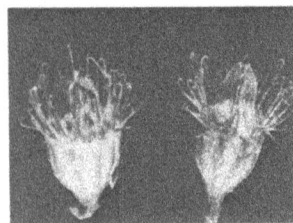
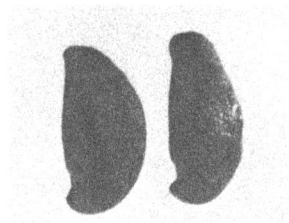

RHAMNACEAE

1 **Rhamnus alnifolius** L.'Her. (x4.5) Alder Buckthorn. Stone 4.5 x 3.4 x 1.7 mm, obovate 48–49 in l.s., rather abruptly pointed at the base, elliptic 9 to oblong 21 in c.s.; inner face with a prominent, narrow ridge, outer surface with a broader, rounded ridge, brown, and roughened.

2 **Rhamnus catharticus** L. (x4.5) Common Buckthorn. Stone 5.1 x 3.2 x 3.0 mm, obliquely obovate 47–48 in l.s., pointed at the base, obovate 49–50 in c.s.; inner face with a prominent ridge and the outer with a conspicuous sulcus, surface green to black.

3 **Rhamnus frangula** L. (x4) European Buckthorn. Stone 4.0 x 3.7 x 2.0 mm, elliptic 5–6 in l.s., obliquely elliptic 8–9 in c.s., more or less plano-convex; surface smooth, green to light brown, hilum region with a corneous ridge.

ROSACEAE

4 **Agrimonia gryposepala** Wallr. (x2) Agrimony. Fruits and seeds enclosed in the erect, persistent hypanthium tube crowned by numerous, spreading, glandular, hooked bristles, hypanthium 7.0 x 7.0 x 7.0 mm, obtriangular 90 in l.s., elliptic 6 in c.s.; surface ridged and sulcate, glandular strigose at the base; achenes oblong 18 in l.s., elliptic 6 in c.s., longitudinally ridged and sulcate.

5 **Agrimonia striata** Michx. (x3) Similar to the previous species in general appearance; hypanthium 3.0 x 3.0 x 3.0 mm, strongly reflexed at maturity, bristles glabrous, erect and connivent, tube strigose and glandular; achenes similar to those of *A. gryposepala.*

6 **Amelanchier arborea** (Michx. f.) Fern. (x6.5) Seeds 3.5 x 1.7 x 1.4 mm, obliquely elliptic 2–3 in l.s., slightly curved at the hilar end, elliptic 7–8 in c.s.; surface smooth and dark brown.

1 **Amelanchier laevis** Wieg. (x5.5) Juneberry. Seeds 4.0 x 2.0 x 1.3 mm, features similar to the previous species; surface brown smooth or faintly striate and puncticulate.

2 **Amelanchier sanguinea** (Pursh) DC. (x5.5) Seeds 4.0 x 2.3 x 1.5 mm, in general appearance similar to *A. arborea*; surface faintly striate and puncticulate.

3 **Amelanchier spicata** (Lam.) K. Koch. (x6) Seeds 3.5 x 1.8 x 1.3 mm, not obviously different in appearance from *A. arborea*.

4 **Aronia arbutifolia** (L.) Ell. (x10) Red Chokeberry. Seeds 2.3 x 1.3 x 1.0 mm, obliquely elliptic 3-4 in l.s., slightly curved at the hilum end, elliptic 7-8 in c.s.; surface rugulose and puncticulate.

5 **Aronia melanocarpa** (Michx.) Ell. (x8) Black Chokeberry. Seeds 2.5 x 1.5 x 1.0 mm, similar in appearance to the previous species.

6 **Aronia prunifolia** (Marsh.) Rehder (x7) Purple Chokeberry. Seeds 2.9 x 1.6 x 1.0 mm, similar in appearance to those of *A. arbutifolia*.

7 **Crataegus monogyna** Jacq. (x3) English Hawthorn. Stones 6.5 x 4.6 x 4.6 mm, obovate 48-49 in l.s., elliptic 6 in c.s.; surface rough, gray, or light brown.

8 **Crataegus pruinosa** (Wendl.) K. Koch (x4.5) Stones 5.5 x 3.6 x 3.2 mm, obliquely elliptic 3-4 in l.s., obliquely obtriangular 89-90 in c.s.; surface ridged, rough, and light brown.

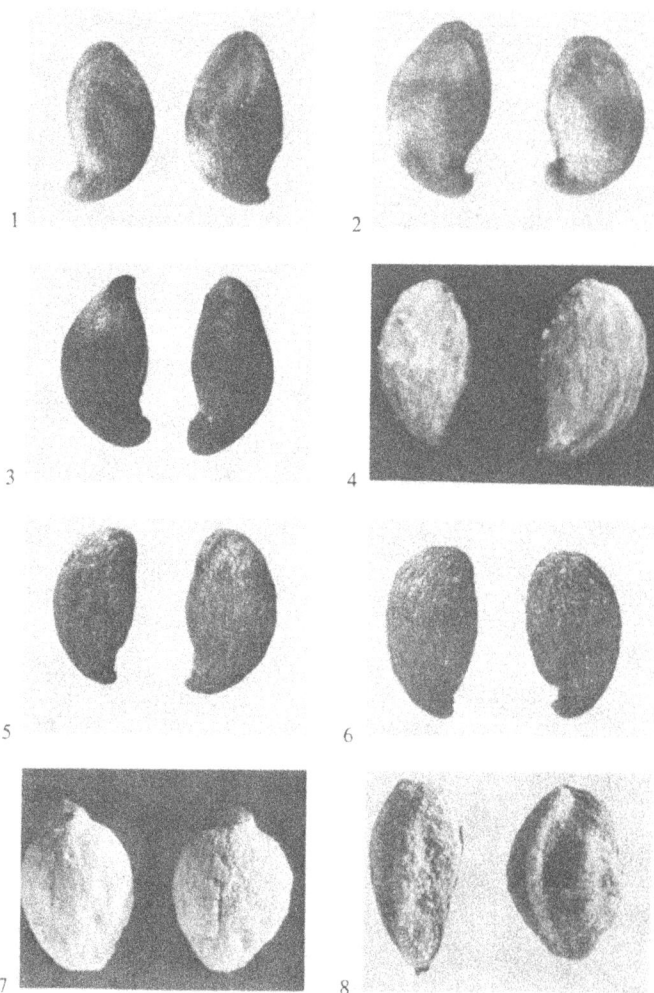

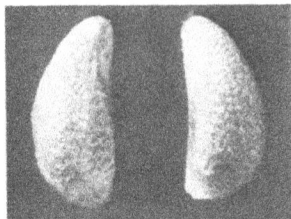
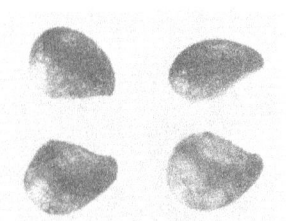
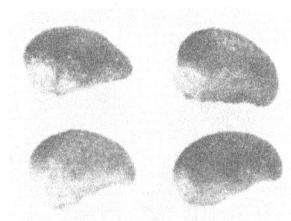
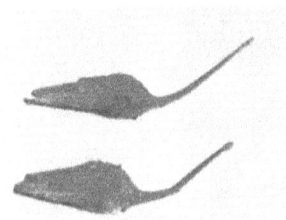
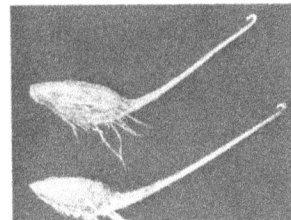
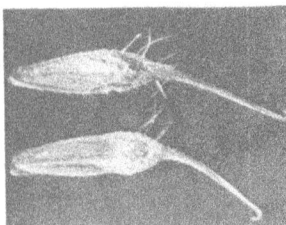

1 **Crataegus punctata** Jacq. (x3) Stones 7.5 x 5.3 x 4.5 mm, obliquely elliptic 4-5 in l.s., obliquely obtriangular 89-90 in c.s.; surface ridged, smooth, light brown.

2 **Dalibarda repens** L. (x7.5) False Violet. Achenes 3.4 x 1.5 x 1.3 mm, obliquely ovate 37-38 in l.s., obovate 49-50 in c.s.; surface with an acute longitudinal ridge, rugulose, pubescent, and brown.

3 **Fragaria vesca** L. (x14) Strawberry. Achenes 1.2 x 1.0 x 0.6 mm, obliquely ovate 40 in l.s., rather acute at the apex, obovate 47-48 in c.s.; inner margin ridged, surface smooth or faintly reticulately veined.

4 **Fragaria virginiana** Duchesne (x14) Achenes 1.4 x 1.0 x 0.7 mm, in appearance not obviously different from the previous species.

Most achenes of the genus *Geum* have a long, persistent style, hooked at the apex. The surface is usually pubescent to some degree particularly at the apex.

5 **Geum aleppicum** Jacq. var. **strictum** (Ait.) Fern. (x5.5) Yellow Avens. Achenes 3.0 x 1.1 x 0.5 mm, obovate 46-47 in l.s., tapering to the style, elliptic 9-10 in c.s.; margin slightly ridged, surface appressed pubescent, apex with a number of stiff hairs.

6 **Geum canadense** Jacq. (x4) Canada Avens. Achenes 3.0 x 1.2 x 0.6 mm, obovate 46-47 in l.s. tapering to the style, elliptic 9 in c.s.; margin slightly ridged, surface with some pubescence, inner margin with stiff hairs or bristles.

7 **Geum laciniatum** Murr. (x4.5) Cut-leaved Avens. Achenes 4.5 x 1.5 x 0.8 mm, obovate 46 in l.s. tapering to the style, elliptic 8-9 in c.s. with a marginal ridge; surface glabrous except at the slightly hispid apex.

1 **Geum macrophyllum** Willd. (x5) Achenes 4.0 x 1.5 x 0.8 mm, obovate 46–47 in l.s., tapering to the style, elliptic 8–9 in c.s., margin slightly ridged; surface finely pubescent, strigose and bristly at the apex, base of the style minutely glandular.

Geum rivale L. (x4) Purple Avens. Achenes 3.3 x 1.0 x 0.8 mm, obovate 45–46 in l.s., elliptic 7–8 in c.s. with a marginal ridge; surface hirsute and the base of the style glandular pubescent.

Geum triflorum Pursh (x5) Old Man's Whiskers. Achenes 3.0 x 1.0 x 0.6 mm, obovate 46 in l.s., tapering to a long, plumose style (3–5 cm long), elliptic 8–9 in c.s.; surface silky-pubescent.

Physocarpus opulifolius (L.) Maxim. (x11) Ninebark. Seeds 1.7 x 1.1 x 0.9 mm, obliquely obovate 47–48 in l.s., obliquely obovate 48–49 in c.s.; hilum lateral at the base of the seed, surface puncticulate, ivory-coloured and glossy.

The achenes of *Potentilla* species are most frequently ovate in outline, and laterally compressed with a distinct marginal ridge. The position of attachment of the style is usually indicated by a scar on the inner margin. The surface may be smooth or irregularly ribbed or reticulate.

Potentilla anserina L. (x13) Silverweed. Achenes 1.5 x 1.1 x 0.9 mm, obliquely ovate 39–40 in l.s., elliptic 7–8 in c.s.; dorsal surface with a sulcus, base rather corky, style lateral, surface rugulose or obscurely ribbed and areolate.

Potentilla argentea L. (x16) Silvery Cinquefoil. Achenes 1.0 x 0.6 x 0.4 mm, obliquely ovate 38–39 in l.s., elliptic 8 in c.s.; style subterminal, surface ribbed with the ribs varying from obvious to obscure, often branching.

1

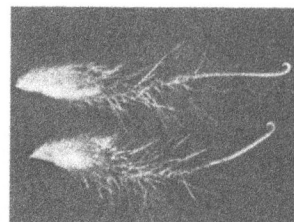

2

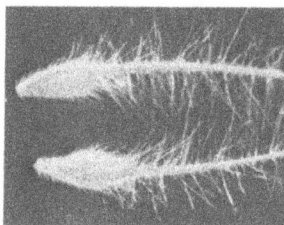

3

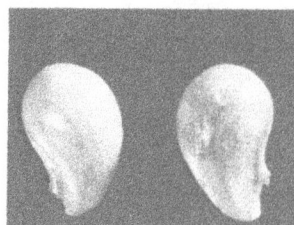

4

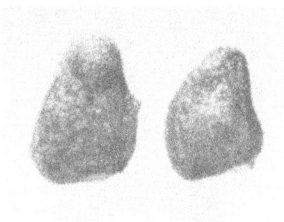

5

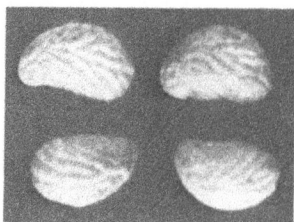

6

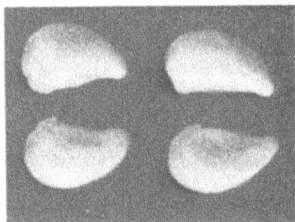

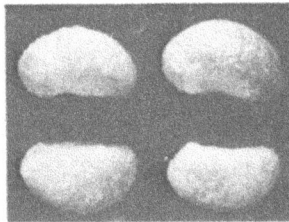

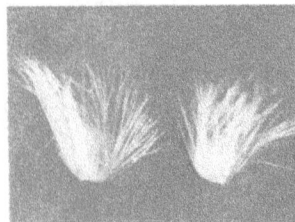

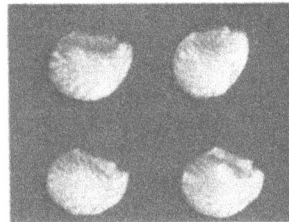

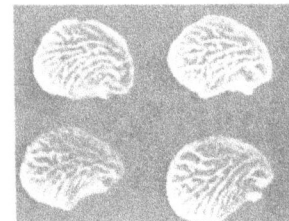

1 **Potentilla arguta** Pursh (x16) Tall Cinquefoil. Achenes 1.0 x 0.6 x 0.4 mm, obliquely ovate 38–39 in l.s., elliptic 8 in c.s.; style sub-basal, surface brown and faintly ribbed or rugulose.

2 **Potentilla canadensis** L. (x15) Achenes 1.0 x 0.6 x 0.4 mm, obliquely ovate 38–39 in l.s., elliptic 8 in c.s.; style sub-terminal, surface irregularly ribbed or rugulose and obscurely areolate.

3 **Potentilla fruticosa** L. (x6) Shrubby Cinquefoil. Achenes 1.2 x 0.8 x 0.6 mm, obliquely ovate 39 in l.s., elliptic 7–8 in c.s.; style sub-basal, surface densely, white, silky-pubescent.

4 **Potentilla intermedia** L. (x13) Achenes 1.0 x 0.7 x 0.5 mm, obliquely ovate 39–40 in l.s., slightly curved at the apex, elliptic 7–8 in c.s.; surface with longitudinal, branching ribs or veins and faintly areolate, style lateral near the apex.

5 **Potentilla norvegica** L. (x13) Rough-fruited Cinquefoil. Achenes 0.9 x 0.7 x 0.4 mm, obliquely ovate 39–40 in l.s., elliptic 8–9 in c.s.; style sub-terminal, surface with prominent, longitudinal, branching ribs and faintly areolate.

6 **Potentilla palustris** (L.) Scop. (x8.5) Marsh Cinquefoil. Achenes 1.2 x 1.2 x 0.8 mm, obliquely elliptic 6 in l.s., elliptic 8 in c.s.; style sub-basal, surface smooth or faintly areolate, greenish with brown or reddish dots.

7 **Potentilla pensylvanica** L. (x12) Achenes 1.2 x 0.9 x 0.6 mm, obliquely ovate 39–40 in l.s., elliptic 8 in c.s.; style sub-terminal, surface irregularly ribbed.

8 **Potentilla recta** L. (x12) Rough Cinquefoil. Achenes 1.2 x 0.9 x 0.5 mm, obliquely ovate 39–40 in l.s., elliptic 8–9 in c.s., margin slightly winged; style sub-terminal, surface with sharp, curved, branching ribs, ribs light brown and the intervals dark coloured, finely areolate or punticulate.

1 **Potentilla simplex** Michx. (x12) Achenes 1.2 x 0.8 x 0.6 mm, obliquely ovate 39 in l.s., elliptic 7–8 in c.s.; style lateral, surface with inconspicuous ribs and finely areolate.

2 **Potentilla tridentata** Soland. (x12) Three-toothed Cinquefoil. Achenes 1.1 x 0.9 x 0.6 mm, obliquely ovate 39–40 in l.s., elliptic 8 in c.s.; style sub-basal, surface white-hirsute at the base, ribs or veins inconspicuous.

3 **Prunus americana** Marsh. (x1.8) Wild Plum. Stone 11.0 x 7.0 x 4.9 mm, elliptic 4–5 in l.s., elliptic 7–8 in c.s.; 1/2 of the margin ridged, the other sulcate, surface veined or slightly rugulose.

4 **Prunus nigra** Ait. (x1.8) Canada Plum. Stone 14.0 x 11.0 x 7.3 mm, elliptic 4–5 in l.s., elliptic 8–9 in c.s.; 1/2 of the margin ridged, the other sulcate, surface smooth or rugulose.

5 **Prunus pensylvanicus** L. f (x3.5) Pin Cherry. Stone 5.5 x 3.9 x 3.8 mm, elliptic 4–5 in l.s., elliptic 6–7 in c.s.; 1/2 the margin ridged, the other sulcate, surface slightly rugulose.

6 **Prunus pumila** L. (x2.5) Sand Cherry. Stone 7.7 x 6.0 x 5.5 mm, elliptic 4–5 in l.s., elliptic 6–7 in c.s.; margin with a ridge and sulcus, surface slightly rugulose.

7 **Prunus serotina** Ehrh. (x2.5) Black Cherry. Stone 7.0 x 6.0 x 5.0 mm, elliptic 5–6 in l.s., elliptic 7 in c.s.; margin ridged on one-half, sulcate on the other, surface smooth or rugulose.

8 **Prunus virginiana** L. (x3) Choke Cherry. Stone 6.5 x 5.2 x 4.8 mm, elliptic 5–6 in l.s., elliptic 6–7 in c.s.; margin as in the previous species, surface smooth or slightly rugulose.

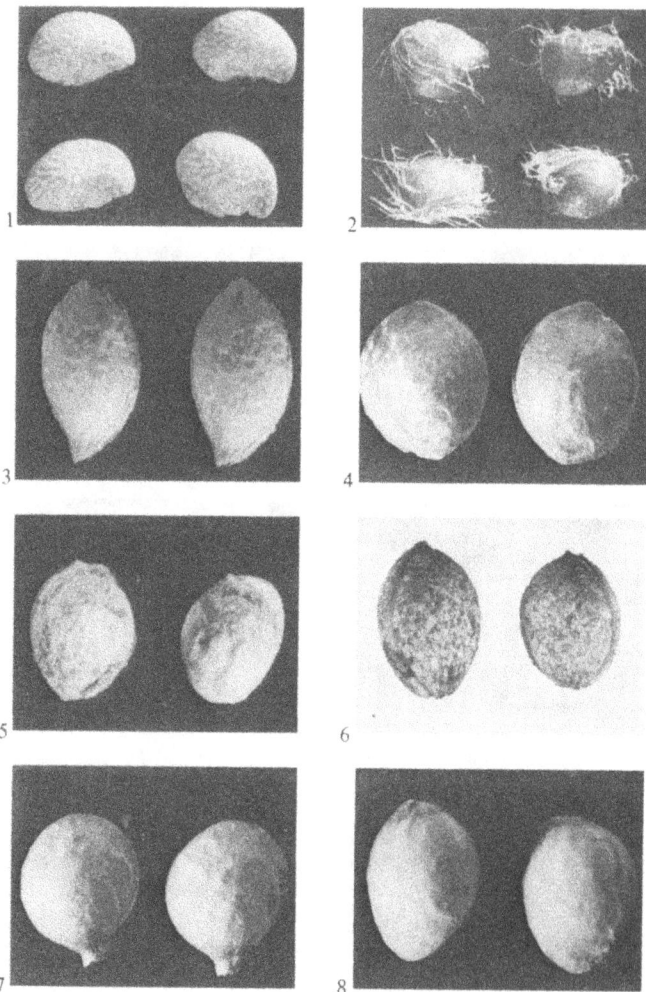

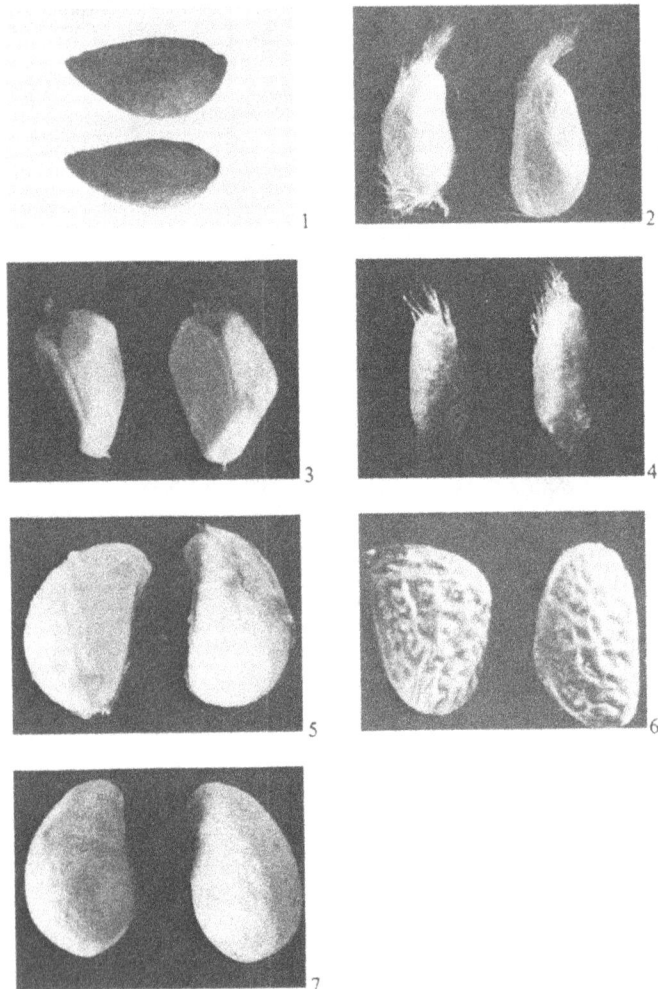

1 **Pyrus coronaria** L. (x2.7) Wild Crab-apple. Seeds 7.7 x 4.8 x 2.7 mm, obovate 47–48 in l.s., acuminate at the base, elliptic 8–9 in c.s.; surface dark brown and finely areolate.

2 **Rosa acicularis** Lindl. (x5) Prickly Rose. Achenes 4.2 x 2.1 x 2.2 mm, obliquely ovate 38 in l.s., obliquely elliptic 5–6 in c.s., tending to be somewhat angular; ridged on one margin and sulcate on the other, surface glabrous except for a tuft of hairs at the apex.

3 **Rosa blanda** Ait. (x5) Achenes 4.0 x 2.7 x 2.7 mm, obliquely ovate 39–40 in l.s., obliquely elliptic 6 to slightly angular in c.s.; surface smooth, glabrous or with a tuft of hairs at the base or apex, glossy and brown.

4 **Rosa carolina** L. (x7) Carolina Rose. Achenes 4.3 x 2.1 x 1.7 mm, obliquely ovate 37–38 in l.s., elliptic 7–8 to obtriangular 88–89 in c.s., sulcate on one angle; surface glabrous except for a few hairs around the apex, light brown and glossy.

The fruit of species of *Rubus* is an aggregate of druplets on a fleshy or dry receptacle. The stones or 'pits' are obliquely elliptic, ovate, or obovate in l.s., the margins are ridged and the surface is rugulose, coarsely reticulate or alveolate.

5 **Rubus acaulis** Michx. (x10) Stone 2.3 x 1.5 x 1.1 mm, obliquely ovate 38–39 in l.s., elliptic 7–8 in c.s.; surface puncticulate or smooth.

6 **Rubus allegheniensis** Porter (x13) Common Blackberry. Stone 2.0 x 1.5 x 1.2 mm, obliquely ovate 39–40 in l.s., elliptic 7–8 in c.s.; margin ridged, surface grayish and coarsely reticulate or alveolate.

7 **Rubus chamaemorus** L. (x7) Baked Apple. Stones 3.7 x 2.5 x 2.1 mm, obliquely ovate 39–40 in l.s. elliptic 6–7 in c.s.; surface gray, dull, faintly reticulately veined.

1 **Rubus flagellaris** L. (x9.5) Northern Dewberry. Stones 2.5 x 2.0 x 1.2 mm, obliquely ovate 39–40 in l.s., elliptic 8–9 in c.s.; margin ridged, surface coarsely reticulate or alveolate.

2 **Rubus idaeus** L. (x10) Common Raspberry. Stones 2.3 x 1.5 x 1.2 mm, obliquely elliptic 3–4 in l.s., elliptic 7–8 in c.s.; margin ridged, surface reticulate or alveolate, gray or pink.

3 **Rubus occidentalis** L. (x10) Black Raspberry. Stones 2.2 x 1.3 x 1.0 mm, obliquely elliptic 3–4 in l.s. elliptic 7–8 in c.s.; margin ridged, surface coarsely reticulate or alveolate, gray or pink.

4 **Rubus odoratus** L. (x10) Flowering Raspberry. Stones 2.1 x 1.3 x 1.1 mm, obliquely elliptic 3–4 in l.s., elliptic 6–7 in c.s.; margin ridged, surface coarsely reticulate or alveolate.

5 **Rubus parviflorus** Nutt. (x10) Thimbleberry. Stones 2.2 x 1.2 x 1.0 mm, obliquely elliptic 3–4 in l.s., elliptic 7 in c.s.; margin ridged, surface coarsely reticulate or alveolate, gray.

6 **Rubus pubescens** Raf. (x10) Dwarf Raspberry. Stones 2.3 x 1.5 x 1.0 mm, obliquely ovate 38–39 in l.s., elliptic 8 in c.s.; margin ridged, surface alveolate or rugulose, gray to pink.

7 **Sanguisorba canadensis** L. (x8.5) American Burnet. Achenes 2.5 x 1.5 x 1.5 mm, obovate 47–48 in l.s., elliptic 6 in c.s. with 4 herbaceous wings; surface faintly alveolate, green or brown.

8 **Sanguisorba minor** Scop. (x5.5) Salad Burnet. Achenes 3.8 x 2.0 x 2.0 mm, elliptic 3–4 in l.s., oblong 18 to elliptic 6 in c.s., with the angles prominently ridged or winged and the faces slightly convex; surface coarsely reticulate or alveolate and brown.

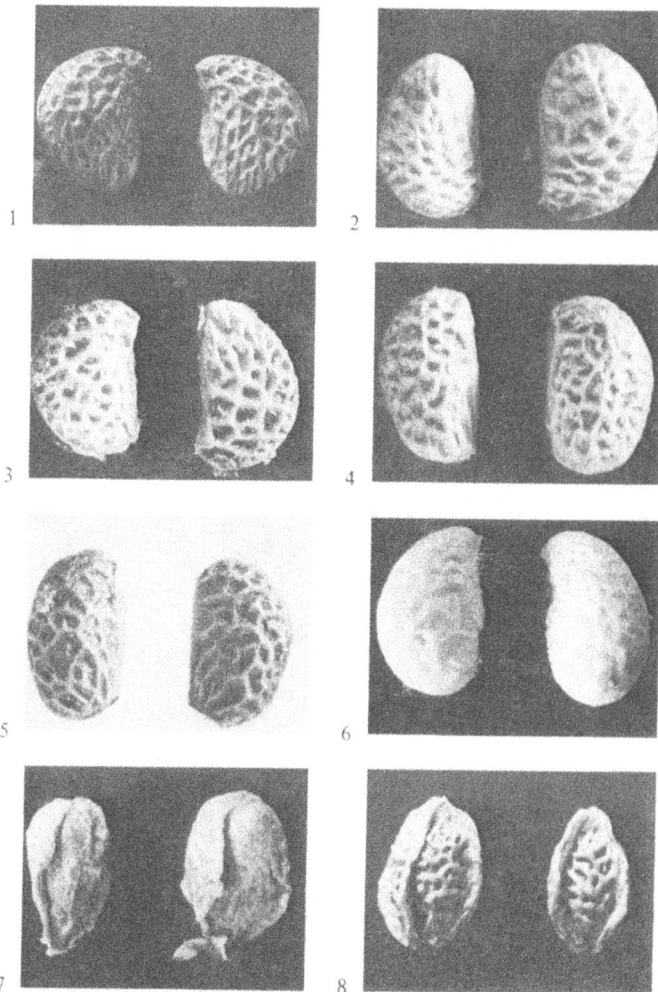

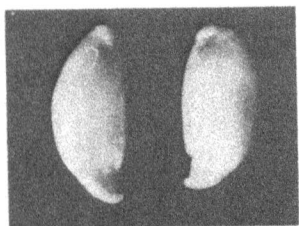
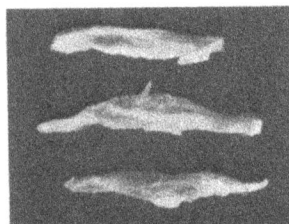
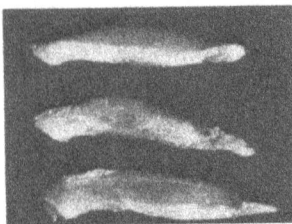

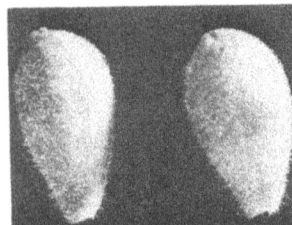
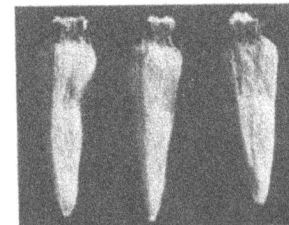

ROSACEAE

1 **Sorbus americana** Marsh. (x7) American Mountain Ash. Seeds 3.4 x 1.7 x 1.2 mm, obliquely elliptic 3 in l.s., pointed at the base, elliptic 7–8 in c.s.; margin with a ridge, surface faintly scalariform and brown.

2 **Sorbus aucuparia** L. (x6) European Mountain Ash. Seeds 4.0 x 1.7 x 1.0 mm, obliquely elliptic 2–3 in l.s., pointed at the base, elliptic 8–9 in c.s.; margin ridged, surface faintly scalariform and brown.

3 **Sorbus decora** (Sarg.) C.K. Schneid. (x5) Seeds 4.2 x 1.0 x 1.3 mm, shape and surface features not unlike those of *S. americana.*

4 **Spiraea alba** DuRoi (x14) Meadowsweet. Seeds 2.0 x 0.2 x 0.2 mm, irregular in form or oblong 12–13 in l.s., irregularly angular in c.s.; surface faintly areolate. Follicles elliptic 2–3 in l.s., elliptic 8 in c.s., one margin ribbed, sulcate at the base where they contact other follicles.

5 **Spiraea latifolia** (Ait.) Borkh. (x14) Seeds 2.0 x 0.3 x 0.2 mm, not dissimilar to those of *S. alba.*

6 **Spiraea tomentosa** L. (x16) Steeplebush. Seeds 1.5 x 0.3 x 0.2 mm, very similar to those of *S. alba* in surface characteristics. Follicles with a white, woolly pubescence.

7 **Waldsteinia fragarioides** (Michx.) Tratt. (x9) Barren Strawberry. Achenes 2.7 x 1.4 x 1.4 mm, obovate 47–48 in l.s., elliptic 6 in c.s.; surface green to brown and densely white-pubescent.

RUBIACEAE

8 **Cephalanthus occidentalis** L. (x4.5) Buttonbush. Achenes 5.5 x 1.7 x 1.3 mm, obtriangular 85–86 in l.s., rhombic 31–32 in c.s. and 2–4 celled; surface grayish and slightly pubescent.

1 **Galium aparine** L. (x5) Spring Bedstraw. Fruit 3.0 x 3.0 x 3.0 mm, elliptic 6 in l.s. and in c.s., surface uncinate hispid and verrucose; one of the pairs of fruits is usually aborted.

2 **Galium boreale** L. (x10) Northern Bedstraw. Fruit 1.4 x 1.0 x 1.0 mm, obliquely elliptic 3–4 in l.s., elliptic 6 in c.s.; surface rugulose, black and with a sparse, straight pubescence.

3 **Galium circaezans** Michx. (x6) Wild Licorice. Fruits 2.5 x 2.5 x 2.5 mm, elliptic 6 in l.s. and in c.s.; surface uncinate hispid and rugulose; one of the pairs of fruits often aborted.

4 **Galium lanceolatum** Torr. (x5) Fruits 2.9 x 2.9 x 2.9 mm, similar in characteristics to the previous species.

5 **Galium mollugo** L. (x13) Smooth Bedstraw. Fruits 1.3 x 0.8 x 0.9 mm, obliquely elliptic 3–4 in l.s., elliptic 5–6 in c.s.; surface gray to black and verrucose.

6 **Galium verum** L. (x15) Yellow Bedstraw. Fruits 1.3 x 1.0 x 1.0 mm, obliquely elliptic 4–5 in l.s., elliptic 6 in c.s.; surface black and rugulose.

7 **Houstonia caerulea** L. (x15) Bluets. Seeds 0.7 x 0.6 x 0.3, elliptic 5–6 in l.s. tending to be concave-convex with the hilum in the concavity, elliptic 8–9 in c.s.; surface faintly pitted or areolate.

8 **Houstonia canadense** Willd. (x14) Seeds 0.2 x 1.0 x 0.8 mm, irregular in form or obliquely oblong 22–23 in l.s., rather concave-convex with the hilum in the concavity, elliptic 7–8 in c.s.; surface reticulate, black, and shiny.

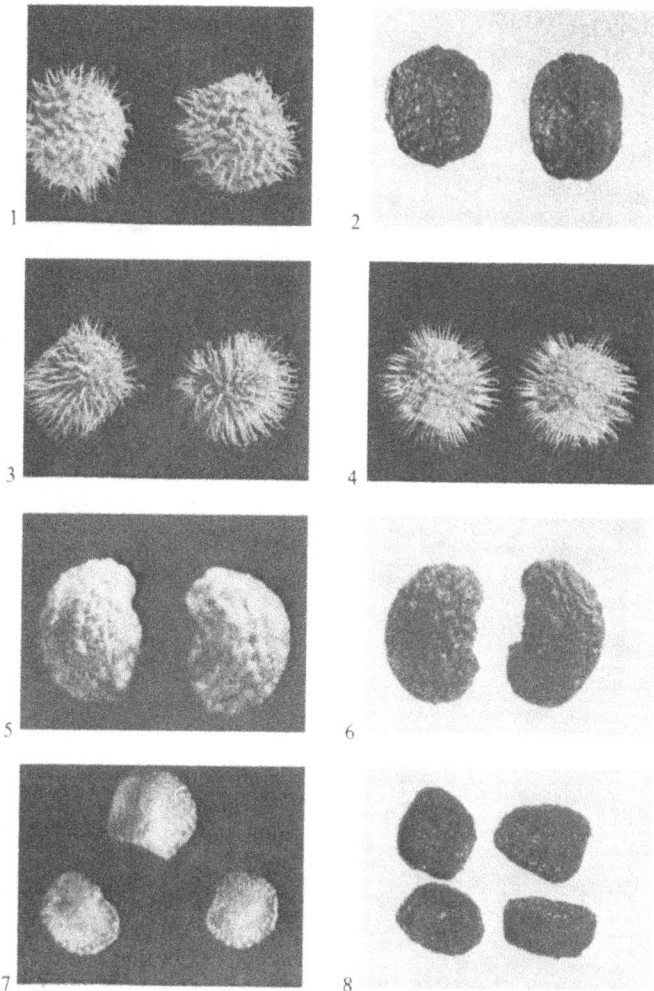

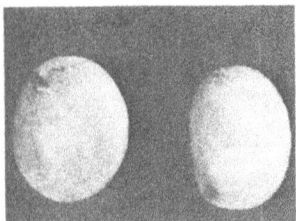
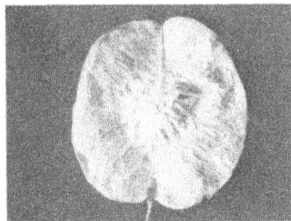
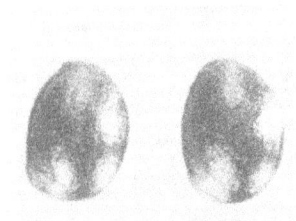
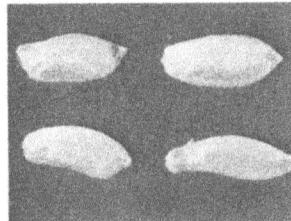
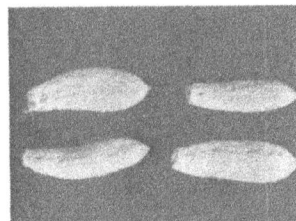
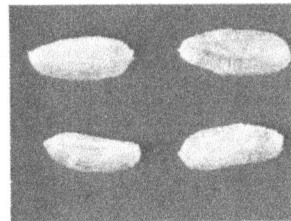
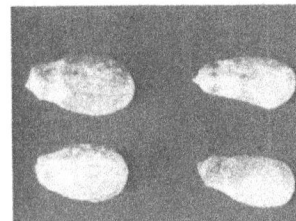
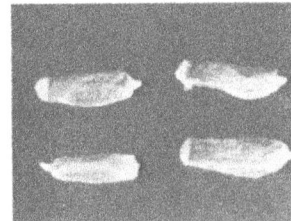

RUBIACEAE

1 **Mitchella repens** L. (x9.5) Partridge Berry. Seeds 2.1 x 1.7 x 1.0 mm, elliptic 4–5 in l.s., elliptic 8–9 in c.s.; surface gray and rugulose.

RUTACEAE

2 **Ptelea trifoliata** L. (x1) Hop Tree. Fruit an indehiscent, thin, disk-like, orbicular samara, 17 x 17 x 1.7 mm, apex retuse or subulate-tipped, wing reticulately veined; seeds elliptic 2–3 in l.s., elliptic 8–9 in c.s.; surface black, shiny, and reticulate.

3 **Zanthoxylum americanum** Mill. (x5) Prickly Ash. Seeds 3.8 x 2.8 x 2.6 mm elliptic 4–5 in l.s., elliptic 6–7 in c.s.; surface black, glossy, finely reticulate.

SALICACEAE

4 **Populus balsamifera** L. (x9.5) Balsam Poplar. Seeds 1.7 x 0.6 x 0.5 mm, elliptic 2–3 in l.s., elliptic 7 in c.s.; surface smooth, base with a silky or cobwebby coma.

5 **Populus deltoides** Marsh. (x5.5) Cottonwood. Seeds 2.9 x 1.0 x 0.6 mm, not obviously different from the previous species.

6 **Populus grandidentata** Michx. (x13) Large-toothed Aspen. Seeds 1.1 x 0.6 x 0.3 mm, similar to *P. balsamifera*.

7 **Populus tremuloides** Michx. (x13) Quaking Aspen. Seeds 1.0 x 0.5 x 0.3 mm, similar in appearance to *P. balsamifera*.

8 **Salix amygdaloides** Anderss. (x16) Peach-leaf Willow. Seeds 0.9 x 0.4 x 0.3 mm, obliquely oblong 14–15 in l.s., elliptic 7–8 in c.s.; surface faintly ribbed and gray, base with a silky coma.

1 **Salix bebbiana** Sarg. (x13) Long-beaked Willow. Seeds 1.1 x 0.4 x 0.3 mm, not obviously different from the previous species.

2 **Salix lucida** Muhl. (x15) Shining Willow. Seeds 1.1 x 0.4 x 0.3 mm, surface features similar to *S. amygdaloides*.

SANTALACEAE

3 **Comandra livida** Richards. (x3.2) Northern Comandra. Stone 4.5 x 4.5 x 4.5 mm, elliptic 6 in l.s. and in c.s.; surface gray, veined and with small sulci.

4 **Comandra umbellata** (L.) Nutt. (x7.5) Bastard Toadflax. Stone 2.0 x 2.0 x 2.0 mm, elliptic 6 in l.s. and in c.s.; surface rugulose and wrinkled, dull and gray.

SARRACENIACEAE

5 **Sarracenia purpurea** L. (x11) Pitcher Plant. Seeds 1.9 x 1.0 x 0.6 mm, obovate 47–48 in l.s., obliquely obovate 47–48 in c.s. with a narrow wing-like ridge on one margin; surface rugulose, granular, dull, and brownish.

SAURURACEAE

6 **Saururus cernuus** L. (x6) Lizard's-tail. Fruits 1.0 x 0.7 x 0.7 mm, consisting of 3–4 indehiscent carpels, elliptic 4–5 in l.s., elliptic 6 in c.s.; surface verrucose.

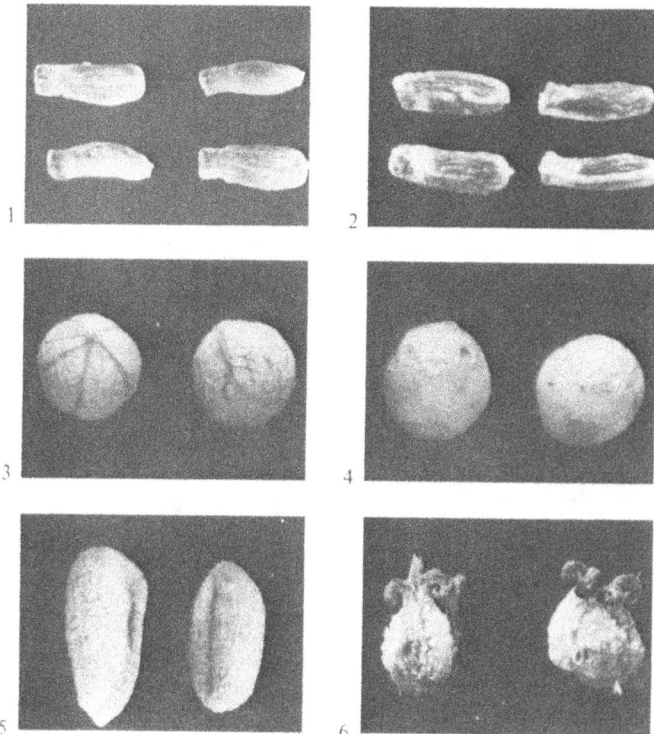

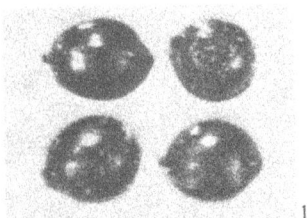

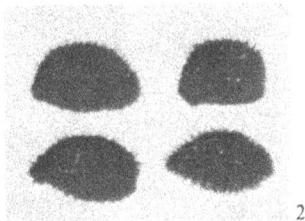

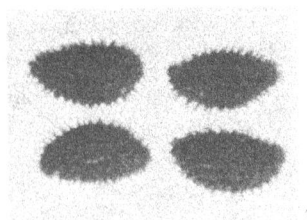

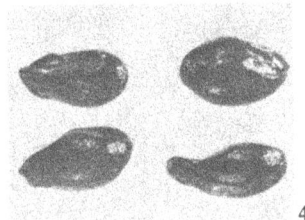

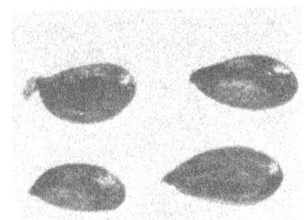

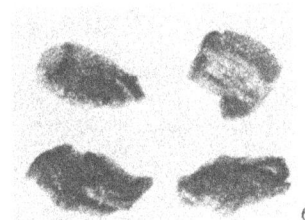

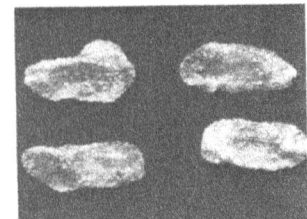

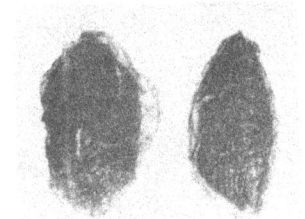

1 **Chrysosplenium americanum** Schwein. (x17) Golden Saxifrage. Seeds 0.7 x 0.6 x 0.5 mm, elliptic 5–6 in l.s., elliptic 7 in c.s.; margin ridged, surface black and with a glandular pubescence.

2 **Heuchera americana** L. (x30) Rock Geranium. Seeds 0.6 x 0.3 x 0.3 mm, elliptic 3 in l.s., elliptic 6 in c.s., surface black and with numerous, minute, gland-tipped spines.

3 **Heuchera richardsonii** R. Br. (x22) Seeds 0.7 x 0.4 x 0.4 mm, elliptic 3–4 in l.s., elliptic 6 in c.s.; surface black and covered with gland-tipped spines.

4 **Mitella diphylla** L. (x9) Miterwort. Seeds 1.7 x 0.8 x 0.8 mm, obovate 46–47 in l.s., elliptic 6 in c.s.; surface with a conspicuous ridge, glossy, black, and finely areolate.

5 **Mitella nuda** L. (x12) Seeds 1.2 x 0.6 x 0.6 mm, obovate 47 in l.s., elliptic 6 in c.s.; surface with a prominent ridge, black, glossy, and scalariform.

6 **Parnassia glauca** Raf. (x10) Grass-of-Parnassus. Seeds 1.4 x 0.8 x 0.3 mm, oblong to irregularly angular in l.s., very flattened in c.s. with a spongy, cellular wing; surface areolate or reticulate, light brown.

7 **Parnassia palustris** L. (x20) Seeds 0.8 x 0.3 x 0.1 mm, oblong to angular in l.s., flattened in c.s. with a spongy wing; surface brown and reticulate.

In general the seeds of *Ribes* species are irregular in longitudinal and in cross section. Tissues of the ovary often remain firmly attached to the seeds and may appear somewhat wing-like.

8 **Ribes americanum** Mill. (x12) Wild Black Currant. Seeds 2.0 x 1.0 x 0.6 mm, irregularly or elliptic 3 in l.s., elliptic 8–9 in c.s.; surface black, dull, and rugulose.

1 **Ribes cynosbati** L. (x9) Prickly Gooseberry. Seeds 2.2 x 1.4 x 1.0 mm, irregular or elliptic 3-4 in l.s., elliptic 7-8 in c.s.; surface brown, dull, and rugulose.

2 **Ribes glandulosum** Grauer (x8) Skunk Currant. Seeds 2.4 x 1.6 x 1.0 mm, irregular or obliquely elliptic 4 in l.s., obliquely elliptic 8-9 in c.s.; margin narrowly winged, surface rugulose and reticulate, brown.

3 **Ribes hirtellum** Michx. (x10) Seeds 2.3 x 1.1 x 0.8 mm, irregular or elliptic 2-3 in l.s., elliptic 7-8 in c.s.; surface dark brown and rugulose.

4 **Ribes hudsonianum** Richards. (x11) Seeds 1.8 x 0.8 x 0.6 mm, irregular or obliquely elliptic 2-3 in l.s., irregularly angular in c.s.; surface black and rugulose.

5 **Ribes lacustre** (Pers.) Poir. (x6) Swamp Currant. Seeds 2.5 x 1.9 x 0.9 mm, irregular or elliptic 4-5 in l.s., elliptic 9-10 in c.s.; surface dark brown and rugulose.

6 **Ribes oxyacanthoides** L. (x8) Seeds 2.5 x 1.3 x 1.0 mm, irregular or obliquely elliptic 3-4 in l.s., elliptic 7-8 in c.s.; margin slightly winged, surface brown and rugulose.

7 **Ribes triste** Pall. (x10) Red Currant. Seeds 2.1 x 2.1 x 1.1 mm, irregular or obliquely elliptic 6 in l.s., elliptic 8-9 in c.s.; margin slightly winged, surface light brown and slightly rugulose.

8 **Saxifraga virginiensis** Michx. (x22) Early Saxifrage. Seeds 0.4 x 0.3 x 0.3 mm, elliptic 4-5 in l.s., obliquely elliptic 6 in c.s.; surface longitudinally ribbed with the ribs finely muricate.

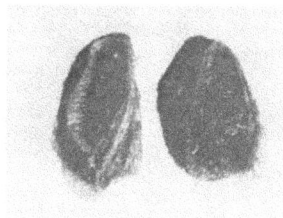

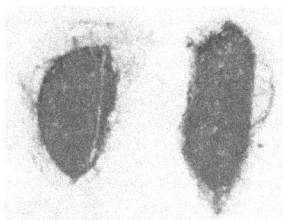
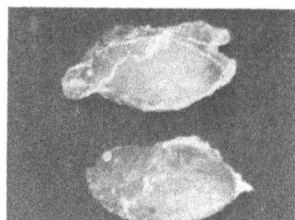
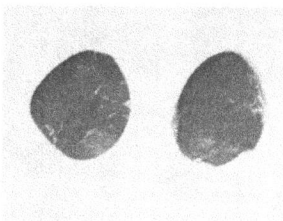
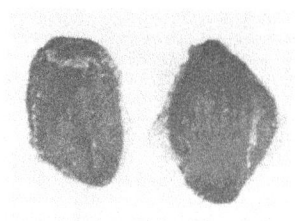

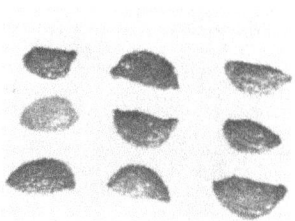

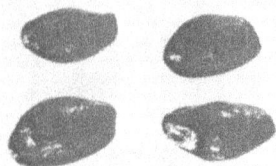

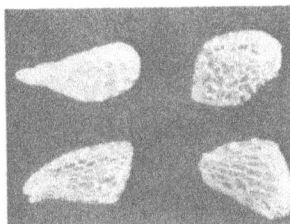

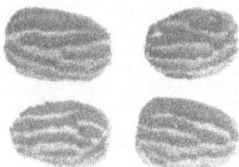
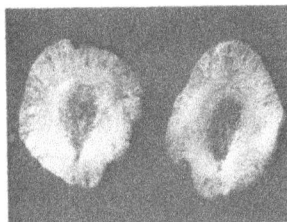

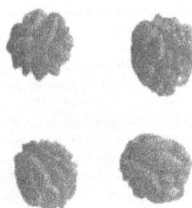
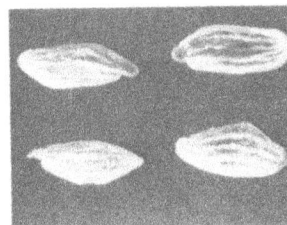

SAXIFRAGACEAE

1 **Tiarella cordifolia** L. (x13) Foamflower. Seeds 1.1 x 0.6 x 0.5 mm, elliptic 3–4 in l.s., elliptic 7 in c.s. with a prominent longitudinal ridge; surface black, glossy, and faintly areolate.

SCROPHULARIACEAE

2 **Aureolaria pedicularia** (L.) Raf. (x22) Seeds 0.9 x 0.9 x 0.9 mm, irregularly angular in form; surface reticulate, gray to black.

3 **Buchnera americana** L. (x25) Bluehearts. Seeds 0.6 x 0.25 x 0.25 mm, irregularly oblong and broadened or slightly L-shaped at the base; surface black, shiny, and reticulate.

4 **Castilleja coccinea** (L.) Spreng. (x12) Indian Paintbrush. Seeds 1.2 x 0.7 x 0.6 mm, obliquely obovate 47–48 in l.s. or irregular, irregularly angular in c.s.; embryo loosely enclosed in a white, net-like seed coat.

5 **Chaenorrhinum minus** (L.) Lange (x25) Dwarf Snapdragon. Seeds 0.6 x 0.4 x 0.4 mm, elliptic 4 in l.s., elliptic 6 in c.s.; surface prominently and irregularly longitudinally ridged, dull, and brown to black.

6 **Chelone glabra** L. (x6) Turtlehead. Seeds 3.2 x 2.7 x 0.4 mm, elliptic 5–6 in l.s., oblong 24 in c.s., distinctly winged; surface with scalariform markings radiating from the centre; wing gray and corky or spongy, seed body black.

7 **Cymbalaria muralis** (Gaertn.) Meyer & Scherb. (x12) Kenilworth Ivy. Seeds 0.8 x 0.8 x 0.8 mm, elliptic 6 in l.s. and in c.s.; surface dull, black, with coarse, irregular ridges, faintly reticulate.

8 **Euphrasia officinalis** L. (x11) Seeds 1.3 x 0.5 x 0.4 mm, obliquely elliptic 2–3 in l.s., elliptic 7–8 in c.s.; surface with white, longitudinal ribs, sulci transversely rugulose.

1 **Gerardia purpurea** L. (x13) Seeds 1.1 x 0.6 x 0.4 mm, irregularly angular in l.s. and in c.s.; surface coarsely reticulate and finely areolate, brown to black.

2 **Gerardia tenuifolia** Vahl (x17) Seeds 0.9 x 0.6 x 0.5 mm, irregularly angular in l.s. and in c.s., often broadened or L-shaped at the base; surface coarsely reticulate and finely areolate, gray to brown.

3 **Gratiola neglecta** Torr. (x25) Hedge Hyssop. Seeds 0.6 x 0.2 x 0.2 mm, obovate 46 in l.s., elliptic 6 in c.s.; surface coarsely reticulate and finely areolate.

4 **Linaria canadensis** (L.) Dum. (x20) Old-field Toadflax. Seeds 0.4 x 0.3 x 0.3 mm, obliquely oblong 16–17 in l.s., oblong 18 in c.s.; surface rugulose and reticulate, dull and black.

5 **Linaria dalmatica** (L.) Mill. (x14) Giant Toadflax. Seeds 1.4 x 0.9 x 0.5 mm, irregular angular with the angles ridged to almost winged; surface black, dull, rugulose and faintly areolate.

6 **Linaria vulgaris** Hill (x15) Toadflax. Seeds 1.7 x 1.7 x 0.3 mm, elliptic 6 in l.s., elliptic 10–11 in c.s.; margin with a broad, thin wing, surface black, dull, the central seed body dotted with black papillae.

7 **Melampyrum lineare** Desr. (x9) Cow Wheat. Seeds 3.5 x 1.4 x 0.9 mm, elliptic 2–3 to oblong 14–15 in l.s., elliptic 8–9 in c.s.; base with a prominent caruncle, surface brown and faintly areolate.

8 **Mimulus ringens** L. (x30) Monkey Flower. Seeds 0.5 x 0.2 x 0.2 mm, obliquely oblong 14–15 in l.s., elliptic 6 in c.s.; surface with brown, papillose ribs, intermediate sulci light brown, apex and base tipped with black.

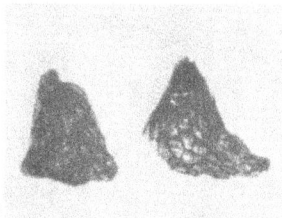

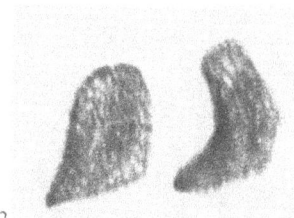

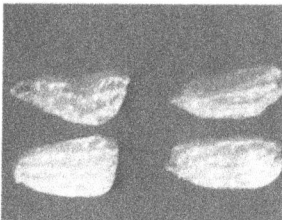

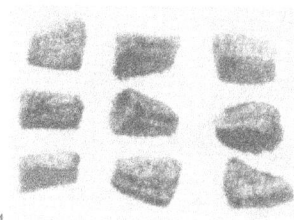

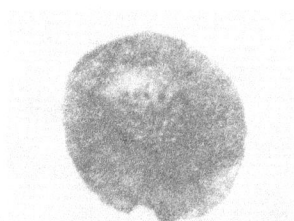

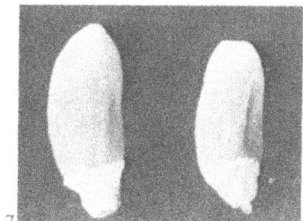

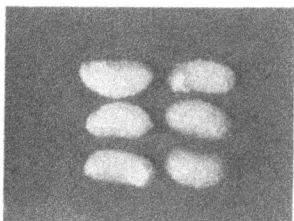

SCROPHULARIACEAE

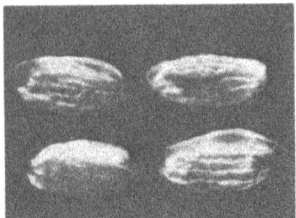
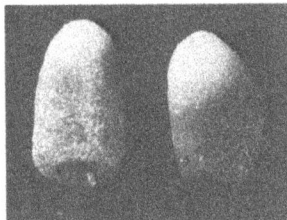
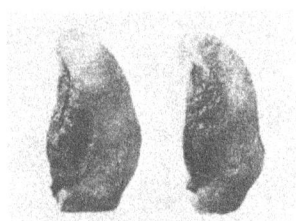
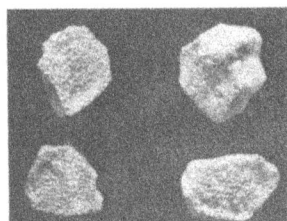
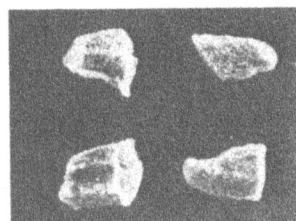
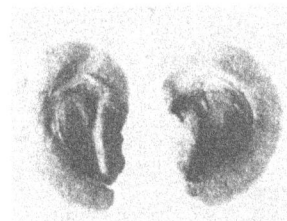
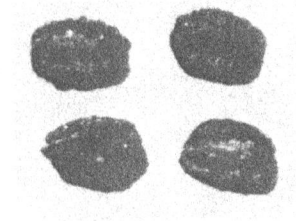
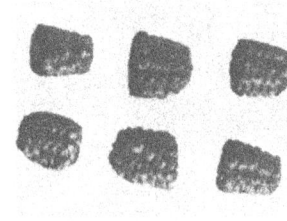

1 **Odontites serotina** (Lam.) Dum. (x11) Red Bartsia. Seeds 1.3 x 0.6 x 0.6 mm, obliquely elliptic 2-3 in l.s., elliptic 6 in c.s.; surface with longitudinal ribs, coarsely reticulate or transversely rugulose.

2 **Pedicularis canadensis** L. (x11) Wood Betony. Seeds 1.9 x 1.2 x 1.0 mm, obliquely oblong 15-16 in l.s., rather pointed at the apex and truncate and concave at the base, elliptic 7 in c.s.; surface with a single longitudinal ridge, rugulose and reticulate, brown.

3 **Pedicularis lanceolata** Michx. (x8) Seeds 3.1 x 1.5 x 0.5 mm, obliquely elliptic 2-3 in l.s., oblong 22 in c.s.; seed body lateral to a broad, spongy wing, body black, reticulate and areolate.

4 **Penstemon digitalis** Nutt. (x10) Seeds 1.2 x 1.0 x 0.5 mm, irregularly angular in l.s. and in c.s.; surface grayish and areolate.

5 **Penstemon hirsutus** (L.) Willd. (x10) Seeds 0.8 x 0.5 x 0.5 mm, irregularly angular in l.s. and in c.s., angles ridged or slightly winged and light brown; surface black and finely areolate or reticulate.

6 **Rhinanthus crista-galli** L. (x4) Yellow Rattle. Seeds 3.0 x 4.5 x 0.5 mm, obliquely elliptic 8 in l.s., very flattened in c.s.; seed body surrounded by a broad, spongy wing, scalariform and light brown; body dark brown, rugulose or veined and areolate.

7 **Scrophularia lanceolata** Pursh (x16) Seeds 0.8 x 0.6 x 0.6 mm, irregular in form, strongly longitudinally ridged or rugose, ridges coarsely reticulate, black.

8 **Verbascum blattaria** L. (x10) Moth Mullein. Seeds 0.8 x 0.6 x 0.6 mm, oblong 16-17 in l.s., obliquely elliptic 6 in c.s.; surface with prominent longitudinal ridges, transversely rugose and finely areolate.

1 **Verbascum thapsus** L. (x10) Common Mullein. Seeds 0.8 x 0.5 x 0.5 mm, scarcely distinguishable from the previous species.

2 **Veronica agrestis** L. (x35) Field Speedwell. Seeds 1.3 x 0.9 x 0.5 mm, elliptic 4–5 to obovate 48–49 in l.s. elliptic 8–9 in c.s.; slightly concave on one surface and light brown in colour.

3 **Veronica americana** (Raf.) Schw. (x30) American Brooklime. Seeds 0.6 x 0.4 x 0.2 mm, elliptic 4 in l.s., elliptic 9 in c.s.; slightly concave on one surface, margin slightly winged, surface inconspicuously rugulose.

4 **Veronica anagallis-aquatica** L. (x30) Water Speedwell. Seeds 0.5 x 0.4 x 0.2 mm, elliptic 4–5 in l.s., elliptic 9 in c.s.; surface slightly granular, light brown.

5 **Veronica arvensis** L. (x20) Corn Speedwell. Seeds 1.0 x 0.6 x 0.3 mm, elliptic 3–4 to obovate 47–48 in l.s., elliptic 9 in c.s.; one surface flat or slightly concave, surfaces granular or rugulose.

6 **Veronica officinalis** L. (x17) Common Speedwell. Seeds 0.9 x 0.7 x 0.3 mm, elliptic 4–5 in l.s., elliptic 9–10 in c.s. tending to be plano-convex; surface obscurely rugulose.

7 **Veronica peregrina** L. (x18) Purslane Speedwell. Seeds 0.7 x 0.4 x 0.1 mm, elliptic 3–4 in l.s., elliptic 10–11 in c.s.; surface faintly rugulose and light brown.

8 **Veronica scutellata** L. (x12) Marsh Speedwell. Seeds 1.2 x 1.0 x 0.2 mm, elliptic 5 in l.s., elliptic 10–11 in c.s.; surface inconspicuously areolate or rugulose, light brown.

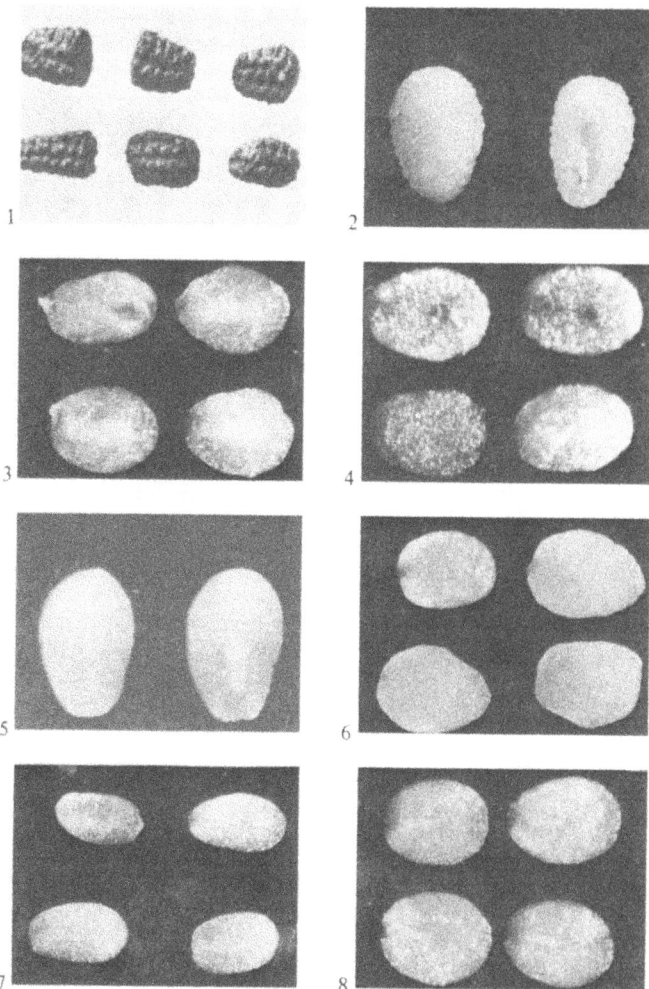

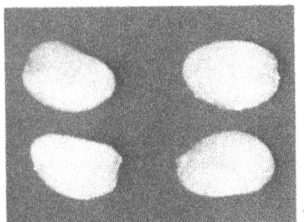

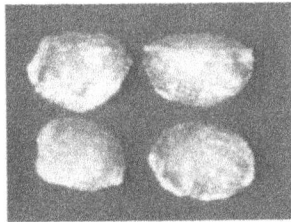

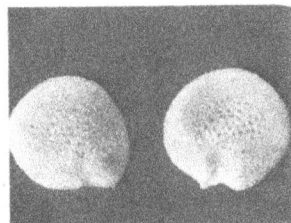

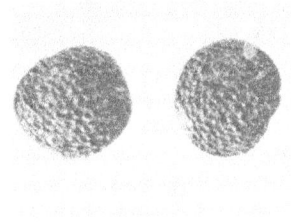

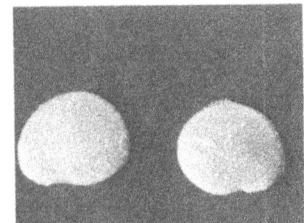

SCROPHULARIACEAE

1 **Veronica serpyllifolia** L. (x18) Thyme-leaved Speedwell. Seeds 0.7 x 0.5 x 0.1 mm, elliptic 4–5 in l.s., elliptic 10–11 in c.s.; surface faintly reticulate or rugulose and light brown.

2 **Veronicastrum virginicum** (L.) Farw. (x30) Culver's-physic. Seeds 0.5 x 0.3 x 0.2 mm, elliptic 3–4 in l.s., elliptic 8 in c.s.; surface rugulose and areolate or reticulate, light brown.

SOLANACEAE

The seeds of this family have the embryos strongly curved and the hilum is usually depressed, giving the usually elliptic seeds a rather reniform appearance. Most of the seeds are reticulate or areolate.

3 **Datura stramonium** L. (x9) Jimsonweed. Seeds 2.8 x 3.5 x 1.4 mm, obliquely elliptic 7–8 in l.s., obliquely oblong 21–22 in c.s.; surface rugulose, reticulate, dull, and black.

4 **Lycium halimifolium** Mill. (x6) Matrimony-vine. Seeds 2.0 x 2.4 x 0.8 mm, elliptic 7 in l.s., oblong 22 in c.s.; surface puncticulate and light brown.

5 **Nicandra physalodes** (L.) Gaertn. (x8) Apple-of-Peru. Seeds 1.5 x 1.8 x 0.6 mm, elliptic 7 in l.s., oblong 22 in c.s.; surface reticulate and dark brown.

6 **Physalis grandiflora** Hook. (x9) White-flowered Ground Cherry. Seeds 1.7 x 1.7 x 0.5 mm, elliptic 6 in l.s., elliptic 10–11 in c.s.; surface reticulate, gray or brownish.

7 **Physalis heterophylla** Nees. (x6) Ground Cherry. Seeds 1.5 x 2.0 x 0.6 mm, elliptic 7–8 in l.s., elliptic 10–11 in c.s.; surface finely reticulate and light brown.

1 **Physalis virginiana** Mill. (x8) Seeds 1.8 x 2.1 x 0.8 mm, elliptic 6–7 in l.s., elliptic 9–10 in c.s.; surface areolate and light brown.

2 **Solanum carolinense** L. (x7) Horse Nettle. Seeds 2.0 x 2.2 x 0.7 mm, obliquely elliptic 6–7 in l.s., elliptic 10–11 in c.s.; surface smooth or very faintly areolate, slightly glossy and yellowish.

3 **Solanum dulcamara** L. (x7.5) Climbing Nightshade. Seeds 2.1 x 2.3 x 0.7 mm, elliptic 6–7 in l.s., elliptic 10–11 in c.s.; surface finely areolate, slightly glossy and yellowish.

4 **Solanum nigrum** L. (x12) Black Nightshade. Seeds 1.5 x 1.2 x 0.5 mm, elliptic 4–5 in l.s., elliptic 9–10 in c.s.; surface finely areolate and light brown.

5 **Solanum rostratum** Dunal (x7) Buffalo Bur. Seeds 2.0 x 2.5 x 1.0 mm, obliquely elliptic 7–8 in l.s., obliquely oblong 21–22 in c.s.; surface undulate, reticulate with the areolae lustrous, brown.

6 **Solanum triflorum** Nutt. (x8) Seeds 2.3 x 1.9 x 0.6 mm, obliquely elliptic 4–5 in l.s., elliptic 9–10 in c.s.; margin acute, surface areolate and light brown.

SPARGANIACEAE

7 **Sparganium americanum** Nutt. (x4.5) Achenes forming a globose head; individual achenes 4.0 x 2.5 x 2.5 mm, elliptic 3–4 in l.s. with a short basal stipe and tapering at the apex to a style about 3.0 mm long, elliptic 6 in c.s.; surface smooth and brown.

8 **Sparganium eurycarpum** Engelm. (x3) Bur-reed. Achenes forming a globose head; individual achenes 7.5 x 5.0 x 5.0 mm, obtriangular 88 in l.s., truncate and depressed at the apex, irregularly 3–5-angled in c.s. with the angles rounded; surface smooth, slightly glossy and light brown.

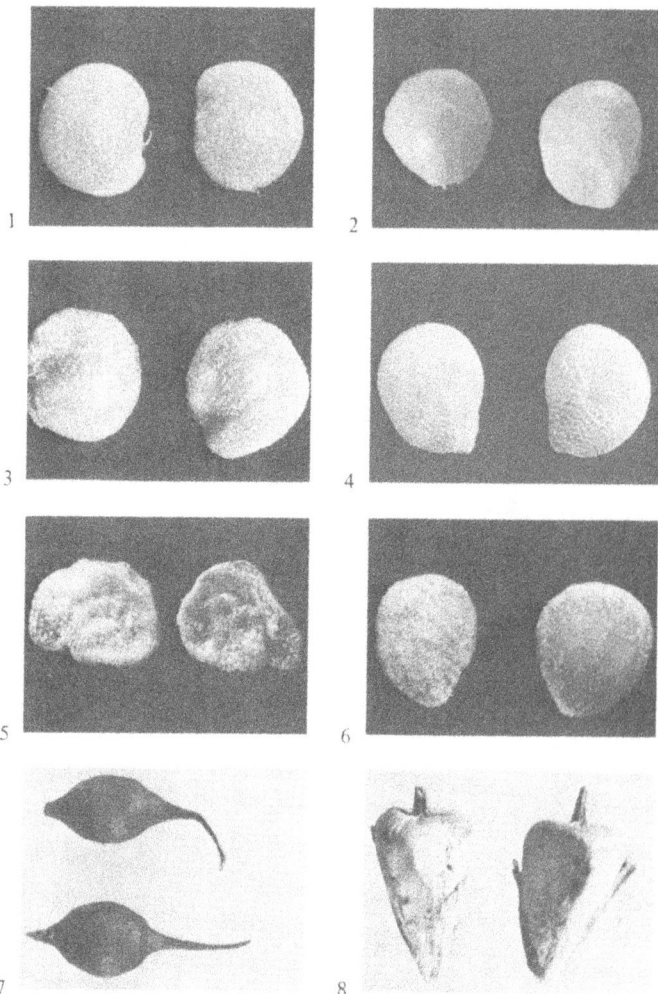

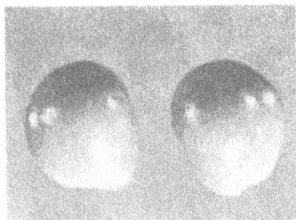

STAPHYLEACEAE

1 **Staphylea trifolia** L. (x3.3) Bladder-nut. Seeds enclosed in a large, inflated, thin-walled, triangular pericarp; seeds 5.5 x 5.2 x 4.1 mm, elliptic 5--6 to obovate 49-50 in l.s., elliptic 7-8 in c.s.; surface smooth, semiglossy and light brown.

TAXACEAE

2 **Taxus canadensis** Marsh. (x4) Ground Hemlock, Yew. Seeds with an outer, red, fleshy, aril-like structure, seed 4.8 x 3.8 x 3.8 mm, elliptic 4-5 in l.s., truncate at the base and abruptly pointed at the apex, elliptic 6 in c.s.; surface finely scalariform and somewhat glossy.

THYMELAEACEAE

3 **Dirca palustris** L. (x3) Leatherwood. Stone 6.8 x 4.3 x 4.3 mm, ovate 38-39 in l.s., elliptic 6 in c.s.; surface rugose and brown.

TILIACEAE

4 **Tilia americana** L. (x3.3) Basswood. Fruit a nut, 6.0 x 6.0 x 6.0 mm, elliptic 6 in l.s. and in c.s., abruptly contracting to a short style; surface velvety pubescent with short, gray, stellate hairs occurring singly or in small clusters.

TYPHACEAE

5 **Typha angustifolia** L. (x13) Narrow-leaved Cattail. Achenes 1.0 x 0.3 x 0.3 mm, irregular in form or elliptic 1-2 in l.s., tapering at apex to a long style and at the base to a long stipe; elliptic 6 in c.s.; surface grayish and finely areolate.

1 **Typha latifolia** L. (x17) Common Cattail. Achenes 1.0 x 0.3 x 0.3 mm, irregular or elliptic 1–2 in l.s., elliptic 6 in c.s., surface not unlike that of the previous species.

ULMACEAE

2 **Celtis occidentalis** L. (x3.4) Hackberry. Fruit a small drupe with a thin, sweet, outer pulp; stones 6.5 x 5.5 x 5.0 mm elliptic 5–6 in l.s., elliptic 6–7 in c.s.; surface with a single ridge, reticulately veined and grayish.

3 **Ulmus americana** L. (x1.8) White or American Elm. Samaras 11.0 x 7.5 x 1.3 mm, elliptic 4–5 in l.s., very flattened in c.s.; wing comparatively narrow, surface reticulately veined, margin densely silky-pubescent.

4 **Ulmus rubra** Muhl. (x1.6) Slippery Elm. Samaras 15.0 x 15.0 x 1.3 mm, elliptic 6 in l.s., very flattened in c.s.; broadly winged, wing reticulately veined, seed body pubescent but not the wing.

5 **Ulmus thomasi** Sarg. (x1.6) Rock Elm. Samaras 18.0 x 13.0 x 2.0 mm, elliptic 4–5 in l.s., very flattened in c.s.; broadly winged, the wing reticulately veined and both the wing and the seed body pubescent, margins ciliate.

UMBELLIFERAE

The fruits of this family are schizocarps. They consist of two carpels (mericarps) fused along a middle line but, with few exceptions, separating at maturity. Each mericarp has 5 primary ribs and often 4 secondary ribs. Oil tubes are frequently evident between the ribs and on the commissures. The styles are usually persistent on their enlarged bases or stylopodia.

6 **Anethum graveolens** L. (x6) Dill. Fruits dorso-ventrally compressed, 3.8 x 2.1 x 1.5 mm, elliptic 3–4 in l.s., elliptic 7–8 in c.s.; all ribs prominent but the lateral ones winged, ribs and wings light brown and the intervals darker brown.

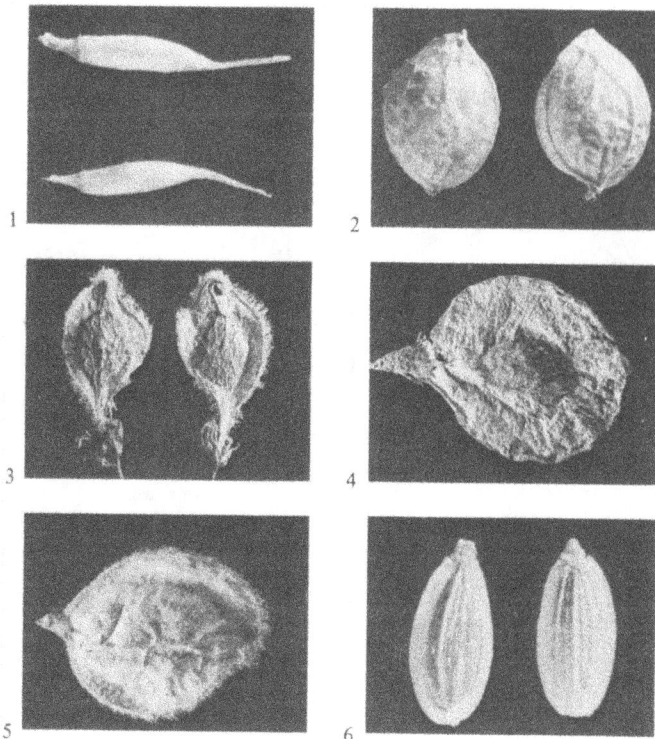

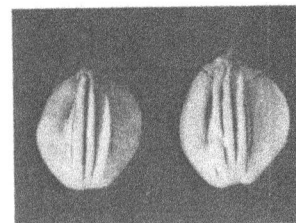

1 **Angelica atropurpurea** L. (x4.5) Angelica. Fruits dorso-ventrally compressed, 5.2 x 4.3 x 3.2 mm, oblong 16–17 in l.s., elliptic 7–8 in c.s.; dorsal ribs filiform to narrowly winged, oil tubes not prominent.

2 **Carum carvi** L. (x6) Caraway. Fruits 3.5 x 2.3 x 1.0 mm, elliptic 3–4 in l.s., flattened laterally and more or less oblong 21–22 in c.s.; ribs prominent, narrow and light-coloured, intervals dark and faintly areolate; oil tubes evident in the intervals and two apparent on the commissures.

3 **Cicuta bulbifera** L. (x8) Fruits 1.6 x 1.5 x 0.8 mm, obliquely elliptic 5–6 in l.s., elliptic 8–9 in c.s.; each mericarp with 3, low, dorsal ribs and the two lateral ribs reduced or absent. More frequently seen in this species are the sessile, vegetative bulbils occurring in the axils of the upper leaves. They are more or less ovate and surrounded by scarious bracts.

4 **Cicuta maculata** L. (x6) Spotted Cowbane. Fruits 3.5 x 2.5 x 1.8 mm, ovate 39–40 in l.s., elliptic 7–8 in c.s.; each mericarp with rounded, light brown, corky ribs and dark brown intervals, lateral ribs of each mericarp contiguous; two oil tubes evident on the commissures.

5 **Conioselinum chinense** (L.) BSP. (x4) Hemlock Parsley. Fruits 4.3 x 3.9 x 2.0 mm, elliptic 5–6 in l.s., elliptic 8–9 in c.s.; each mericarp with 3 thin, narrow, dorsal ribs and 2 thin, broad, lateral wings; oil tubes evident in the intervals, commissures with 4–5 purplish oil tubes.

6 **Conium maculatum** L. (x6) Poison Hemlock. Fruits 3.0 x 2.8 x 1.5 mm, elliptic 5–6 to ovate 40–41 in l.s., slightly laterally compressed, elliptic 8–9 in c.s.; ribs prominent, undulate, crenate and pale brown, intervals usually a darker brown, oil tubes not evident.

7 **Cryptotaenia canadensis** (L.) DC. (x3) Honewort. Fruits 6.5 x 1.5 x 0.5 mm, elliptic 1–2 in l.s., tapering to slender stylopodia, slightly laterally compressed, elliptic 10 in c.s.; ribs low, rounded and lighter brown than the dark intervals.

1 **Daucus carota** L. (x6) Wild Carrot. Fruits 3.0 x 1.3 x 1.0 mm, elliptic 2–3 in l.s., elliptic 7–8 in c.s.; the primary ribs of each mericarp much reduced and the bristles short, the secondary ribs are prominently winged and with rows of long spines or bristles, body brown but the spines grayish.

2 **Erigenia bulbosa** (Michx.) Nutt. (x9) Harbinger-of-spring. Fruits 2.3 x 3.0 x 1.5 mm, elliptic 7–8 in l.s., elliptic 9 in c.s.; ribs low, rounded and with very narrow, corky wings; oil tubes 1–3 in the intervals and several indistinct ones on the commissures.

3 **Heracleum lanatum** Michx. (x2.3) Cow Parsnip. Fruits dorso-ventrally compressed, obovate 48–49 in l.s., elliptic 10–11 in c.s.; dorsal ribs not prominent but the two lateral ones broadly winged; oil tubes appearing as 2–4 brown, clavate rays extending downward from the apex; surface usually pubescent.

4 **Hydrocotyle americana** L. (x12) Water Pennywort. Fruits 1.5 x 1.5 x 0.5 mm, elliptic 6 in l.s., oblong 22 in c.s.; laterally compressed; ribs evident but not prominent.

5 **Ligusticum scothicum** L. (x3) Scotch Lovage. Fruits 7.5 x 2.5 x 2.5 mm, oblong 14 in l.s., elliptic 6 in c.s.; slightly laterally compressed, ribs prominent, narrow and winged, surface brown and areolate; oil tubes 2 or 3 in the intervals and 6 on the commissures.

6 **Osmorhiza claytoni** (Michx.) Clarke (x1.8) Sweet Cicely. Fruits 15.0 x 2.2 x 1.4 mm, oblong 12–13 in l.s., oblong 20–21 in c.s.; slightly laterally compressed, each mericarp slightly 5-angled, styles short, 0.7–1.5 mm long, bases of the mericarps long-tailed, faces broadly sulcate, surface black and reticulate and with a few appressed, stiff hairs.

7 **Osmorhiza longistylis** (Torr.) DC. (x2.3) Anise-root. Fruits 14.0 x 2.2 x 1.4 mm, similar in general structure to the previous species but the styles long and slender, 2.0–4.0 mm long.

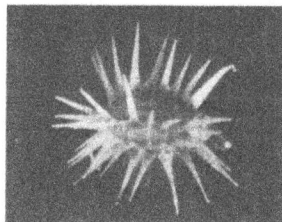

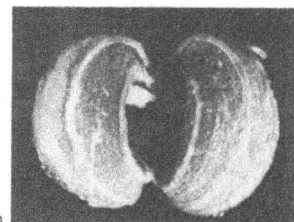

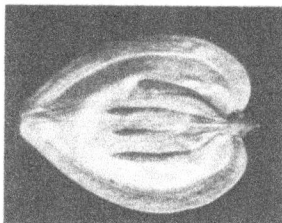

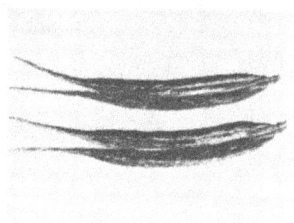

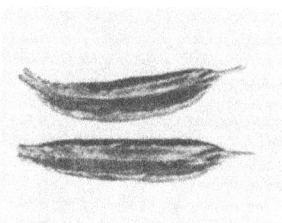

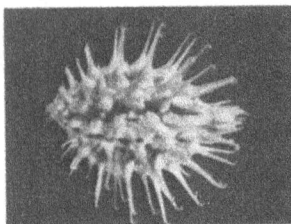

1 **Pastinaca sativa** L. (x4) Wild Parsnip. Fruits dorso-ventrally compressed, 6.5 x 4.5 x 1.5 mm, elliptic 4–5 in l.s., elliptic 10 in c.s.; dorsal ribs merely rounded ridges, lateral ribs narrowly winged; oil tubes reddish and between the dorsal ribs and evident the full length of the mericarp.

2 **Sanicula marilandica** L. (x4.5) Black Snakeroot. Fruits 5.0 x 5.0 x 5.0 mm, elliptic 6 in l.s., and in c.s.; surface covered with hooked bristles, styles slightly longer than the bristles.

3 **Sanicula trifoliata** Bickn. (x4.5) Fruits 5.0 x 3.0 x 3.0 mm, elliptic 3–4 in l.s., elliptic 6 in c.s.; surface covered with hooked bristles, styles about as long as the bristles.

4 **Sium suave** Walt. (x5) Water Parsnip. Fruits 2.7 x 2.5 x 1.6 mm, elliptic 5–6 in l.s., elliptic 8–9 in c.s.; ribs prominent, corky and light-coloured, intervals brown or black.

5 **Taenidia integerrima** (L.) Drude (x5) Yellow Pimpernel. Fruits slightly laterally compressed, 4.0 x 2.6 x 1.4 mm, elliptic 3–4 in l.s., oblong 20–21 in c.s., ribs narrow, surface rugulose, brown or greenish.

6 **Torilis japonica** (Houtt.) DC. (x7) Hedge Parsley. Fruits 3.3 x 1.8 x 1.8 mm, elliptic 3–4 in l.s., elliptic 6 in c.s., prinary ribs inconspicuous, secondary ribs prominent and covered with incurved, hooked bristles.

7 **Zizia aurea** (L.) W.J.D. Koch (x6) Fruits laterally compressed, 3.5 x 2.0 x 1.0 mm, elliptic 3–4 in l.s., oblong 21 in c.s.; ribs narrow, low, lighter in colour than the intervals; oil tubes solitary in the intervals and 2 evident on the commissures.

Boehmeria cylindrica (L.) Sw. (x14) False Nettle. Fruit an achene partially surrounded by the rather spongy calyx, 0.9 x 0.7 x 0.5 mm, ovate 39–40 in l.s., elliptic 7–8 in c.s.; surface light brown, faintly dotted, puberulent, and areolate.

Laportea canadensis (L.) Wedd. (x7) Wood Nettle. Achenes 2.5 x 2.5 x 0.7 mm, elliptic 6 in l.s., elliptic 10–11 in c.s.; embryo coiled so that the hilum appears on a lateral projection, margin slightly winged, surface black and faintly rugulose.

Pilea pumila (L.) Gray (x12) Clearweed. Achenes 1.5 x 1.2 x 0.5 mm, ovate 39–40 in l.s., acute at the apex, elliptic 9–10 in c.s., narrowly wing-margined; surface rugulose and faintly areolate.

Urtica dioica L. var. **procera** (Muhl.) Wedd. (x20) Stinging Nettle. Achenes 0.9 x 0.6 x 0.3 mm, enclosed in the persistent calyx, elliptic 4 to ovate 39 in l.s., elliptic 9 in c.s.; surface gray, dull, and faintly areolate.

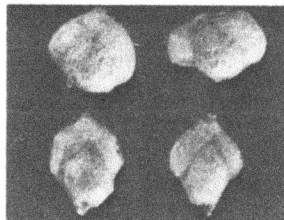

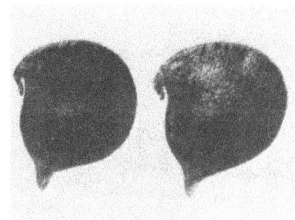

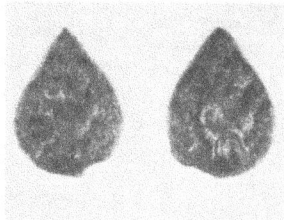

VERBENACEAE

The ovary of the genus *Verbena* is 4-lobed and forms 4 oblong nutlets at maturity. This results in the nutlets being obtriangular in c.s. when the outer surface is considered the apex of the c.s.

Verbena bracteata Lag. & Rodr. (x9) Nutlets 2.3 x 0.6 x 0.6 mm, oblong 13–14 in l.s., obtriangular 90 in c.s. with the dorsal surface rounded; surface 2–4-ribbed on the lower half and reticulate on the upper half, inner faces papillose.

Verbena hastata L. (x11) Nutlets 2.0 x 0.6 x 0.5 mm, similar to the previous species, margins ridged, dorsal surface longitudinally 3–5-ribbed on the lower half and reticulate near the apex, inner faces papillose.

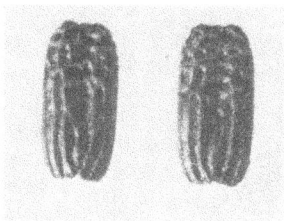

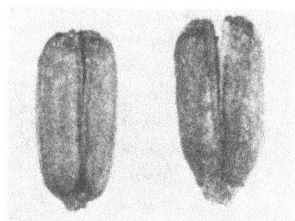

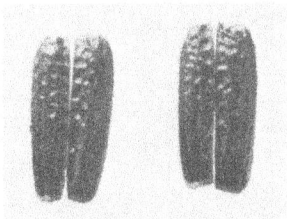

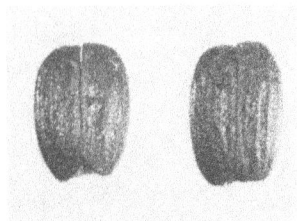

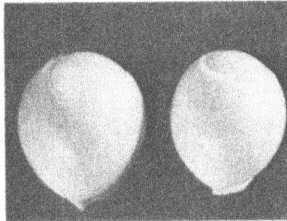

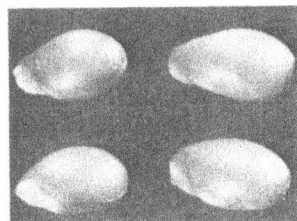

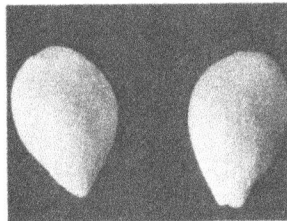

VERBENACEAE

1 **Verbena simplex** Lehm. (x8) Nutlets as in *V. bracteata*, 2.5 x 0.8 x 0.6 mm.

2 **Verbena stricta** Vent. (x8) Hoary Vervain. Nutlets 2.8 x 0.7 x 0.6 mm, form as in *V. bracteata*, most of the dorsal surface coarsely reticulate, inner surfaces with a white pubescence or papillose.

3 **Verbena urticifolia** L. (x10) White Vervain. Nutlets 1.8 x 0.7 x 0.6 mm, rounded dorsal surface obscurely veined and finely reticulate.

VIOLACEAE

4 **Cubelium concolor** (Forst.) Raf. (x4.5) Green Violet. Seeds 4.4 x 4.4 x 4.4 mm, elliptic 6 in l.s. and in c.s.; hilar area with a caruncle and an obvious raphe ridge, surface light brown with a darker area at the apex.

5 **Viola arvensis** Murr. (x11) Wild Pansy. Seeds 1.5 x 0.8 x 0.8 mm, obovate 47–48 in l.s., elliptic 6 in c.s.; base with an obvious caruncle and raphe ridge, surface glossy, light brown, and scalariform.

6 **Viola canadensis** L. (x10) Canada Violet. Seeds 2.0 x 1.4 x 1.4 mm, characteristics similar to those of the previous species.

7 **Viola cucullata** Ait. (x10) Hooded Violet. Seeds 2.0 x 1.5 x 1.5 mm, characteristics as in *V. arvensis*, surface faintly areolate.

8 **Viola eriocarpa** Schw. (x10) Smooth Yellow Violet. Seeds 2.5 x 1.5 x 1.5 mm, features not unlike those of *V. arvensis*.

1 **Viola fimbriatula** Sm. (x10) Seeds 1.6 x 1.0 x 1.0 mm, surface mottled in colour, otherwise as in *V. arvensis*.

2 **Viola pedata** L. (x10) Bird's-foot Violet. Seeds 1.8 x 1.0 x 1.0 mm, resembling very closely the features of *V. arvensis*.

3 **Viola pubescens** Ait. (x10) Downy Yellow Violet. Seeds 2.6 x 1.7 x 1.7 mm, surface areolate but other features as in *V. arvensis*.

4 **Viola rostrata** Pursh (x10) Long-spurred Violet. Seeds 1.8 x 1.2 x 1.2 mm, characteristics similar to those of *V. arvensis*.

VITACEAE

5 **Parthenocissus vitacea** (Knerr) Hitchc. (x4) Virginia Creeper. Seeds 4.2 x 3.7 x 2.7 mm, obovate 49–50 in l.s., obtriangular 88–89 in c.s.; rounded on the dorsal surface and with a dorsal sulcus ending in a circular depression in the middle of the surface or at the chalaza, inner angle with a central ridge and a sulcus on each side of it; surface rugulose and brown.

6 **Vitis riparia** Michx. (x4) Wild Grape. Seeds 5.2 x 3.8 x 2.9 mm, obovate 48–49 in l.s., obovate 51–52 in c.s.; dorsal surface with a sulcus from the middle to the chalaza and a ridge from the middle to the base, inner surface with a raphe ridge and a sulcus on each side of it, surface smooth and brown.

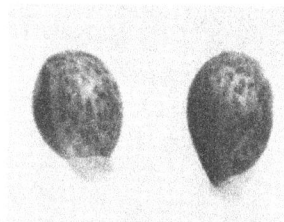

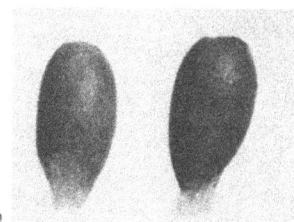

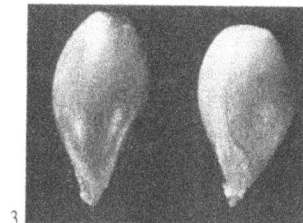

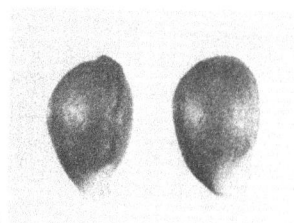

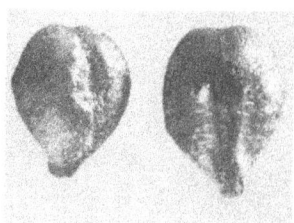

Glossary of terms

ACCUMBENT Cotyledons placed edgewise to the radicle

ACHENE A dry, indehiscent, one-seeded fruit with the seed free from the ovary wall

ACUMINATE Tapering to a long point, or attenuate

ALVEOLATE With pits or depressions giving the appearance of a honeycomb

AMPHITROPOUS The ovule of the seed half inverted so that the hilum becomes lateral or between the micropyle and the chalaza

ANATROPOUS The ovule completely inverted so that the hilum is basal with the micropyle adjacent to it and the chalaza at the opposite end

APICULATE Minutely pointed at the apex

AREOLATE Having distinct but very fine network of spaces

AREOLE A space outlined on a surface

ARIL An appendage or an outer covering of a seed growing out from the hilum or funiculus

ARILLATE Having an aril

ARTICLE A segment of a constricted pod or fruit as in *Desmodium* spp.

ARTICULATE Jointed; with a place where separation may occur

BARBED With stiff points or short bristles, usually reflexed like the barb of a fish-hook

BARBELLATE With very fine barbs

BLOOM A white, waxy, powdery covering on many fruits, leaves or stems

BRACT A small, modified, leaf-like structure

BRISTLE A stiff hair

CALLUS A hard extension or swelling at the base of the grass lemma or palea

CANESCENT Hoary or whitish, usually with a gray pubescence

CAPILLARY Hair-like or slender

CAPITATE Terminating with a head-like swelling

CARUNCLE An excrescence or appendage at or about the hilum of some seeds; an aril

CAUDATE With a tail-like appendage

CHAFFY With small, membranous scales

CHALAZA That part of the ovule that joins the seed coats; actually the base of the ovule

CHARTACEOUS Having the texture of paper

CILIATE Referring to a margin fringed with hairs

COLLICULOSE Covered with small, rounded elevations or hillocks

COMA A tuft of hairs at the base or apex of some seeds

COMMISSURE The inner face of carpels joining each other as in the Umbelliferae

COMOSE Bearing a tuft of hairs

CONDUPLICATE Cotyledons folded lengthwise and around the radicle

CORDIFORM Heart-shaped

CORIACEOUS Leather-like

CORNEOUS Hard and with the texture of horny material

DECIDUOUS The falling-off of plant or flower parts

DEHISCENT A flower part that breaks open is dehiscent

EMARGINATE With a shallow notch at the apex

FARINACEOUS With a mealy or flour-like material

FARINOSE Covered with a mealy material, frequently waxy

FILIFORM Thread-like

FIMBRIATE With a hairy fringe

FLEXUOUS Bent alternately in opposite directions

FLORET A small flower as in the Compositae or Gramineae

FRUIT A mature ovary or ovaries with or without other adhering flower parts

FUNICULUS The connective of the ovule or seed to the placenta, a connecting thread

GENICULATE Abruptly bent like a knee joint

GLABROUS Smooth, without any pubescence

GLANDULAR Having glands

GLAUCOUS Covered with a white, usually waxy, coating that will rub off as on a plum or cabbage leaf

GRANULAR Covered with small grains

HERBACEOUS Having the texture of an herb

HILUM The scar or mark on a seed indicating the point of attachment of the funiculus

HIRSUTE With stiff hairs

HIRTELLOUS Minutely hirsute

HISPID Covered with rough hairs or bristles

HISPIDULOUS Minutely hispid

HOARY With a close, white pubescence

HYALINE Thin and translucent

INCUMBENT The flat surface of the cotyledons against the radicle

INDEHISCENT Not opening

INDURATE Hard

INVOLUCRE The bracts subtending a flower or inflorescence

LACERATE Appearing as if torn

LACINIATE Deeply cut and with very narrow segments

LANATE Covered with a wool-like pubescence

LIGULATE Strap-shaped

LOMENT A legume with constrictions between the seeds

LUNATE Crescent-shaped

MEMBRANOUS Thin, flexible and translucent like a thin membrane

MERICARP One of the two seed-like carpels of an umbelliferous fruit

MUCILAGINOUS Slimy, or like mucilage

MUCRONATE Bristle-tipped

MURICATE Surface features with short, sharp points

NODULOSE With small knobs or knots

NUT A dry, indehiscent, commonly one-seeded fruit, usually derived from a compound ovary, with an indurated wall and often partially or wholly enclosed in an involucre or husk.

NUTLET The diminutive of a nut

OBLIQUE The bending of the axis, or, the form asymmetrical in outline

OCELLATE With eye-like depressions or markings

PAPILLA A small, nipple-shaped projection

PAPILLOSE or PAPILLATE Bearing papillae

PAPPUS The name for the various terminal structures on the achenes of the *Compositae*

PEDICEL The stalk of the individual flower of an inflorescence

PELTATE Flat, shield-shaped with the hilum on one flat surface

PERICARP The matured ovary wall

PLUMOSE Hairs branching like a feather

PUBERULENT Minutely pubescent

PUBESCENCE Hairiness

PUBESCENT Covered with short rather soft hairs

PUNCTATE Marked with dots, depressions, or glands

PUNCTICULATE Finely punctate

PUSTULOSE Having blister-like swellings particularly at the base of the hairs

RAPHE A ridge of vascular tissue connecting the nucellus with the placenta

RENIFORM Kidney-shaped

RETICULATE With the appearance of a network, and here, considered to be much more distinct than areolate

RETRORSÉ Directed backward or downward

RETUSE Shallowly notched at a rounded apex

RUGOSE Coarsely wrinkled

RUGULOSE With very fine wrinkles

SAMARA A winged fruit

SCABROUS With short, stiff hairs and rough to the touch

SCALARIFORM With markings suggesting a ladder-like appearance

SCARIOUS Thin, dry and membranous

SCHIZOCARP A fruit which splits into one-seeded portions or mericarps

SCURFY With small, bran-like scales

SERRATE With sharp, forward-pointing teeth

SETOSE Bristly

SPICULOSE Covered with fine points

STELLATE Star-like

STIPE A short stalk

STIPITATE With a short stalk or stipe

STONE The pit of a drupe or drupe-like fruit like the plum or cherry

STRIATE With fine grooves, ridges, or lines

STRIGOSE Having sharp-pointed, appressed, stiff hairs or bristles

STROPHIOLE The caruncle as found in the Caryophyllaceae
SUBULATE Awl-like
SULCATE Grooved or furrowed lengthwise
SULCUS A groove or furrow

TESTA The outer seed coat, usually hard and brittle
TOMENTOSE Densely woolly, or pubescent with soft, matted, wool-like pubescence
TRUNCATE Ending abruptly as if cut off
TUBERCLE A small, swollen structure of different texture and appearance than the organ on which it occurs
TUBERCULATE Having tubercles

UMBELLATE Like an umbel
UNCINATE Hooked at the apex
UNDULATE Wavy

VELUTINOUS With a coating of fine, soft hairs, velvet-like
VERRUCOSE Covered with wart-like elevations
VILLOSE With long, silky, straight hairs

Bibliography

1 Babcock, E.B., and G.L. Stebbins Jr. 1938. The American species of *Crepis*. Carnegie Inst. Washington, Pub. 504

2 Barkley, Fred. 1937. A monographic study of *Rhus* and its immediate allies in North and Central America including the West Indies. Ann. Missouri Bot. Gard. 20: 265-498

3 Barneby, Rupert C. 1964. Atlas of North American *Astragalus*. Mem. New York Bot. Gard. 13: vols. 1 & 2

4 Barnhart, J.H. 1916. Lentibulariaceae. Mem. New York Bot. Gard. 6: 39-64

5 Barton, Lela Viola. 1967. *Bibliography of Seeds*. Columbia University Press, New York

6 Beetle, A.A. 1943. A key to the North American species of the genus *Scirpus* based on achene characters. Amer. Midland Naturalist 29: 533-8

7 Beijerink, W. 1947. *Zadenatlas der nederlandsche Flora*. Wageningen: H. Vennman & Zonen

8 Benson, Lyman. 1942. North American *Ranunculi* I-V. Bull. Torrey Bot. Club 68: 157-72, 477-90, 640-59; (1941) 69: 298-316, 373-86

9 – 1948. Treatise on the North American *Ranunculi*. American Midland Naturalist 40: 1-261

10 Berggren, Greta. 1969. Atlas of Seeds and Small Fruits of Northwest-European Plant Species with Morphological Descriptions, Part 2, Cyperaceae. Stockholm: Swedish Natural Science Research Council

11 Blake, Anita Mary. 1928. Akenes of some Compositae. North Dakota Agriculture College Bull., 218

12 Boivin, Bernard. 1944. American *Thalictra* and their old world allies. Rhodora 46: 337-77, 391-445, 453-87

13 Brackett, A. 1923. A revision of the species of *Hypoxis*. Rhodora 25: 120-47, 151-63

210

14 Clapham, A.R., T.G. Tutin, and E.F. Warburg. 1962. *Flora of the British Isles.* Cambridge University Press

15 Cronquist, Arthur. 1947. Revision of the North American species of *Erigeron* north of Mexico. Brittonia 6 (2): 121-302

16 Daoud, H.S., and Robert L. Wilbur. 1965. A revision of the North American species of *Helianthemum* (Cistaceae). Rhodora 67: 63-82, 201-16, 255-80

17 Delorit, Richard J. 1970. *An Illustrated Taxonomy Manual of Weed Seeds.* River Falls, Wisc.: Agronomy Publications

18 Dickson, C.A. 1970. The study of plant macrofossils in British Quaternary deposits. In D. Walker and R.W. West, eds., Cambridge University Press: *Studies in the Vegetational History of the British Isles*, pp. 233-54

19 Fassett, N.C. 1951. *Callitriche* in the New World. Rhodora 53: 137-55, 161-82, 185-94, 209-22

20 Featherly, H.I. 1965. *Taxonomic Terminology of the Higher Plants.* New York: Hafner Publishing Co.

21 Fernald, M.L. 1932. The linear-leaved North American species of *Potamogeton* section *Axillares*. Mem. Amer. Acad. Arts and Sci. 17 (1): 1-183

22 - 1950. *Gray's Manual of Botany*, 8th ed. New York: American Book Co.

23 Gaertner, Erica E. 1953. Key to the seeds of the fleshy fruits. Trans. Roy. Can. Inst. xxx (1): 33-43

24 - 1954. *Two Keys for the Identification of Seeds, Based on the Works of W. Beijerink and J. Scurti.* Toronto: Research Council of Ontario

25 Gale, Shirley. 1944. *Rhynchospora* section *Eurhynchospora* in Canada, the United States and the West Indies. Rhodora 46: 89-134, 159-97, 207-49, 255-78

26 Gillett, John M. 1963. The Gentians of Canada, Alaska and Greenland. Pub. 1180, Biosystematic Research Institute, Central Experimental Farm, Ottawa

27 Gleason, H.A. 1952. *The New Britton and Brown Illustrated Flora of Northeastern United States and Adjacent Canada.* Lancaster, Pa.: Lancaster Press

28 Gleason, H.A., and Arthur Cronquist. 1963. *Manual of Vascular Plants of Northeastern United States and Adjacent Canada.* Princeton, N.J.: D. Van Nostrand Co., Inc.

29 Goodwin, H. 1956. *The History of the British Flora.* Cambridge University Press

30 Greenman, J.M. Monograph of the North and Central American species of the genus *Senecio*, part II. Ann. Missouri Bot. Gard. 2: 573-626 (1915); 3: 85-194 (1916); 4: 15-36 (1917); 5: 37-107 (1918)

31 Herman, F.J. 1936. Diagnostic characteristics in *Lycopus*. Rhodora 38: 373-5

32 Hillman, F.H., and H.H. Henry. 1935. The More Important Forage Plant Seeds and Incidental Seeds Found with Them. USDA Bureau of Plant Industry, Seeds Investigation

33 Hitchcock, A.S. 1950. Manual of Grasses of the United States. USDA Misc. Pub. 200, Washington, DC

34 Hodgdon, Albion R. 1938. A Taxonomic study of *Lechea*. Rhodora 40: 29-69, 87-131

35 Hopkins, Milton. 1937. *Arabis* in Eastern and Central North America. Rhodora 39: 63-98, 106-48, 155-86

36 Hosie, R.C. 1969. *Native Trees of Canada*, 7th ed. Canadian Forestry Service, Dept. of Fisheries and Forestry. Ottawa: Queen's Printer for Canada

37 Hui-Lin Li. 1952. The genus *Tovara* (Polygonaceae). Rhodora 54: 19-25

38 Isely, Duane. 1947. Seeds of weedy thistles. Proc. Assoc. Official Seed Analysts 37

39 - 1947. Investigations in seed classification by family characteristics. Research Bull. 351, Agriculture Experimental Station, Iowa State College, Ames, Iowa

0 – 1955. The Leguminosae of the North Central United States II, Hedysareae. Iowa State Coll. J. Sci. 30 (1): 33-118

1 Jones, G.N. 1968. *Taxonomy of American Species of Linden (Tilia)*. Urbana, Ill.: University of Illinois Press

2 Jones, G.N., and F.F. Jones. 1943. A revision of the perennial species of *Geranium* of the United States and Canada. Rhodora 45: 5-26, 32-53

3 Katz, N.J., S.V. Katz, and M.G. Kipriani. 1965. *Atlas and Keys of Fruits and Seeds Occurring in the Quaternary Deposits of the USSR* (in Russian but partly translated by J.C. Ritchie). Moscow: Nauka

4 Kelly, William R. 1953. Study of seed identification and seed germination of *Potentilla* spp. and *Veronica* spp. Cornell University Agriculture Experiment Station, Ithaca, N.Y., Mem. 317

5 Martin, Alexander C., and W.D. Barkley. 1961. *Seed Identification Manual*. University of California Press

6 McGivney, Sister M. Vincent de Paul. 1938. A revision of the subgenus *Eucyperus* found in the United States. Catholic University America Pub., Biol. Ser. 26

7 Murley, Margaret R. 1944. A key to 14 species of Geraniaceae. Iowa Acad. Sci. Proc. 51: 241-6

8 – 1945. Distribution of Euphorbiaceae in Iowa with seed keys. Iowa State Coll. J. Sci. 19: 415-27

9 – 1946. Fruit key to the Umbelliferae in Iowa with plant distribution records. Iowa State Coll. J. Sci. 20 (3): 349-64

0 – 1951. Seeds of the Cruciferae of Northeastern North America. Amer. Midland Naturalist 46 (1): 1-81

1 New York Botanical Garden. 1905 et seq. North American Flora. New York Bot. Gard., New York

2 Norton, J.B.S. 1900. A revision of the American species of *Euphorbia* of the section *Tithymalus* occurring north of Mexico. Missouri Bot. Gard. Ann. Rept. XI: 85-144

53 Ogden, Eugene C. 1943. The broad-leaved species of *Potamogeton* of North America north of Mexico. Rhodora 45: 57-105, 119-63, 171-214

54 Olmsted, F.L., F.V. Coville, and H.P. Kelsey, 1924. Standardized Plant Names. American Joint Committee on Horticultural Nomenclature, Salem, Mass.

55 Perry, Lily May. 1933. A revision of the North American species of *Verbena*. Ann. Missouri Bot. Gard. 20 (2): 239-358

56 Pennell, F.W. 1935. The Scrophulariaceae of Eastern Temperate North America. Academy Natural Science Philadelphia Monograph

57 Porter, R.H. 1949. Recent developments in seed technology. Bot. Rev. 15 (4): 221-82; (5) 283-334

58 Ritchie, J.C., and B. de Vries. 1964. Contributions to the Holocene paleoecology of west-central Canada; a late-glacial deposit from the Missouri Coteau. Can. J. Bot. 42: 677-92

59 Rock, Howard F.L. 1957. A revision of the vernal species of *Helenium* (Compositae). Rhodora 59: 101-16, 128-58, 168-78, 203-16

60 Rosendahl, C.O. 1948. A contribution to the knowledge of the Pleistocene flora of Minnesota. Ecology 29: 284-315

61 St John, Harold. 1916. A revision of the North American species of *Potamogeton* of the section *Coleophylli*. Rhodora 18: 121-38

62 Stanford, E.E. 1925. The amphibious group of *Polygonum*, subgenus *Persicaria*. Rhodora 27: 156-66

63 – 1927. *Polygonum hydropiper* in Europe and North America. Rhodora 29: 77-87

64 Svenson, H.K. Monographic studies in the genus *Eleocharis*. Rhodora 31 (1929); 34 (1932); 36 (1934); 39 (1937); 41 (1939)

65 Systematics Association, Committee for Descriptive Biological Terminology II. 1962. Terminology of simple symmetrical plane shapes. Taxon XI (5): 145-56

66 Van der Pijl, L. 1972. *Principles of Dispersal in Higher Plants*, 2nd ed. New York: Springer-Verlag

67 Wahl, H.A. 1954. A preliminary study in the genus *Chenopodium* in North America. Bartonia 27: 1-46

68 Watts, W.A., and T.C. Winter. 1966. Plant macrofossils from Kirchner Marsh, Minnesota – a paleoecological study. Bull. Geol. Soc. Am. 77: 1339-60

69 Wheeler, L.C. 1941. *Euphorbia* sub-genus *Chamaesyce* in Canada and the United States. Rhodora 43: 97-154, 168-205, 223-86

70 Wilbur, Robert L. 1963. A revision of the North American genus *Uvularia* (Liliaceae). Rhodora 65: 158-88

71 Wood, Carroll E. 1949. The American barbistyled species of *Tephrosia* (Leguminosae). Rhodora 51: 193-231, 233-302, 305-64, 369-84

72 Woodson, R.E. 1930. Studies in Apocynaceae I. Ann. Missouri Bot. Gard. 17: 1-212

73 Wright, W.H. *Weed Seeds.* Canada Department of Agriculture. Ottawa: Queen's Printer for Canada

Index

www.ingramcontent.com/pod-product-compliance
Lightning Source LLC
Chambersburg PA
CBHW080556030426
42336CB00019B/3215